# CAMBRIDGE LIBRARY COLLECTION

*Books of enduring scholarly value*

## Earth Sciences

In the nineteenth century, geology emerged as a distinct academic discipline. It pointed the way towards the theory of evolution, as scientists including Gideon Mantell, Adam Sedgwick, Charles Lyell and Roderick Murchison began to use the evidence of minerals, rock formations and fossils to demonstrate that the earth was older by millions of years than the conventional, Bible-based wisdom had supposed. They argued convincingly that the climate, flora and fauna of the distant past could be deduced from geological evidence. Volcanic activity, the formation of mountains, and the action of glaciers and rivers, tides and ocean currents also became better understood. This series includes landmark publications by pioneers of the modern earth sciences, who advanced the scientific understanding of our planet and the processes by which it is constantly re-shaped.

## Views of Nature

Alexander von Humboldt (1769–1859) was an intellectual giant: an explorer who helped lay the foundations of biogeography, a naturalist who influenced Charles Darwin, and a botanist who developed a model of the Earth's climate zones. He travelled extensively in Europe, carried out scientific explorations across the Russian Empire and in Latin America, and devoted much energy to seeking a unified view of the different branches of scientific knowledge. *Ansichten der Natur*, published in 1808 with a second edition in 1826, aimed to 'engage the imagination' as well as to communicate new ideas, and was translated into many European languages. This authorised translation of the third and final 1849 edition, dating from Humboldt's eightieth year, was published in 1850, though another English translation (by Mrs Sabine) had appeared the previous year. The wide coverage, including geology, geography and biology, is typical of Humboldt, as is the precise and engaging style.

Cambridge University Press has long been a pioneer in the reissuing of out-of-print titles from its own backlist, producing digital reprints of books that are still sought after by scholars and students but could not be reprinted economically using traditional technology. The Cambridge Library Collection extends this activity to a wider range of books which are still of importance to researchers and professionals, either for the source material they contain, or as landmarks in the history of their academic discipline.

Drawing from the world-renowned collections in the Cambridge University Library, and guided by the advice of experts in each subject area, Cambridge University Press is using state-of-the-art scanning machines in its own Printing House to capture the content of each book selected for inclusion. The files are processed to give a consistently clear, crisp image, and the books finished to the high quality standard for which the Press is recognised around the world. The latest print-on-demand technology ensures that the books will remain available indefinitely, and that orders for single or multiple copies can quickly be supplied.

The Cambridge Library Collection will bring back to life books of enduring scholarly value (including out-of-copyright works originally issued by other publishers) across a wide range of disciplines in the humanities and social sciences and in science and technology.

# Views of Nature

*Or Contemplations on the Sublime Phenomena of Creation*

Alexander von Humboldt

CAMBRIDGE UNIVERSITY PRESS

Cambridge, New York, Melbourne, Madrid, Cape Town,
Singapore, São Paolo, Delhi, Tokyo, Mexico City

Published in the United States of America by Cambridge University Press, New York

www.cambridge.org
Information on this title: www.cambridge.org/9781108037358



This edition first published 1850
This digitally printed version 2011

ISBN 978-1-108-03735-8 Paperback



The original edition of this book contains a number of colour plates, which have been reproduced in black and white. Colour versions of these images can be found online at www.cambridge.org/9781108037358

# VIEWS OF NATURE:

OR CONTEMPLATIONS ON

## THE SUBLIME PHENOMENA OF CREATION;

WITH

SCIENTIFIC ILLUSTRATIONS.

BY

ALEXANDER VON HUMBOLDT.

TRANSLATED FROM THE GERMAN

BY E. C. OTTE, AND HENRY G. BOHN.

WITH A FRONTISPIECE FROM A SKETCH BY THE AUTHOR, A FAC-SIMILE OF HIS HAND-WRITING, AND A COMPREHENSIVE INDEX.

LONDON:
HENRY G. BOHN, YORK STREET, COVENT GARDEN.
1850.

## PREFACE BY THE PUBLISHER.

GREAT pains have been taken with the present translation, as well in regard to fidelity and style, as in what may be termed the accessories. In addition to all that is contained in the original work, it comprises an interesting view of Chimborazo, from a sketch by Humboldt himself; a fac-simile of the author's handwriting; head lines of contents; translations of the principal Latin, French, and Spanish quotations;* a very complete index; and a conversion of all the foreign measurements. It was at first intended to give both the foreign and English measurements, in juxta-position; but this plan was abandoned on perceiving that the pages would become overloaded with figures, and present a perplexing and somewhat appalling aspect, without affording any equivalent advantage to the English reader. In some few instances, however, where it seemed desirable, and in all the parallel tables, duplicate measurements have been inserted. The French *toises* are converted into their relative number of English feet; and German miles, whether simple or square, are reduced to our own. The longitudes have been calculated from Greenwich, conformably to English maps, in lieu of those given by Humboldt, which are calculated from Paris. The degrees of temperature, instead of Reaumur's, are Fahrenheit's, as now the most generally recognised.

It here becomes necessary to say something of the trans-

* To instance a few, see pp. 241, 245, 255, 259, 304, 320, 325, 326, 886, 422, 424.

lators, and the cause of so much unexpected delay in producing this volume; the more so as many of the subscribers to the Scientific Library have expressed an interest in the subject, owing, in some measure, to a controversy which arose out of my previous publication of *Cosmos*. The translation was originally entrusted to E. C. Otté, with an agreement as to time, according to which I had every reason to expect that I should fulfil my engagement to publish it in October last, or at latest in November; but, after much of the manuscript was prepared, the translator's indisposition and subsequent absence from London, occasioned a serious suspension. In this dilemma I found it necessary to call in aid, as well as to assist personally. The result of this "co-operation of forces" will no doubt prove satisfactory to the reader, inasmuch as every sheet has been at least trebly revised, and it is hoped proportionably improved. In addition to the responsible translator, my principal collaborateur has been Mr. R. H. Whitelocke, a gentleman well qualified for the task.

All the measurements are calculated by the scientific friend, who fulfilled this department so satisfactorily in my edition of *Cosmos*.

The translation of the pretty poem, *The Parrot of Atures*, (page 189,) now first given in English, is contributed by Mr. Edgar A. Bowring.

For the additional notes subscribed " Ed." I am myself, in most instances, responsible.

Much has been said, pro and con, about the sanction of the Author to the several translations of his works. My answer has, I believe, been generally considered satisfactory and conclusive. I have now only to add, that when I wrote to Baron Humboldt, more than a year and a-half ago, presenting him with my then unpublished edition of *Cosmos*, I announced my intention of proceeding with his other works, and consulted him on the subject. He replied in the kindest spirit, without intimating any previous engagement, and honoured me with several valuable suggestions. A portion of one of his letters is annexed in facsimile. In consequence of what I then presumed to be his recommendation, I determined to make the *Ansichten* my next volume, and announced it, long before any one else, though not at first by its English name. At that time I had reason to hope that I should receive the new German edition at least as early as any one, but was disappointed. This circumstance, added to the delay already alluded to, has brought me late into the field. In now, however, presenting my subscribers with what I have taken every available means to render a perfect book, I hope I shall afford them ample atonement.

A few words respecting the work itself. The first edition was published forty-three years ago, the second in 1826, and the third, of which the present volume is a translation, in August last. The difference between the three editions in respect to the text (if I may so distinguish the more entertaining part of the work from the scientific "Illustrations") is not material, excepting that each has one or more new chapters. Thus to the second edition was added the *Essay on Volcanos* and the curious allegory on vital force, entitled *The Rhodian Genius*, and to the third *The Plateau of Caxamarca*.

The additions to the "Illustrations" however in the third edition are considerable, and comprise a rapid sketch of whatever has been contributed by modern science in illustration of the Author's favourite subjects.

No intellectual reader can peruse this masterly work

without intense interest and considerable instruction. After feasting on the highly wrought and, it may be said, poetical descriptions, written in the Author's earlier years, he will turn with increased zest to the elaborate illustrations, which, in a separate form, are brought to bear on every subject of the text. This scientific portion, although not at first the most attractive, presents many delightful episodes, which will amply repay the perusal of even those who merely read for amusement.

HENRY G. BOHN.

*York Street, January,* 1850.

---

FAC-SIMILE OF THE HANDWRITING
OF BARON HUMBOLDT.

EXTRACTS OF A LETTER TO THE PUBLISHER.

Vous me faites l'honneur de
me demander, Monsieur, lesquelles
de mes publications d'un intérêt
plus général et plus littéraire
n'ont pas été traduites en anglois.
Les auteurs ne sont pas bons à con-
sulter à ce sujet.

Je mets dans ce moment la derni-
ère main à une nouvelle édition
augmentée et entièrement refondue
des "Ansichten der Natur" Views
into Nature. C'est un de mes ou-
vrages les plus lus en Allemagne et en
France.

Veuillez bien excuser, Monsieur, l'écriture
oblique et illisible de cette lettre (j'ai
un bras malade depuis le voyage dans
les forêts de l'Orinoque!) et agréer
l'expression de ma haute considération.

V. t. h. et t. o. ser.

Alexandre Humboldt
au Château de Sanssouci
ce 20 Octobr 1848

# AUTHOR'S PREFACE,

## TO THE FIRST EDITION.

---

With some diffidence, I here present to the public a series of papers which originated in the presence of the noblest objects of nature,—on the Ocean,—in the forests of the Orinoco,—in the Savannahs of Venezuela,—and in the solitudes of the Peruvian and Mexican Mountains. Several detached fragments, written on the spot, have since been wrought into a whole. A survey of nature at large,—proofs of the co-operation of forces,—and a renewal of the enjoyment which the immediate aspect of the tropical countries affords to the susceptible beholder,—are the objects at which I aim. Each Essay was designed to be complete in itself; and one and the same tendency pervades the whole. This æsthetic mode of treating subjects of Natural History is fraught with great difficulties in the execution, notwithstanding the marvellous vigour and flexibility of my native language. The wonderful luxuriance of nature presents an accumulation of separate images, and accumulation disturbs the harmony and effect of a picture. When the feelings and the imagination are excited, the style is apt to stray into poetical prose. But these ideas require no amplification here, for the following pages afford but too abundant examples of such deviations and of such want of unity.

Notwithstanding these defects, which I can more easily

perceive than amend, let me hope that these "Views" may afford the reader, at least some portion of that enjoyment which a sensitive mind receives from the immediate contemplation of nature. As this enjoyment is heightened by an insight into the connection of the occult forces, I have subjoined to each treatise scientific illustrations and additions.

Everywhere the reader's attention is directed to the perpetual influence which physical nature exercises on the moral condition and on the destiny of man. It is to minds oppressed with care that these pages are especially consecrated. He who has escaped from the stormy waves of life will joyfully follow me into the depths of the forests, over the boundless steppes and prairies, and to the lofty summits of the Andes. To him are addressed the words of the chorus who preside over the destinies of mankind:

On the mountains is freedom! the breath of decay
  Never sullies the fresh flowing air;
Oh! nature is perfect wherever we stray;
  'Tis man that deforms it with care.*

* These lines are from Schiller's *Bride of Messina*, as translated by A. Lodge, Esq See Schiller's works (Bohn's ed.) vol. iii. p. 509.

# AUTHOR'S PREFACE,

## TO THE SECOND AND THIRD EDITIONS.

---

THE twofold object of this work,—an anxious endeavour to heighten the enjoyment of nature by vivid representations, and at the same time to increase, according to the present state of science, the reader's insight into the harmonious co-operation of forces,—was pointed out by me in the preface to the first edition, nearly half a century ago. I there alluded to the several obstacles which oppose themselves to the æsthetic treatment of the grand scenes of nature. The combination of a literary and a purely scientific aim, the desire to engage the imagination, and at the same time to enrich life with new ideas by the increase of knowledge, render the due arrangement of the separate parts, and what is required as unity of composition, difficult of attainment. Notwithstanding these disadvantages, however, the public have continued to receive with indulgent partiality, my imperfect performance.

The second edition of the *Views of Nature*, was published by me in Paris in 1826. Two papers were then added, one, "An inquiry into the structure and mode of action of Volcanos in different regions of the earth;" the other, "Vital Force, or The Rhodian Genius." Schiller, in remembrance of his youthful medical studies, loved to converse with me, during my long stay at Jena, on physiological subjects. The inquiries in which I was then engaged, in preparing my work "On the condition of the fibres of muscles and nerves, when irritated by contact with substances chemically opposed," often imparted a more serious direction to our conversation. It was at this period that I wrote the little allegory on Vital Force, called The Rhodian Genius. The predilection which Schiller entertained for this piece, and which he admitted into his periodical, *Die Horen*, gave me courage to introduce it here. My brother, in a letter which has recently been published (William von Humboldt's Letters to a Female Friend, vol. ii. p. 39), delicately alludes to the subject, but at the same time very justly adds; "The development of a physiological idea is exclusively the object of the essay. Such semi-poetical clothings of grave truths were more in vogue at the time this was written than they are at present."

In my eightieth year I have still the gratification of completing a third edition of my work, and entirely remoulding it to meet the demands of the age. Almost all the scientific illustrations are either enlarged or replaced by new and more comprehensive ones.

I have indulged a hope of stimulating the study of nature, by compressing into the smallest possible compass, the numerous results of careful investigation on a variety of interesting subjects, with a view of shewing the importance of accurate numerical data, and the necessity of comparing them with each other, as well as to check the dogmatic smattering and fashionable scepticism which have too long prevailed in the so-called higher circles of society.

My expedition into northern Asia (to the Ural, the Altai,

and the shores of the Caspian Sea) in the year 1829, with Ehrenberg and Gustavus Rose, at the command of the Emperor of Russia, took place between the second and third editions of my work. This expedition has essentially contributed to the enlargement of my views in all that concerns the formation of the earth's surface, the direction of mountain-chains, the connexion of the Steppes and Deserts, and the geographical distribution of plants according to ascertained influences of temperature. The ignorance which has so long existed respecting the two great snow-covered mountain-chains, the Thian-schan and the Kuen-lün, situated between the Altai and Himalaya, has (owing to the injudicious neglect of Chinese sources of information) obscured the geography of Central Asia, and propagated fancies instead of facts, in works of extensive circulation. Within the last few months the hypsometric comparisons of the culminating points of both continents have unexpectedly received important and corrective illustration, of which I am the first to avail myself in the following pages. The measurement (now divested of former errors) of the altitude of the two mountains, Sorata and Illimani, in the eastern chain of the Andes of Bolivia, has not yet, with certainty, restored the Chimborazo to its ancient pre-eminence among the snowy mountains of the new world. In the Himalaya the recent barometric measurement of the Kinchin-jinga (26,438 Parisian, or 28,178 English feet) places it next in height to the Dhawalagiri, which has also been trigonometrically measured with greater accuracy.

To preserve uniformity with the two former editions of the *Views of Nature*, the calculations of temperature, unless where the contrary is stated, are given according to the eighty degrees thermometer of Reamur. The lineal measurement is the old French, in which the *toise* is equivalent to six Parisian feet. The miles are geographical, fifteen to a degree of the equator. The longitudes are calculated from the first meridian of the Parisian Observatory.

*Berlin, March,* 1849.

BAXTER'S PATENT OIL COLOR PRINTINO,
XI, NORTHAMPTON SQUARE.

CHIMBORAZO.

# CONTENTS.

# SUMMARY OF CONTENTS.

Modern views on the mountain systems of the two American peninsulas. Chains, which have a direction from S.W. to N.E., in Brazil and in the Atlantic portion of the United States of North America. Depression of the Province of Chiquitos; ridges as water-marks between the Guaporé and Aguapehi in 15° and 17° south lat., and between the fluvial districts of the Orinoco and Rio Negro in 2° and 3° north lat.—pp. 29–31.

Continuation of the Andes-chain north of the isthmus of Panamá through the territory of the Aztecs, (where the Popocatepetl, recently ascended by Capt. Stone, rises to an altitude of 17,720 feet,) and through the Crane and Rocky Mountains. Valuable scientific investigations of Capt. Frémont. The greatest barometric levelling ever accomplished, representing a profile of the ground over 28° of longitude. Culminating point of the route from the coast of the Atlantic to the South Sea. The South Pass southward of the Wind-River Mountains. Swelling of the ground in the Great Basin. Long disputed existence of Lake Timpanogos. Coast-chain, Maritime Alps, Sierra Nevada of California. Volcanic eruptions. Cataracts of the Columbia River—pp. 31–38.

General considerations on the contrast between the configuration of the territorial spaces, presented by the two diverging coast-chains, east and west of the central chain, called the Rocky Mountains. Hypsometric constitution of the Eastern Lowland, which is only from 400 to somewhat more than 600 feet above the level of the sea, and of the arid uninhabited plateau of the Great Basin, from 5000 to more than 6000 feet high. Sources of the Mississippi in Lake Istaca according to Nicollet, whose labours are most meritorious. Native land of the Bisons; their ancient domestication in Northern Mexico asserted by Gomara—pp. 38–42.

Retrospective view of the entire Andes-chain from the cliff of Diego Ramirez to Behring's Straits. Long prevalent errors concerning the height of the eastern Andes-chain of Bolivia, especially of the Sorata and Illimani. Four summits of the western chain, which, according to Pentland's latest determinations, surpass the Chimborazo in height, but not the still-active volcano, Aconcagua, measured by Fitz-Roy—pp. 42–44.

The African mountain range of Harudje-el-Abiad. Oases of vegetation, abounding in springs—pp. 44–46.

Westerly winds on the borders of the desert Sahara. Accumulation of sea-weed; present and former position of the great fucus-bank, from the time of Scylax of Caryanda to that of Columbus and to the present period—pp. 46–50.

Tibbos and Tuaryks. The camel and its distribution—pp. 50–53.

Mountain-systems of Central Asia between Northern Siberia and India, between the Altai and the Himalaya, which latter range is aggregated with the Kuen-lün. Erroneous opinion as to the existence of one immense plateau, the so-called "Plateau de la Tartarie"—pp. 53–56.

Chinese literature a rich source of orographic knowledge. Gradations of the High Lands. Gobi and its direction. Probable mean height of Thibet—pp. 56–63.

General review of the mountain systems of Asia. Meridian chains: the Ural, which separates lower Europe from lower Asia or the Scythian Europe of Pherecydes of Syros and Herodotus. Bolor, Khingan, and the Chinese chains, which at the great bend of the Thibetan and Assam-Burmese river, Dzangbo-tschu, stretch from north to south. The meridian elevations alternate between the parallels of 66° and 77° east long. from Cape Comorin to the Frozen Ocean, like displaced veins. Thus the Ghauts, the Soliman chain, the Paralasa, the Bolor, and the Ural follow from south to north. The Bolor gave rise, among the ancients, to the idea respecting the Imaus, which Agathodæmon considered to be prolonged northwards as far as the lowland or basin of the lower Irtysch. Parallel chains, running east and west, the Altai, Thian-schan with its active volcanos, which lie 1528 miles from the frozen ocean at the mouth of the Obi, and 1512 from the Indian Ocean at the mouth of the Ganges; Kuen-lün, already recognized by Eratosthenes, Marinus of Tyre, Ptolemy, and Cosmas Indicopleustes, as the greatest axis of elevation in the Old World, between 35½° and 36° lat. in the direction of the diaphragm of Dicæarchus. Himalaya. The Kuen-lün may be traced, when considered as an axis of elevation, from the Chinese wall near Lung-tscheu, through the somewhat more northerly chains of Nan-schan and Kilian-schan, through the mountain node of the "Starry Sea," the Hindoo Cush (the Paropanisus and Indian Caucasus of the ancients), and, lastly, through the chain of the Demavend and Persian Elburz, as far as the Taurus in Lycia. Not far from the intersection of the Kuen-lün by the Bolor, the corresponding direction of the axes of elevation (inclining from east to west in the Kuen-lün and Hindoo Cush, and on the other hand south-east and north-west in the Himalaya) proves, that the Hindoo Cush is a prolongation of the Kuen-lün, and not of the Himalaya which is associated to the latter in the manner of a gang or vein. The point where the Himalaya changes its direction, that is to say, where it leaves its former east-westerly direction, lies not far from 81° east long. The Djawahir is not, as has hitherto been supposed, the next in altitude to the Dhawalagiri, which is the highest summit of the Himalaya; for, according to Joseph Hooker, this rank is due to a mountain lying in the meridian of Sikhim between Butan and Nepaul, called the Kinchinjinga or Kintschin-Dschunga. This mountain (Kinchinjinga) measured by Col. Waugh, Director of the Trigonometrical Survey of India, has for its western summit an altitude of 28,178 feet, and for its eastern 27,826 feet, according to the *Journal of the Asiatic Soc. of Bengal*, November, 1848. The mountain, now considered higher than the Dhawalagiri, is represented in the engraving to the title-page of Joseph Hooker's splendid work, *The Rhododendrons of Sikkim Himalaya*, 1849. Determination of the snow-limits on the northern and southern slopes of the Himalaya; the former lies in the mean about 3620 up to 4900 feet higher. New statements of Hodgson. But for the remarkable distribution of heat in the upper strata of the air, the table-land of western Thibet would be uninhabitable to millions of human beings—pp. 63–80.

The Hiong-nu, whom Deguignes and John Müller considered to be a tribe of Huns, appear rather to be one of the widely spread Turkish races of the Altai and Tangnu mountains. The Huns, whose name was known even to Dionysius Periegetes, and who are described by Ptolemy as Chuns (hence the later territorial name of Chunigard!) are a Finnish tribe, from the Ural mountains, which separate the two continents—pp. 80–81.

Representations of the sun, animals, and characters, graven on rocks at Sierra Parime, as well as in North America, have frequently been regarded as writing—p. 82.

Description of the cold mountain regions between 11,000 and 13,000 Parisian, or 11,720 and 13,850 English feet in height, which have been designated Paramos. Character of their vegetation—p. 83.

Orographic remarks on the two mountain clusters (Pacaraima and Sierra de Chiquitos) which separate the three plains of the lower Orinoco, the Amazon, and La Plata rivers from each other—p. 84.

Concerning the Dogs of the New Continent, the aboriginal as well as those from Europe, which have become wild. Sufferings of Cats at heights surpassing 13,854 feet—pp. 85–88.

The Low Land of the Sahara and its relations to the Atlas range, according to the latest reports of Daumas, Carette, and Renou. The barometric measurements of Fournel render it very probable, that part of the north African desert lies below the level of the sea. Oasis of Biscara. Abundance of rock-salt in regions which extend from S.W. to N.E. Causes of nocturnal cold in the desert, according to Melloni—pp. 88–92. Information respecting the River Wadi Dra (one-sixth longer than the Rhine), which is dry during a great part of the year. Some account of the territory of the Sheikh Beirouk, who is independent of the Emperor of Morocco, according to manuscript communications of Capt. Count Bouet Villaumez, of the French Marine. The mountains north of Cape Nun (an Edrisian name, in which by a play of words a negation has been assumed since the 15th century) attain an altitude of 9186 feet—pp. 92–94.

Gramineous vegetation of the American Llanos between the tropics, compared with the herbaceous vegetation of the Steppes in Northern Asia. In these, especially in the most fertile of them, a pleasing effect is afforded in spring by the small snow-white and red flowering Rosaceæ, Amygdaleæ, the species of Astragalus, Crown-imperial, Cypripedias, and Tulips. Contrast with the desert of the salt-steppes full of Chenopodiæ, and of species of Salsola and Atriplex. Numerical considerations with respect to the predominant families. The plains which skirt the Frozen Ocean (north of what Admiral Wrangel has described as the boundary of Coniferæ and Amentaceæ), are the domain of cryptogamic plants. Physiognomy of the Tundra on an ever-frozen soil, covered with a thick coating of Sphagnum and other foliaceous mosses, or with the snow-white Cenomyce and Stereocaulon paschale—pp. 94–96.

Chief causes of the very unequal distribution of heat in the European and American continents. Direction and inflection of the isothermal lines (equal mean heat of the year, in winter and summer)—pp. 96–105.

Is there reason to believe that America emerged later from the chaotic covering of waters?—pp. 105–107. Thermal comparison between the northern and southern hemispheres in high latitudes—pp. 107–109. Apparent connexion between the sand-seas of Africa, Persia, Kerman, Beloochistan, and Central Asia. On the western portion of the Atlas, and the connection of purely mythical ideas, with geographical legends. Indefinite allusions to fiery eruptions. Triton Lake. Crater forms, south of Hanno's "Bay of the Gorilla Apes." Singular description of the Hollow Atlas, from the Dialexes of Maximus Tyrius—pp. 110–11.

Explanations of the Mountains of the Moon (Djebel-al-Komr) in the interior of Africa, according to Reinaud, Beke, and Ayrton. Werne's instructive report of the second expedition, which was undertaken by command of Mehemet Ali. The Abyssinian high mountain chain, which, according to Rüppell, attains nearly the height of Mont Blanc. The earliest account of the snow between the tropics is contained in the inscription of Adulis, which is of a somewhat later date than Juba. Lofty mountains, which between 6° and 4°, and even more southerly, approach the Bahr-el-Abiad. A considerable rise of ground separates the White Nile from the basin of the Goschop. Line of separation between the waters which flow towards the Mediterranean and Indian seas, according to Carl Zimmermann's map. Lupata chain, according to the instructive researches of Wilhelm Peters—pp. 114–120.

Oceanic currents. In the northern part of the Atlantic the waters are agitated in a true rotatory movement. That the first impulse to the Gulf-stream is to be looked for at the southern apex of Africa, was a fact already known to Sir Humphry Gilbert in 1560. Influence of the Gulf-stream on the climate of Scandinavia. How it contributed to the discovery of America. Instances of Esquimaux, who, favoured by north-west winds, have been carried, through the returning easterly inclined portion of the warm gulf-stream, to the European coasts. Information of Cornelius Nepos and Pomponius Mela respecting Indians, whom a King of the Boii sent as a present to the Gallic Proconsul Quintus Metellus Celer; and again of others in the times of the Othos, Frederick Barbarossa, Columbus, and Cardinal Bembo. Again, in the years 1682 and 1684, natives of Greenland appeared at the Orkney Islands—pp. 120–125.

Effects of lichens and other cryptogamia in the frigid and temperate zones, in promoting the growth of the larger phanerogamia. In the tropics the preparatory ground-lichens often find substitutes in the oleaginous plants. Lactiferous animals of the New Continent; the Llama, Alpaca, and Guanaco—pp. 125–128. Culture of farinaceous grasses—pp. 128–131. On the earliest population of America—pp. 131–134.

The coast-tribe the Guaranes (Warraus), and the littoral palm Mauritia, according to Bembo, Raleigh, Hillhouse, Robert and Richard Schomburgk—pp. 134–136.

## NOCTURNAL LIFE OF ANIMALS IN THE PRIMEVAL FOREST . . . . . . . pp. 191–201.

Difference in the richness of languages as regards precise and definite words for characterizing natural phenomena, such as the state of vegetation and the forms of plants, the contour and grouping of clouds, the appearance of the earth's surface, and the shape of mountains. Loss which languages sustain in such expressive words. The misinterpretation of a Spanish word has enlarged mountain-chains on maps, and created new ranges. PRIMEVAL FOREST. Frequent misuse of this term. Want of uniformity in the association of the arboral species is characteristic of the forests within the tropics. Causes of their imperviousness. The Climbing plants (*Lianes*) often form but a very inconsiderable portion of the underwood—pp. 191–196.

Aspect of the Rio Apure in its lower course. Margin of the forest fenced like a garden by a low hedge of Sauso (*Hermesia*). The wild animals of the forest issue with their young through solitary gaps, to approach the river-side. Herds of large Capybaræ, or Cavies. Fresh-water dolphins—pp. 196–199. The cries of wild animals resound through the forest. Cause of the nocturnal noises — pp. 199–200. Contrast to the repose which reigns at noontide on very hot days within the tropics. Description of the rocky narrows of the Orinoco at the Baraguan. Buzzing and humming of insects; in every shrub, in the cracked bark of trees, in the perforated earth, furrowed by hymenopterous insects, life is audible and manifest—pp. 200–201.

SCIENTIFIC ILLUSTRATIONS AND ADDITIONS . . pp. 202–203.

Characteristic denominations of the surface of the earth (Steppes, Savannahs, Prairies, Deserts) in the Arabic and Persian. Richness of the dialects of Old Castile for designating the forms of mountains. Fresh-water rays and fresh-water dolphins. In the giant streams of both continents some organic sea-forms are repeated. American nocturnal apes with cat's eyes; the tricoloured striped Douroucoali of the Cassiquiare—pp. 202–203.

HYPSOMETRIC ADDENDA. . . . . . . pp. 204–209.

Pentland's measurements in the eastern mountain-chain of Bolivia. Volcano of Aconcagua, according to Fitz-Roy and Darwin. Western mountain-chain of Bolivia—pp. 204–205. Mountain systems of North America. Rocky Mountains and snowy chain of California. Laguna de Timpanogos—pp. 205–207. Hypsometric profile of the Highland of Mexico as far as Santa Fé—pp. 207–209.

## IDEAS FOR A PHYSIOGNOMY OF PLANTS . pp. 210–231.

Universal profusion of life on the slopes of the highest mountain summits, in the ocean and in the atmosphere. Subterranean Flora. Siliceous-shelled polygastrica in masses of ice at the pole. Podurellæ in the ice tubules of the glaciers of the Alps; the glacier-flea (*Desoria glacialis*). Minute organisms of the dust fogs—pp. 210–213.

History of the vegetable covering. Gradual extension of vegetation over the naked crust of rock. Lichens, mosses, oleaginous plants. Cause of the present absence of vegetation in certain districts.—pp. 213–220.

Each zone has its peculiar character. All animal and vegetable conformation is bound to fixed and ever-recurring types. Physiognomy of Nature. Analysis of the combined effect produced by a region. The individual elements of this impression. Outline of the mountain ranges; azure of the sky; shape of the clouds. That which chiefly determines the character is the vegetable covering. Animal organizations are deficient in mass; the mobility of individual species, and often their diminutiveness, conceals them from view—pp. 220–223.

Enumeration of the forms of Plants which principally determine the physiognomy of Nature, and which increase or diminish from the equator towards the Pole, in obedience to established laws—

| | Text. | Illustrations. |
|---|---|---|
| Palms | pp. 223–224 | pp. 296–304 |
| Banana form | p. 224 | p. 305 |
| Malvaceæ | p. 224 | pp. 305–307 |
| Mimosæ | p. 225 | pp. 307–308 |
| Ericeæ | p. 225 | pp. 308–310 |
| Cactus form | p. 226 | pp. 310–312 |
| Form of Orchideæ | p. 226 | pp. 312–313 |
| Casuarinæ | p. 226 | pp. 313–314 |
| Acicular-leaved Trees | p. 227 | pp. 314–329 |
| Pothos form, and that of the Aroideæ | p. 227 | pp. 329–331 |
| Lianes and Climbing plants | pp. 227–228 | pp. 331–332 |
| Aloes | p. 228 | pp. 332–334 |
| Grass form | p. 228 | pp. 334–337 |
| Ferns | p. 229 | pp. 337–341 |
| Lilies | p. 229 | pp. 341–343 |
| Willow form | p. 229 | p. 343 |
| Myrtles | p. 229 | pp. 343–346 |
| Melastomaceæ | p. 229 | p. 346 |
| Laurel form | p. 229 | p. 346 |

Enjoyment resulting from the natural grouping and contrasts of these plant-forms. Importance of the physiognomical study of plants to the landscape-painter—pp. 229–231.

SCIENTIFIC ILLUSTRATIONS AND ADDITIONS . . . . pp. 232–352.

Organisms, both animal and vegetable, in the highest Alpine regions, near the line of eternal snow, in the Andes chain, and the Alps; insects are carried up involuntarily by the ascending current of air. The small field-mouse (*Hypudæus nivalis*) of the Swiss Alps. On the real height to which the Chinchilla laniger mounts in Chili—pp. 232–233.

Lecideæ, Parmeliæ on rocks not entirely covered with snow; but certain phanerogamic plants also stray in the Cordilleras beyond the boundary of perpetual snow, thus Saxifraga Boussingaulti to 15,773 feet above the level of the sea. Groups of phanerogamic Alpine plants in the Andes chain at from 13,700 to nearly 15,000 feet high. Species of Culcitium, Espeletia, Ranunculus, and small moss-like umbellifera, Myrrhis andicola, and Fragosa arctioides—pp. 233–234. Measurement of Chimborazo, and etymology of the name—p. 234–236. On the greatest absolute height to which men in both continents, in the Cordilleras and the Himalaya,—on the Chimborazo and Tarhigang—have as yet ascended —p. 236.

Economy, habitat, and singular mode of capturing the Condor (*Cuntur*, in the Inca language) by means of palisades—pp. 237–239. Use of the Gallinazos (*Cathartes urubu* and *C. aura*) in the economy of nature, for purifying of the air in the neighbourhood of human dwellings; their domestication—pp. 239–240.

On the so-called revivification of the rotifera, according to Ehrenberg and Doyère; according to Payen, germs of Cryptogamia retain their power of reproduction in the highest temperature—pp. 240–241.

Diminution, if not total suspension, of organic functions in the winter-sleep of the higher classes of animals—p. 242. Summer-sleep of animals in the tropics. Drought acts like the cold of winter. Tenrecs, Crocodiles, Tortoises, and East-African Lepidosirens—pp. 242–244.

Pollen, Fructification of Plants. The experience of many years concerning the Cœlebogyne; it brings forth mature seeds in England without a trace of male organs—pp. 244–245.

The phosphorescence of the Ocean through luminous animals as well as organic fibres and membranes of the decomposing animalculæ. Acalephæ and siliceous-shelled luminous infusoria. Influence of nervous irritability on the coruscation—pp. 245–250.

Pentastoma, inhabiting the lungs of the rattle-snake of Cumana—p. 251.

Rock-constructing Coral animals. The structure surviving the architects. More correct views of the present period. Coast-reefs, Reefs surrounding islands and Lagoon-islands. Atolls, Coral walls inclosing a lagoon. The royal gardens of Christopher Columbus, The Coral Islands south of Cuba. The living gelatinous coating of the calcareous fabric of the coral-stems allures fishes in quest of food, and also turtles. Singular mode of fishing with the Remora, *Echeneis Naucrates* (the little angling fish)—pp. 251–258.

Probable depth of the coralline structures—pp. 258–260. Besides a great quantity of carbonate of lime and magnesia, the madrepores and Astreæ contain also some fluoric and phosphoric acid—pp. 260–261. Oscillating state of the sea-bottom according to Darwin—pp. 261–262.

Irruptions of the sea. Mediterranean Sea. Sluice-theory of Strato. Samothracian legends. The Myth of Lyctonia and the submerged Atlantis—pp. 262–266. Concerning the precipitation of clouds—

p. 266. The indurating crust of the earth while giving out caloric. Heated currents of air, which in the primordial period, during the frequent corrugations of the mountainous strata, and the upheaval of lands, have poured into the atmosphere through temporary fissures and chasms—pp. 266-268.

Colossal size and great age of certain genera of trees, *e. g.*, the dragon-tree of Orotava of 13, the Adansonia digitata (Baobab) of 33 feet in diameter. Carved characters of the 15th century. Adanson assigns to certain Baobab-stems of Senegambia an age of from 5000 to 6000 years—pp. 268-273.

According to an estimate based on the number of the annual rings, there are yews (Taxus baccata) of from 2600 to 3000 years old. Whether in the temperate northern zone that part of a tree which faces the north has narrower rings, as Michael Montaigne asserted in 1581? Gigantic trees, of which some individuals attain a diameter of above 20 feet and an age of several centuries, belong to the most opposite natural families —pp. 273-274.

Diameter of the Mexican Schubertia disticha of Santa Maria del Tule 43, of the oak near Saintes (Dep. de la Charente inf.) 30 feet. The age of this oak considered by its annual rings to be from 1800 to 2000 years. The main stem of the rose-tree (27 feet high) at the crypt of the church of Hildesheim is 800 years old. A species of fucus, Macrocystis pyrifera, attains a length of more than 350 feet, and therefore exceeds all the conifera in length, not excepting the Sequoia gigantea itself—pp. 274-276.

Investigations into the supposed number of the phanerogamic species of plants, which have hitherto been described or are preserved in herbariums. Numerical ratios of plant-forms. Discovered laws of the geographical distribution of the families. Ratios of the great divisions: of the Cryptogamia to the Cotyledons, and of the Monocotyledons to the Dicotyledons, in the torrid, temperate, and frigid zones. Outlines of arithmetical botany. Number of the individuals, predominance of social plants. The forms of organic beings stand in mutual dependence on each other. If once the number of species in one of the great families of the Glumaceæ, Leguminosæ, or Compositæ, on any one point of the earth, be known, an approximative conclusion may be arrived at not only as to the number of all the phanerogamia, but also of the species of all remaining plant-families growing there. Connection of the numerical ratios here treated on in the geographical distribution of the families, with the direction of the isothermal lines. Primitive mystery in the distribution of types. Absence of Roses in the southern, and of Calceolarias in the northern zone. Why has our heath (Calluna vulgaris), and why have our Oaks not progressed eastwards across the Ural into Asia? The vegetation-cycle of each species requires a certain minimum heat for its due organic development—pp. 273-287.

Analogy with the numeric laws in the distribution of animal forms. If more than 35,000 species of phanerogamia are now cultivated in

Europe, and if from 160,000 to 212,000 phanerogamia are now contained, described and undescribed, in our herbariums; it is probable that the number of collected insects scarcely equals that number of phanerogamia; whereas in individual European districts the insects collected preponderate in a threefold ratio over the phanerogamia—pp. 287–291.

Considerations on the proportion borne by the number of the phanerogamia actually ascertained, to the entire number existing on the globe—pp. 291–295.

Influence of the pressure of atmospheric strata on the form and life of plants, with reference to Alpine vegetation—pp. 295–296.

Specialities on the plant-forms already enumerated. Physiognomy of plants discussed from three different points of view: the absolute difference of the forms, their local preponderance in the sum total of the phanerogamic Floras, and their geographical as well as climatic dispersion—pp. 296–346. Greatest height of arboral plants; examples of 223 to 246 feet in Pinus Lambertiana and P. Douglasii, of 266 in P. Strobus, of 300 feet in Sequoia gigantea and Pinus trigona. All these examples are from the north-western part of the New Continent. The Araucaria excelsa of Norfolk Island, accurately measured, rises only from 182 to 223 feet; the Alpine palms of the Cordilleras (Ceroxylon andicola), only 190 feet—pp. 322–324. A contrast to these gigantic vegetable forms, presented not merely by the stem of the arctic willow (Salix arctica, two inches in height,) stunted by cold and exposure on the mountains, but also in the tropical plains by the Tristicha hypnoides, a phanerogamic plant which is hardly three French lines (quarter of an inch) in height, when fully developed—pp. 324–325.

Bursting forth of blossoms from the rough bark of the Crescentia Cujete, of the Gustavia augusta, from the roots of the Cacao tree. The largest blossoms borne by the Rafflesia Arnoldi, Aristolochia cordata, Magnolia, Helianthus annuus—p. 348.

The different forms of plants determine the scenic character of vegetation in the different zones. Physiognomic classification, or distribution of the groups according to external facies, is from its basis of arrangement entirely different from the classification according to the system of natural families. The physiognomy of plants is based principally on the so-called organs of vegetation, on which the preservation of the individual depends; systematic botany bases the classification of the natural families on the consideration of the organs of reproduction, on which the preservation of the species depends—pp. 348–352.

## ON THE STRUCTURE AND MODE OF ACTION OF VOLCANOS IN DIFFERENT PARTS OF THE EARTH—pp. 353–375.

Influence of travels in distant lands on the generalization of our ideas and on the progress of physical orology. Influence of the conformation of the Mediterranean on the earliest ideas respecting volcanic phenomena.—COMPARATIVE GEOLOGY OF VOLCANOS. Periodical return of certain revolutions in nature, the cause of which lies deep in the interior of the globe. Proportion of the height of volcanos to that of their cone of ashes in the Pichincha, Peak of Teneriffe, and Vesuvius. Changes in the height of volcanic mountain summits. Measurements of the margins of the crater of Vesuvius from 1773 to 1822; the author's measurements embrace the period from 1805 to 1822—pp. 353–365. Circumstantial description of the eruption in the night between the 24th and 25th of October, 1822. Falling in of a cone of ashes more than 400 feet high, which stood in the interior of the crater. The eruption of ashes from the 24th to the 28th of October, was the most memorable among those, of which authentic accounts are possessed, since the time of the elder Pliny—pp. 365–371.

Difference between volcanos that are of very diverse forms, with permanent craters, and the phenomena more rarely observed in historic times, in which trachytic mountains suddenly open, eject lava and ashes, and reclose, perhaps for ever. The latter phenomena are peculiarly instructive for geognosy, because they remind us of the earliest revolutions that occurred in the oscillating, upheaved, fissured surface of the earth. In ancient times they led to the notion of the Pyriphlegethon. Volcanos are intermittent earth-springs, the result of a permanent or transitory connection between the interior and exterior of our planet, the result of a reaction of the still fluid interior against the crust of the earth; hence the question is useless, as to what chemical substance burns in the volcanos, and furnishes the material for combustion—pp. 371–373. The primary cause of subterranean heat is, as in all planets, the formative process itself, the separation of the conglomerating mass from a cosmic vaporous fluid. Power and influence of the calorific radiation from numerous deep fissures, unfilled veins in the primordial world. Great independence, at that period, of the climate (atmospheric temperature) in respect to geographical latitude, the position of the planet towards the central body, the sun. Organisms of the present tropical world buried in the icy north—pp. 373–375.

SCIENTIFIC ILLUSTRATIONS AND ADDITIONS . . . . pp. 376–379.

Barometric measurements on Vesuvius, comparison of the two crater-margins and the Rocca del Palo—pp. 376–379. Increase of temperature with depth, being 1° of Fahrenheit for every 54 feet. Temperature of the Artesian well in Oeynhausen's Bath (New Salt-works near Minden), at the greatest depth yet reached below the level of the sea. As early as the third century the thermal springs near Carthage led Patricius, Bishop of Pertusa, to form correct suppositions respecting the cause of calorific increase in the interior of the earth—p. 379.

VITAL FORCE, OR THE RHODIAN GENIUS; AN ALLEGORY. pp. 380–385.

ILLUSTRATIONS AND NOTE . . . . . . . pp. 386–389.

The Rhodian Genius is the development of a physiological idea in a mythical garb. Difference of views concerning the necessity and non-necessity for the assumption of peculiar vital forces—pp. 386–387. The difficulty of satisfactorily reducing the vital phenomena of the organism to physical and chemical laws is, principally, based on the complexity of the phenomena, on the multiplicity of forces acting simultaneously, as well as on the varying conditions of their activity. Definition of the expressions, *animate* and *inanimate* matter. Criteria of the miscent state ensuing upon separation, are the simple enunciation of a fact—pp. 387–389.

## THE PLATEAU OF CAXAMARCA, THE ANCIENT CAPITAL OF THE INCA ATAHUALLPA, AND FIRST VIEW OF THE PACIFIC FROM THE RIDGE OF THE ANDES. pp. 390–420.

Cinchona, or Quina-woods in the valleys of Loxa. First use of the fever-bark in Europe; the Vice-Queen Countess of Chinchon—pp. 390–392.

Alpine vegetation of the Paramos. Ruins of ancient Peruvian causeways; they rise in the Paramo del Assuay almost to the height of Mont Blanc—p. 394. Singular mode of communication, by a swimming courier—p. 399.

Descent to the Amazon River. Vegetation around Chamaya and Tomependa; red groves of Bougainvillæa. Rocky ridges which cross the Amazon River. Cataracts. Narrows of the Pongo de Manseriche, in which the mighty stream, measured by La Condamine, is hardly 160 feet broad. Fall of the rocky dam of Rentema, which for several hours, laid bare the bed of the river, to the terror of the inhabitants on its banks—p. 401.

Passage across the Andes chain, where it is intersected by the magnetic equator. Ammonites of nearly 15 inches, Echini and Isocardia of the chalk-formation, collected between Guambos and Montan, nearly 12,800 feet above the sea. Rich silver-mines of Chota. The picturesque, tower-like Cerro de Gualgayoc. An enormous mass of filamentous virgin silver in the Pampa de Navar. A treasure of virgin gold, twined round with filamentous silver, in the shell-field (Choropampa), so named on account of the numerous fossils. Outbursts of silver and gold ores in the chalk-formations. The little mountain-town of Micuipampa lies 11,873 feet above the sea—pp. 402–405.

Across the mountain wilderness of the Paramo de Yanaguanga the traveller descends into the beautiful embosomed valley or rather Plateau of Caxamarca (almost at an equal altitude with the city of Quito). Warm baths of the Inca. Ruins of Atahuallpa's palace, inhabited by his indigent descendants, the family of Astorpilca. Belief entertained there, in the existence of subterranean golden gardens of the Inca; said to be situated in the lovely valley of Yucay, under the Temple of the Sun at Cuzco, and at many other points. Conversation with the son of the Curaca Astorpilca. The room is still shown in which the unfortunate Atahuallpa was kept prisoner for nine months, from the November of 1532; also the wall on which he made a mark to indicate the height to which he would cause the room to be filled with gold, if his persecutors would set him free. Account of the prince's execution on the 29th of August, 1533, and remarks on the so-called "indelible blood stain" on a stone slab before the altar in the chapel of the city prison —pp. 406–414. How the hope in a restoration of the Inca empire, also indulged in by Raleigh, has been maintained among the natives. Causes of this fanciful belief—p. 414.

Journey from Caxamarca to the sea-coast. Passage across the Cordilleras through the Altos de Guangamarca. The often disappointed hope of enjoying the sight of the Pacific from the crest of the Andes, at last gratified, at a height of 9380 feet—pp. 415–420.

SCIENTIFIC ILLUSTRATIONS AND ADDITIONS . . . pp. 421–436.

On the origin of the name borne by the Andes Chain . . p. 421.
Epoch of the introductiou of Cinchona (Peruvian) bark into Europe —p. 422.

Ruins of the Inca's causeways and fortified dwellings; Aposentos de Mulalo, Fortaleza del Canar, Inti-Guaycu—p. 423.

On the ancient civilization of the Chibchas or Muyscas of New Granada—p. 425. Age of the culture of the potato and banana—p. 427. Etymology of the word Cundinamarca, corrupted from Cundirumarca, and which, in the first years of republican independence, designated the whole country of New Granada—p. 427.

Chronometric connection of the city of Quito with Tomependa, on the upper course of the Amazon River, and the Callao de Lima, the position of which was accurately determined by the transit of Mercury on the 9th of November, 1802—p. 428.

On the tedious court ceremonies of the Incas. Atahuallpa's imprisonment and unavailing ransom—p. 429.

Free-thinking of the Inca Huayna Capac. Philosophical doubts on the official worship of the sun, and obstacles to the diffusion of knowledge among the lower and poorer classes of people, according to the testimony of Padre Blas Valera—p. 431.

Raleigh's project for the restoration of the Inca dynasty under English protection, which should be granted for an annual tribute of several hundred thousand pounds—p. 432.

Columbus' earliest evidence of the existence of the Pacific. It was first seen on the 25th of September, 1513, by Vasco Nunez de Balboa, and first navigated by Alonso Martin de Don Benito—p. 432.

On the possibility of constructing an Oceanic canal through the isthmus of Panama (with fewer locks than the Caledonian Canal). Points, the exploration of which has been hitherto totally neglected—p. 435.

Determination of the longitude of Lima—p. 435.

# ON STEPPES AND DESERTS.

At the foot of the lofty granitic range which, in the early age of our planet, resisted the irruption of the waters on the formation of the Caribbean Gulf, extends a vast and boundless plain. When the traveller turns from the Alpine valleys of Caracas, and the island-studded lake of Tacarigua (1), whose waters reflect the forms of the neighbouring bananas,—when he leaves the fields verdant with the light and tender green of the Tahitian sugar-cane, or the sombre shade of the cacoa groves,—his eye rests in the south on Steppes, whose seeming elevations disappear in the distant horizon.

From the rich luxuriance of organic life the astonished traveller suddenly finds himself on the dreary margin of a treeless waste. Nor hill, nor cliff rears its head, like an island in the ocean, above the boundless plain: only here and there broken strata of floetz, extending over a surface of two hundred square miles, (more than three thousand English square miles*,) appear sensibly higher than the surrounding district. The natives term them *banks* (2), as if the spirit of language would convey some record of that ancient condition of the world, when these elevations formed the shoals, and the Steppes themselves the bottom, of some vast inland sea.

Even now, illusion often recalls, in the obscurity of night, these images of a former age. For when the guiding constellations illumine the margin of the plain with their rapidly rising and setting beams, or when their flickering forms are

* It is not intended in every instance to trouble the reader with duplicate measurements; but they will be introduced occasionally Wherever only one measurement is given, it must be understood as English.—Ed.

reflected in the lower stratum of undulating vapour, a shoreless ocean seems spread before us (3). Like a limitless expanse of waters, the Steppe fills the mind with a sense of the infinite, and the soul, freed from the sensuous impressions of space, expands with spiritual emotions of a higher order. But the aspect of the ocean, its bright surface diversified with rippling or gently swelling waves, is productive of pleasurable sensations,—while the Steppe lies stretched before us, cold and monotonous, like the naked stony crust of some desolate planet (4).

In all latitudes nature presents the phenomenon of these vast plains, and each has some peculiar character or physiognomy, determined by diversity of soil and climate, and by elevation above the level of the sea.

In northern Europe the Heaths which, covered by one sole form of vegetation, to the exclusion of all others, extend from the extremity of Jutland to the mouth of the Scheldt, may be regarded as true Steppes. They are, however, both hilly and of very inconsiderable extent when compared with the Llanos and Pampas of South America, or even with the Prairies on the Missouri (5) and Copper River, the resort of the shaggy Bison and the small Musk Ox.

The plains in the interior of Africa present a grander and more imposing spectacle. Like the wide expanse of the Pacific, they have remained unexplored until recent times. They are portions of a sea of sand, which towards the east separates fruitful regions from each other, or incloses them like islands, as the desert near the basaltic mountains of Harudsch (6), where, in the Oasis of Siwah, rich in date-trees, the ruins of the temple of Ammon indicate the venerable seat of early civilization. Neither dew nor rain refreshes these barren wastes, or unfolds the germs of vegetation within the glowing depths of the earth; for everywhere rising columns of hot air dissolve the vapours and disperse the passing clouds.

Wherever the desert approaches the Atlantic Ocean, as

between Wadi Nun and the White Cape, the moist sea-air rushes in to fill the vacuum caused by these vertically ascending currents of air. The navigator, in steering towards the mouth of the river Gambia, through a sea thickly carpeted with weeds, infers by the sudden cessation of the tropical east wind (7), that he is near the far-spreading and radiating sandy desert.

Flocks of swift-footed ostriches and herds of gazelles wander over this boundless space. With the exception of the newly discovered group of Oases, rich in springs, whose verdant banks are frequented by nomadic tribes of Tibbos and Tuaricks (8), the whole of the African deserts may be regarded as uninhabitable by man. It is only periodically that the neighbouring civilized nations venture to traverse them. On tracks whose undeviating course was determined by commercial intercourse thousands of years ago, the long line of caravans passes from Tafilet to Timbuctoo, or from Mourzouk to Bornou; daring enterprises, the practicability of which depends on the existence of the camel, *the ship of the desert* (9), as it is termed in the ancient legends of the East.

These African plains cover an area which exceeds almost three times that of the neighbouring Mediterranean. They are situated partly within and partly near the tropics, a position on which depends their individual natural character. On the other hand, in the eastern portion of the old continent the same geognostic phenomenon is peculiar to the temperate zone.

On the mountainous range of Central Asia, between the Gold or Altai Mountain and the Kouen-lien (10), from the Chinese wall to the further side of the Celestial Mountains, and towards the Sea of Aral, over a space of several thousand miles, extend, if not the highest, certainly the largest Steppes in the world. I myself enjoyed an opportunity, full thirty years after my South American travels, of visiting that portion of the Steppes which is occupied by Kalmuck-Kirghis tribes, and is situated between the Don, the Volga, the Caspian Sea, and the Chinese Lake of Dsaisang, and which consequently extends over an area of nearly 2,800 geographical miles. The vegetation of the Asiatic Steppes, which are sometimes hilly and interspersed with pine forests, is in its groupings far more varied than that of the Llanos and the Pampas of Caracas and Buenos Ayres. The more beautiful portions of the plains, inhabited by Asiatic pastoral tribes, are adorned with lowly shrubs of luxuriant white-blossomed Rosaceæ, Crown Imperials (Fritillariæ), Cypripedeæ, and Tulips. As the torrid zone is in general distinguished by a tendency in the vegetable forms to become arborescent, so we also find. that some of the Asiatic Steppes of the temperate zone are characterized by the remarkable height to which flowering plants attain; as, for instance, Saussureæ, and other Synanthereæ; all siliquose plants, and particularly numerous species of Astragalus. On crossing the trackless portions of the herb-covered Steppes in the low carriages of the Tartars, it is necessary to stand upright in order to ascertain the direction to be pursued through the copse-like and closely crowded plants that bend under the wheels. Some of these Steppes are covered with grass; others with succulent, evergreen, articulated alkaline plants; while many are radiant with the effulgence of lichen-like tufts of salt, scattered irregularly over the clayey soil like newly fallen snow.

These Mongolian and Tartar Steppes, which are intersected by numerous mountain chains, separate the ancient and long-civilized races of Thibet and Hindostan from the rude nations of Northern Asia. They have also exerted a manifold influence on the changing destinies of mankind. They have inclined the current of population southward, impeded the intercourse of nations more than the Himalayas, or the Snowy Mountains of Sirinagur and Gorka, and placed permanent limits to the progress of civilization and refinement in a northerly direction.

History cannot, however, regard the plains of Central Asia

under the character of obstructive barriers alone. They have frequently proved the means of spreading misery and devastation over the face of the earth. Some of the pastoral tribes inhabiting this Steppe,—the Mongols, Getæ, Alani, and Usüni,—have convulsed the world. If in the course of earlier ages, the dawn of civilization spread like the vivifying light of the sun from east to west; so in subsequent ages and from the same quarter, have barbarism and rudeness threatened to overcloud Europe.

A tawny tribe of herdsmen (11) of Tukiuish *i. e.*, Turkish origin, the Hiongnu, dwelt in tents of skins on the elevated Steppe of Gobi. A portion of this race had been driven southward towards the interior of Asia, after continuing for a long time formidable to the Chinese power. This shock, (dislodgement of the tribes) was communicated uninterruptedly as far as the ancient land of the Fins, near the sources of the Ural.* From thence poured forth bands of Huns, Avars, Chasars, and a numerous admixture of Asiatic races. Warlike bodies of Huns first appeared on the Volga, next in Pannonia, then on the Marne and the banks of the Po, laying waste those richly cultivated tracts, where, since the age of Antenor, man's creative art had piled monument on monument. Thus swept a pestilential breath from the Mongolian deserts over the fair Cisalpine soil, stifling the tender, long-cherished blossoms of art!

From the Salt-steppes of Asia,—from the European Heaths,—smiling in summer with their scarlet, honey-yielding flowers,—and from the barren deserts of Africa, we return to the plains of South America, the picture of which I have already begun to sketch in rude outline.

* The Huns, on being driven from their ancient pastures by the Chinese, traversed Asia, 1300 leagues,) and, swelled by the numerous hordes they conquered *en route*, entered Europe, and gave the first impulse to the great migration of nations. Deguires traces their progress with geographical minuteness, and Gibbon tells their story with his usual eloquence in Chap. XXVI.—Ed.

But the interest yielded by the contemplation of such a picture must arise from a pure love of nature. No Oasis here reminds the traveller of former inhabitants, no hewn stone (12), no fruit-tree once cultivated and now growing wild, bears witness to the industry of past races. As if a stranger to the destinies of mankind, and bound to the present alone, this region of the earth presents a wild domain to the free manifestation of animal and vegetable life.

The Steppe extends from the littoral chain of Caracas to the forests of Guiana, and from the snow-covered mountains of Merida, on whose declivity lies the Natron lake of Urao,—the object of the religious superstition of the natives,—to the vast delta formed by the mouth of the Orinoco. To the south-west it stretches like an arm of the sea (13), beyond the banks of the Meta and of the Vichada, to the unexplored sources of the Guaviare, and to the solitary mountain group to which the vivid imagination of the Spanish warriors gave the name of *Paramo de la Suma Paz*, as though it were the beautiful seat of eternal repose.

This Steppe incloses an area of 256,000 square miles. Owing to inaccurate geographical data, it has often been described as extending in equal breadth to the Straits of Magellan, unmindful that it is intersected by the wooded plain of the Amazon, which is bounded to the north by the grassy Steppes of the Apure, and to the south by those of the Rio de la Plata. The Andes of Cochabamba and the Brazilian mountains approximate each other by means of separate transverse spurs, projecting between the province of Chiquitos and the isthmus of Villabella (14). A narrow plain unites the *Hylæa* of the Amazon with the Pampas of Buenos Ayres. The area of the latter is three times larger than that of the Llanos of Venezuela; indeed so vast in extent, that it is bounded on the north by palms, while its southern extremity is almost covered with perpetual ice. The Tuyu, which resembles the Cassowary, (Struthio Rhea,) is peculiar to these Pampas, as are also those herds of wild dogs (15), which dwell in social community in subterranean caverns, and often ferociously attack man, for whose defence their progenitors fought.

Like the greater part of the desert of Sahara (16), the Llanos, the most northern plains of South America, lie within the torrid zone. Twice in every year they change their whole aspect, during one half of it appearing waste and barren like the Lybian desert; during the other, covered with verdure, like many of the elevated Steppes of Central Asia (17).

The attempt to compare the natural characteristics of remote regions, and to pourtray the results of this comparison in brief outline, though a gratifying, is a somewhat difficult branch of physical geography.

A number of causes, many of them still but little understood (18), diminish the dryness and heat of the New World. Among these are: the narrowness of this extensively indented continent in the northern part of the tropics, where the fluid basis on which the atmosphere rests, occasions the ascent of a less warm current of air; its wide extension towards both the icy poles; a broad ocean swept by cool tropical winds; the flatness of the eastern shores; currents of cold sea-water from the antarctic region, which, at first following a direction from south-west to north-east, strike the coast of Chili below the parallel of 35° south lat., and advance as far north on the coasts of Peru as Cape Parina, where they suddenly diverge towards the west; the numerous mountains abounding in springs, whose snow-crowned summits soar above the strata of clouds, and cause the descent of currents of air down their declivities; the abundance of rivers of enormous breadth, which after many windings invariably seek the most distant coast; Steppes, devoid of sand, and therefore less readily acquiring heat; impenetrable forests, which, protecting the earth from the sun's rays, or radiating heat from the surface of their leaves, cover the richly-watered plains of the Equator, and exhale into the interior of the country, most remote from mountains and the

Ocean, prodigious quantities of moisture, partly absorbed and partly generated—all these causes produce in the flat portions of America a climate which presents a most striking contrast in point of humidity and coolness with that of Africa. On these alone depend the luxuriant and exuberant vegetation and that richness of foliage which are so peculiarly characteristic of the New Continent.

If, therefore, the atmosphere on one side of our planet be more humid than on the other, a consideration of the actual condition of things will be sufficient to solve the problem of this inequality. The natural philosopher need not shroud the explanation of such phenomena in the garb of geological myths. It is not necessary to assume that the destructive conflict of the elements raged at different epochs in the eastern and western hemispheres, during the early condition of our planet; or that America emerged subsequently to the other quarters of the world from the chaotic covering of waters, as a swampy island, the abode of crocodiles and serpents (19).

South America presents indeed a remarkable similarity to the south-western peninsula of the old continent, in the form of its outlines and the direction of its coast-line. But the internal structure of the soil, and its relative position with respect to the contiguous masses of land, occasion in Africa that remarkable aridity which over a vast area checks the development of organic life. Four-fifths of South America lie beyond the Equator, and therefore in a region which, on account of its abundant waters, as well as from many other causes, is cooler and moister than our northern hemisphere (20). To this, nevertheless, the most considerable portion of Africa belongs.

The extent from east to west of the South American Steppes cr Llanos, is only one third that of the African Desert. The former are refreshed by the tropical sea wind, while the latter, situated in the same parallel of latitude as Arabia and Southern Persia, are visited by currents of air which have passed over heat-radiating continents. The venerable father of history, Herodotus, so long insufficiently appreciated, has in the true spirit of a comprehensive observer of nature, described all the deserts of Northern Africa, Yemen, Kerman, and Mekran (the Gedrosia of the Greeks), as far even as Mooltan in Western India, as one sole connected sea of sand (21).

To the action of hot land winds, may be associated in Africa, as far as we know, a deficiency of large rivers, of forests that generate cold by exhaling aqueous vapour, and of lofty mountains. The only spot covered with perpetual snow is the western portion of Mount Atlas (22), whose narrow ridge, seen laterally, appeared to the ancient navigators when coasting the shore, as one solitary and aërial pillar of heaven. This mountain range extends eastward to Dakul, where the famed Carthage, once mistress of the seas, lies in crumbling ruins. This range forms a far extended coast-line or Gætulian rampart, which repels the cool north winds and with them the vapours rising from the Mediterranean.

The Mountains of the Moon, Djebel-al-Komr (23), fabulously represented as forming a mountainous parallel between the elevated plain of Habesch—an African Quito—and the sources of the Senegal, were supposed to rise above the lower sea line. Even the Cordilleras of Lupata, which skirt the eastern coast of Mozambique and Monomotapa, in the same manner as the Andes bound the western shores of Peru, are covered with eternal snow in the gold districts of Machinga and Mocanga. But these mountains, abundantly watered, are situated at a considerable distance from the vast desert which extends from the southern declivity of the chain of Atlas to the Niger, whose waters flow in an easterly direction.

Possibly, these combined causes of aridity and heat would have proved insufficient to convert such large portions of the African plains into a dreary waste, had not some convulsion of nature—as for instance the irruption of the ocean—on some occasion deprived these flat regions of their nutrient soil, as well as of the vegetation which it supported. The epoch when this occurred, and the nature of the forces which determined the irruption, are alike shrouded in the obscurity of the past. Perhaps it may have been the result of the great rotatory current (24), which drives the warmer waters of the Gulf of Mexico over the bank of Newfoundland to the old continent, and by which the cocoa-nut of the West Indies and other tropical fruits have been borne to the shores of Ireland and Norway. One branch of this oceanic current, after it leaves the Azores, has still, at the present time, a south-easterly course, striking the low range of the sandy coasts of Africa with a force that is frequently fraught with danger to the mariner. All sea-coasts—but I refer here more particularly to the Peruvian shore between Amotape and Coquimbo—afford evidence of the hundreds, or even thousands of years, which must pass before the moving sand can yield a firm basis for the roots of herbaceous plants, in those hot and rainless regions where neither Lecideæ nor other lichens can grow (25).

These considerations suffice to explain why, notwithstanding their external similarity of form, the continents of Africa and South America present the most widely different climatic relations and characters of vegetation. Although the South American Steppe is covered with a thin crust of fruitful earth, is periodically refreshed by rains, and adorned with luxuriant herbage, its attractions were not sufficient to induce the neighbouring nations to exchange the beautiful mountain valleys of Caracas, the sea-girt districts, and the richly watered plains of the Orinoco, for this treeless and springless desert. Hence on the arrival of the first European and African settlers, the Steppe was found to be almost without inhabitants.

The Llanos are, it is true, adapted for the breeding of cattle, but the primitive inhabitants of the new continent were almost wholly unacquainted with the management of animals yielding milk (26). Scarcely one of the American tribes knew how to avail themselves of the advantages which nature, in this respect, had placed before them. The American aborigines, who, from 65° north lat. to 55° south lat., constitute (with the exception, perhaps, of the Esquimaux,) but one sole race, passed directly from a hunting to an agricultural life without going through the intermediate stage of a pastoral life. Two species of indigenous horned cattle (the Buffalo and the Musk Ox) graze on the pasture lands of Western Canada and Quivira, as well as in the neighbourhood of the colossal ruins of the Aztek fortress, which rises like some American Palmyra on the desert solitudes of the river Gila. A long-horned *Mouflon*, resembling the so-called progenitor of the sheep, roams over the parched and barren limestone rocks of California; while the camel-like Vicunas, Huanacos, Alpacas, and Llamas, are natives of the southern peninsula. But of these useful animals the two first only (viz. the Buffalo and the Musk Ox) have preserved their natural freedom for thousands of years. The use of milk and cheese, like the possession and culture of farinaceous grasses, is a distinctive characteristic of the nations of the old world (27).

If some few tribes have passed through Northern Asia to the western coast of America, and preferring to keep within a temperate climate, have followed the course of the ridges of the Andes southward (28), such migrations must have been made by routes on which the settlers were unable to transport either flocks or grain. The question here arises, whether on the downfall of the long-declining empire of the Hiongnu, the consequent migration of this powerful race may not have been the means of drawing from the north-east of China and Korea, bands of settlers, by whom Asiatic civilisation was transported to the new continent? If the primitive colonists had been natives of those Steppes in which agriculture was unknown, this bold hypothesis (which as yet is but little warranted by etymological comparisons) would at all events explain the remarkable absence of the Cereals in America. Perhaps contrary winds may have driven to the shores of New California one of those Asiatic Priest-colonies who were instigated by their mystic dreameries to undertake distant voyages, and of which the history of the peopling of Japan, at the time of the *Thsinschihuang-ti*, affords a memorable instance. (29)

If a pastoral life—that beneficent intermediate stage which binds nomadic bands of hunters to fruitful pasture lands, and at the same time promotes agriculture—was unknown to the primitive races of America, it is to the very ignorance of such a mode of life that we must attribute the scantiness of population in the South American Steppes. But this circumstance allowed freer scope for the forces of nature to develop themselves in the most varied forms of animal life; a freedom only circumscribed by themselves, like vegetable life in the forests of the Orinoco, where the Hymenæa and the giant laurel, exempt from the ravages of man, are only in danger of a too luxuriant embrace of the plants which surround them.

Agoutis, small spotted antelopes, the shielded Armadillo, which, rat-like, terrifies the hare in its subterranean retreat; herds of slothful Chiguires, beautifully striped Viverræ, whose pestilential odour infects the air; the great maneless Lion; the variegated Jaguar (commonly known as the tiger), whose strength enables it to drag to the summit of a hill the body of the young bull it has slain—these, and many other forms of animal life (30), roam over the treeless plain.

This region, which may be regarded as peculiarly the habitation of wild animals, would not have been chosen as a place of settlement by nomadic hordes, who like the Indo-Asiatics generally prefer a vegetable diet, had it not possessed some few fan-palms (*Mauritia*) scattered here and there. The beneficent qualities of this tree of life have been universally celebrated (31.) Upon this alone subsist the unsubdued tribe of the Guaranes, at the mouth of the Orinoco northward of the Sierra de Imataca. When they increased in numbers and became over-crowded, it is said that, besides the huts which they built on horizontal platforms supported by the stumps of felled palm-trees, they also ingeniously suspended from stem to stem spreading mats or hammocks woven of the leaf-stalk of the Mauritia, which enabled them, during the rainy season, when the Delta was overflowed, to live in trees in the manner of apes. These pendent huts were partly covered with clay. The women kindled the fire necessary for their culinary occupations on the humid flooring. As the traveller passed by night along the river, his attention was attracted by a long line of flame suspended high in the air, and apparently unconnected with the earth. The Guaranes owe the preservation of their physical, and perhaps even of their moral independence, to the loose marshy soil, over which they move with fleet and buoyant foot, and to their lofty sylvan domiciles; a sanctuary whither religious enthusiasm would hardly lead an American Stylite (32).

The Mauritia not only affords a secure habitation, but likewise yields numerous articles of food. Before the tender spathe unfolds its blossoms on the male palm, and only at that peculiar period of vegetable metamorphosis, the medullary portion of the trunk is found to contain a sago-like meal, which like that of the Jatropha root, is dried in thin bread-like slices. The sap of the tree when fermented constitutes the sweet inebriating palm-wine of the Guaranes. The narrow-scaled fruit, which resembles reddish pine-cones, yields, like the banana and almost all tropical fruits, different articles of food, according to the periods at which it is gathered, whether its saccharine properties are fully matured, or whether it is still in a farinaceous condition. Thus in the lowest grades of man's development, we find the existence of an entire race dependent upon almost a single tree; like certain insects which are confined to particular portions of a flower.

Since the discovery of the new continent, its plains (Llanos) have become habitable to man. Here and there towns (33) have sprung up on the shores of the Steppe-rivers, built to facilitate the intercourse between the coasts and Guiana (the Orinoco district). Everywhere throughout these vast districts the inhabitants have begun to rear cattle. At distances of a day's journey from each other, we see detached huts, woven together with reeds and thongs, and covered with ox-hides. Innumerable herds of oxen, horses, and mules (estimated at the peaceful period of my travels at a million and a half) roam over the Steppe in a state of wildness. The prodigious increase of these animals of the old world is the more remarkable, from the numerous perils with which, in these regions, they have to contend.

When, beneath the vertical rays of the bright and cloudless sun of the tropics, the parched sward crumbles into dust, then the indurated soil cracks and bursts as if rent asunder by some mighty earthquake. And if, at such a time, two opposite currents of air, by conflict moving in rapid gyrations, come in contact with the earth, a singular spectacle presents itself. Like funnel-shaped clouds, their apexes (34) touching the earth, the sands rise in vapoury form through the rarefied air in the electrically-charged centre of the whirling current, sweeping on like the rushing water-spout, which strikes such terror into the heart of the mariner. A dim and sallow light gleams from the lowering sky over the dreary plain. The horizon suddenly contracts, and the heart of the traveller sinks with dismay as the wide Steppe seems to close upon him on all sides. The hot and dusty earth forms a cloudy veil which shrouds the heavens from view, and increases the stifling oppression of the atmosphere (35); while the east wind, when it blows over the long-heated soil, instead of cooling, adds to the burning glow.

Gradually, too, the pools of water, which had been protected from evaporation by the now seared foliage of the fan-palm, disappear. As in the icy north animals become torpid from cold, so here the crocodile and the boa-constrictor lie wrapt in unbroken sleep, deeply buried in the dried soil. Everywhere the drought announces death, yet everywhere the thirsting wanderer is deluded by the phantom of a moving, undulating, watery surface, created by the deceptive play of the reflected rays of light (the mirage, 36). A narrow stratum separates the ground from the distant palm-trees, which seem to hover aloft, owing to the contact of currents of air having different degrees of heat and therefore of density*. Shrouded in dark clouds of dust, and tortured by hunger and burning thirst, oxen and horses scour the plain, the one bellowing dismally, the other with outstretched necks snuffing the wind, in the endeavour to detect, by the moisture in the air, the vicinity of some pool of water not yet wholly evaporated.

The mule, more cautious and cunning, adopts another method of allaying his thirst. There is a globular and articulated plant, the Melocactus (37), which encloses under its prickly integument an aqueous pulp. After carefully striking away the prickles with his forefeet, the mule cautiously ventures to apply his lips to imbibe the cooling thistle juice. But the draught from this living vegetable spring is not always unattended by danger, and these animals are often observed to have been lamed by the puncture of the cactus thorn.

Even if the burning heat of day be succeeded by the cool freshness of the night, here always of equal length, the wearied ox and horse enjoy no repose. Huge bats now attack the animals during sleep, and vampyre-like suck their blood;† or, fastening on their backs, raise festering wounds, in which mosquitoes, hippobosces, and a host of other stinging insects, burrow and nestle. Such is the miserable existence of these

* This effect is well represented in Grindlay's *Scenery of the Western Side of India*, plate 18.—Ed.

† Modern naturalists affirm that *all* bats are insectivorous.—Ed.

poor animals when the heat of the sun has absorbed the waters from the surface of the earth.

When, after a long drought, the genial season of rain arrives, the scene suddenly changes (38). The deep azure of the hitherto cloudless sky assumes a lighter hue. Scarcely can the dark space in the constellation of the Southern Cross be distinguished at night. The mild phosphorescence of the Magellanic clouds fades away. Even the vertical stars of the constellations Aquilà and Ophiuchus shine with a flickering and less planetary light. Like some distant mountain, a single cloud is seen rising perpendicularly on the southern horizon. Misty vapours collect and gradually overspread the heavens, while distant thunder proclaims the approach of the vivifying rain.

Scarcely is the surface of the earth moistened before the teeming Steppe becomes covered with Kyllingiæ, with the many-panicled Paspalum, and a variety of grasses. Excited by the power of light, the herbaceous Mimosa unfolds its dormant, drooping leaves, hailing, as it were, the rising sun in chorus with the matin song of the birds and the opening flowers of aquatics. Horses and oxen, buoyant with life and enjoyment, roam over and crop the plains. The luxuriant grass hides the beautifully spotted Jaguar, who, lurking in safe concealment, and carefully measuring the extent of the leap, darts, like the Asiatic tiger, with a cat-like bound on his passing prey.

At times, according to the account of the natives, the humid clay on the banks of the morasses (39), is seen to rise slowly in broad flakes. Accompanied by a violent noise, as on the eruption of a small mud-volcano, the upheaved earth is hurled high into the air. Those who are familiar with the phenomenon fly from it; for a colossal water-snake or a mailed and scaly crocodile, awakened from its trance by the first fall of rain, is about to burst from his tomb.

When the rivers bounding the plain to the south, as the Arauca, the Apure, and the Payara, gradually overflow their banks, nature compels those creatures to live as amphibious animals, which, during the first half of the year, were perishing with thirst on the waterless and dusty plain. A part of the steppe now presents the appearance of a vast inland sea (40). The mares retreat with their foals to the higher banks, which project, like islands, above the spreading waters. Day by day the dry surface diminishes in extent. The cattle, crowded together, and deprived of pasturage, swim for hours about the inundated plain, seeking a scanty nourishment from the flowering panicles of the grasses which rise above the lurid and bubbling waters. Many foals are drowned, many are seized by crocodiles, crushed by their serrated tails, and devoured. Horses and oxen may not unfrequently be seen which have escaped from the fury of this bloodthirsty and gigantic lizard, bearing on their legs the marks of its pointed teeth.

This spectacle involuntarily reminds the contemplative observer of the adaptability granted by an all-provident nature to certain animals and plants. Like the farinaceous fruits of Ceres, the ox and horse have followed man over the whole surface of the earth—from the Ganges to the Rio de la Plata, and from the sea coast of Africa to the mountainous plain of Antisana, which lies higher than the Peak of Teneriffe (41). in the one region the northern birch, in the other the date-palm, protects the wearied ox from the noonday sun. The same species of animal which contends in eastern Europe with bears and wolves, is exposed, in a different latitude, to the attacks of tigers and crocodiles!

The crocodile and the jaguar are not, however, the only enemies that threaten the South American horse; for even among the fishes it has a dangerous foe. The marshy waters of Bera and Rastro (42) are filled with innumerable electric eels, who can at pleasure discharge from every part of their slimy, yellow-speckled bodies a deadening shock. This species of gymnotus is about five or six feet in length. It is powerful enough to kill the largest animals when it discharges its nervous organs at one shock in a favourable direction. It was once found necessary to change the line of road from Urituсu across the Steppe, owing to the number of horses which, in fording a certain rivulet, annually fell a sacrifice to these gymnoti, which had accumulated there in great numbers. All other species of fish shun the vicinity of these formidable creatures. Even the angler, when fishing from the high bank, is in dread lest an electric shock should be conveyed to him along the moistened line. Thus, in these regions, the electric fire breaks forth from the lowest depths of the waters.

The mode of capturing the gymnotus affords a picturesque spectacle. A number of mules and horses are driven into a swamp, which is closely surrounded by Indians, until the unusual noise excites the daring fish to venture on an attack. Serpent-like they are seen swimming along the surface of the water, striving cunningly to glide under the bellies of the horses. By the force of their invisible blows numbers of the poor animals are suddenly prostrated; others, snorting and panting, their manes erect, their eyes wildly flashing with terror, rush madly from the raging storm; but the Indians, armed with long bamboo staves, drive them back into the midst of the pool.

By degrees the fury of this unequal contest begins to slacken. Like clouds that have discharged their electricity, the wearied eels disperse. They require long rest and nourishing food to repair the galvanic force which they have so lavishly expended. Their shocks gradually become weaker and weaker. Terrified by the noise of the trampling horses, they timidly approach the brink of the morass, where they are wounded by harpoons, and drawn on shore by non-conducting poles of dry wood.

Such is the remarkable contest between horses and fish. That which constitutes the invisible but living weapon of these inhabitants of the water—that, which awakened by the contact of moist and dissimilar particles (43), circulates through all the organs of animals and plants—that which flashing amid the roar of thunder illuminates the wide canopy of heaven—which binds iron to iron, and directs the silent recurring course of the magnetic needle—all, like the varied hues of the refracted ray of light, flow from one common source, and all blend together into one eternal all-pervading power.

I might here close my bold attempt of delineating the natural picture of the Steppe; but, as on the ocean, fancy delights in dwelling on the recollections of distant shores, so will we, ere the vast plain vanishes from our view, cast a rapid glance over the regions by which the Steppe is bounded.

The northern desert of Africa separates two races of men which originally belonged to the same portion of the globe, and whose inextinguishable feuds appear as old as the myth of Osiris and Typhon (44). To the north of Mount Atlas there dwells a race characterised by long and straight hair, a sallow complexion, and Caucasian features; while to the south of Senegal, in the direction of Soudan, we find hordes of Negroes occupying various grades in the scale of civilization. In Central Asia the Mongolian Steppe divides Siberian barbarism from the ancient civilization of the peninsula of Hindostan.

In like manner, the South American Steppes are the boundaries of a European semi-civilization (45). To the north, between the mountain chain of Venezuela and the Caribbean Sea, lie, crowded together, industrial cities, clean and neat villages, and carefully tilled fields. Even a taste for arts, scientific culture, and a noble love of civil freedom, have long since been awakened within these regions.

To the south, a drear and savage wilderness bounds the Steppe. Forests, the growth of thousands of years, in one impenetrable thicket, overspread the marshy region between the rivers Orinoco and Amazon. Huge masses of lead-coloured granite (46) contract the beds of the foaming rivers. Mountains and forests re-echo with the thunder of rushing waters, the roar of the tiger-like jaguar, and the dull rain-foreboding howl of the bearded ape (47).

Where the shallower parts of the river disclose a sandbank, the crocodile may be seen, with open jaws, and motionless as a rock, its uncouth body often covered with birds (48); while the chequered boa-constrictor, its tail lashed round the trunk of a tree, lies coiled in ambush near the bank, ready to dart with certain aim on its prey. Rapidly uncoiling, it stretches forth its body to seize the young bull, or some feebler prey, as it fords the stream, and moistening its victim with a viscid secretion, laboriously forces it down its dilating throat (49).

In this grand and wild condition of nature dwell numerous races of men. Separated by a remarkable diversity of languages, some are nomadic, unacquainted with agriculture, and living on ants, gums, and earth, mere outcasts of humanity (50), such as the Ottomaks and Jarures: others, for instance the Maquiritares and Macos, have settled habitations, live on fruits cultivated by themselves, are intelligent, and of gentler manners. Extensive tracts between the Cassiquiare and the Atabapo are inhabited solely by the Tapir and social apes; not by man. Figures graven on the rocks (51) attest that even these deserts were once the seat of a higher civilization. They bear testimony, as do also the unequally developed and varying languages (which are amongst the oldest and most imperishable of the historical records of man), to the changing destinies of nations.

While on the Steppe tigers and crocodiles contend with horses and cattle, so on the forest borders and in the wilds of Guiana the hand of man is ever raised against his fellow man. With revolting eagerness, some tribes drink the flowing blood of their foes, whilst others, seemingly unarmed, yet prepared for murder, deal certain death with a poisoned thumb-nail (52). The feebler tribes, when they tread the sandy shores, carefully efface with their hands the traces of their trembling steps.

Thus does man, everywhere alike, on the lowest scale of brutish debasement, and in the false glitter of his higher culture, perpetually create for himself a life of care. And thus, too, the traveller wandering over the wide world by sea and land, and the historian who searches the records of bygone ages, are everywhere met by the unvarying and melancholy spectacle of man opposed to man.

He, therefore, who amid the discordant strife of nations, would seek intellectual repose, turns with delight to contemplate the silent life of plants, and to study the hidden forces of nature in her sacred sanctuaries; or yielding to that inherent impulse, which for thousands of years has glowed in the breast of man, directs his mind, by a mysterious presentiment of his destiny, towards the celestial orbs, which, in undisturbed harmony, pursue their ancient and eternal course.*

* Ipsa suæ meminit stirpis, seseque Deisque
Mens fruitur felix, et novit in astra reverti.
*Barclaii Argenis,* lib. v. Ed.

---

## ILLUSTRATIONS AND ADDITIONS.

### (1) p. 1—" *The Lake of Tacarigua.*"

On advancing through the interior of South America, from the coast of Caracas or of Venezuela towards the Brazilian frontier (from the 10th degree of north latitude to the equator), the traveller first passes a lofty chain of mountains (the littoral chain of Caracas) inclining from west to east; next vast treeless Steppes or plains (*Los Llanos*), which extend from the foot of the littoral chain to the left bank of the Orinoco; and, lastly, the mountain range which gives rise to the cataracts of Atures and Maypure. This mountain chain, which I have named the Sierra Parime, passes in an easterly direction between the sources of the Rio Branco and Rio Esquibo, in the direction of Dutch and French Guiana. This region, which is the seat of the marvellous myths of the Dorado, and is composed of a mountain mass, divided into numerous gridiron-like ridges, is bounded on the south by the woody plain through which the Rio Negro and the Amazon have formed themselves a channel. Those who would seek further instruction regarding these geographical relations, may compare the large chart of La Cruz Olmedilla (1775), which has served as the basis of nearly all the more modern maps of South America, with that of Columbia, which I drew up in accordance with my own astronomical determinations of place, and published in the year 1825.

The littoral chain of Venezuela is, geographically considered, a portion of the Peruvian Andes. These are divided at the great mountain node of the sources of the Magdalena (lat. 1° 55′ to 2° 20′) into three chains, running to the south of Popayan, the easternmost of which extends into the snowy mountains of Merida. These mountains gradually decline towards the Paramo de las Rosas into the hilly district of Quibor and Tocuyo, which connects the littoral chain of Venezuela with the Cordilleras of Cundinamarca.

This littoral chain extends murally and uninterruptedly from Portocabello to the promontory of Paria. Its mean elevation is scarcely 750 toises, or 4796 English feet; but some few summits, like the Silla de Caracas (also called the Cerro de Avila), which is adorned with the purple-flowering Befaria (the red-blossomed American Alpine rose), rise 1350 toises, or 8633 English feet above the level of the sea. The coast of the Terra Firma everywhere bears traces of devastation, giving evidence of the action of the great current which runs from east to west, and which, after the disintegration of the Caribbean Islands, formed the present Sea of the Antilles. The tongues of land of Araya and Chuparipari, and more especially the coasts of Cumana and New Barcelona, present to the geologist a remarkable aspect. The rocky islands of Boracha, Caracas, and Chimanas rise like beacon-towers from the sea, affording evidence of the fearful irruption of the waters against the shattered mountain chain. The Sea of the Antilles may once have been an inland sea, like the Mediterranean, which has suddenly been connected with the ocean. The islands of Cuba, Hayti, and Jamaica still exhibit the remains of the mountains of micaceous schist which formed the northern boundary of this lake. It is a remarkable fact that the highest peaks are situated at the very point where these islands approach one another the closest. It may be conjectured that the principal nucleus of the chain was situated between Cape Tiburon and Morant Point. The height of the copper mountains (montañas de cobre) near Saint Iago de Cuba has not yet been measured, but this range is probably higher than the Blue Mountains of Jamaica (1138 toises, or 7277 English feet), whose elevation somewhat exceeds that of the Pass of St. Gothard. I have already expressed my conjectures more fully regarding the valley-like form of the Atlantic Ocean, and the ancient connection of the continents, in a treatise written at Cumana, entitled *Fragment d'un Tableau géologique de l'Amérique méridionale*, which appeared in the *Journal de Physique*, *Messidor*, *an* IX. It is remarkable that Columbus himself makes mention, in his official report, of the connection between the course of the equinoctial current and the form of the coast-line of the Greater Antilles.*

The northern and more cultivated portion of the province of Caracas is a mountainous region. The marginal chain is divided, like that of the Swiss Alps, into many ranges, enclosing longitudinal valleys. The most remarkable among these is

* *Examen critique de l'Hist. de la Géographie*, t. iii., pp. 104—108.

the charming valley of Aragua, which produces an abundance of indigo, sugar, and cotton, and, what is perhaps the most singular of all, even European wheat. The southern margin of this valley is bounded by the beautiful Lake of Valencia, the ancient Indian name of which was Tacarigua. The contrast presented by its opposite shores gives it a striking resemblance to the Lake of Geneva. The barren mountains of Guigue and Guiripa have indeed less grandeur and solemnity of character than the Savoy Alps; but, on the other hand, the opposite shore, which is covered with bananas, mimosæ, and triplaris, far surpasses in picturesque beauty the vineyards of the Pays de Vaud. The lake is 10 leagues, (of which 20 form a degree of the Equator), *i.e.*, about 30 geographical miles, in length, and is thickly studded with small islands, which continually increase in size, owing to the evaporation being greater than the influx of fresh water. Within the last few years several sandbanks have even become true islands, and have acquired the significant name of *Las Aparecidas*, or the "*Newly Appeared.*" On the island of Cura the remarkable species of solanum is cultivated, which has edible fruit, and has been described by Willdenow (in his *Hortus Berolinensis*, 1816, Tab. xxvii.). The elevation of the Lake of Tacarigua above the level of the sea is almost 1400 French feet (according to my measurement, exactly 230 toises, *i.e.*, 1471 English feet) less than the mean height of the valley of Caracas. This lake has several species of fish peculiar to itself,* and ranks among the most beautiful and attractive natural scenes that I am acquainted with in any part of the earth. When bathing, Bonpland and myself were often terrified by the appearance of the *Bava*, a species of crocodile-lizard (*Dragonne?*), hitherto undescribed, from three to four feet in length, of repulsive aspect, but harmless to man. We found in the Lake of Valencia a *Typha*, perfectly identical with the European bulrush, the *Typha angustifolia*—a singular and highly important fact in reference to the geography of plants.

In the valleys of Aragua, skirting the lake, both varieties of the sugar-cane are cultivated, viz., the common *Caña criolla*, and the sp cies newly introduced from the South Sea, the *Caña de Otaheiti.* The latter variety is of a far lighter and

* See my *Observations de Zoologie et d'Anatomie comparée*, t. ii., pp. 179--181.

more beautiful green, and a field of it may be distinguished from the common sugar-cane at a great distance. Cook and George Forster were the first to describe it; but it would appear, from Forster's treatise on the edible plants of the South Sea Islands, that they were but little acquainted with the true value of this important product. Bougainville brought it to the Isle of France, whence it passed to Cayenne and (subsequently to the year 1792) to Martinique, Saint Domingo or Haiti, and many of the Lesser Antilles. The enterprising but unfortunate Captain Bligh transported it, together with the bread-fruit tree, to Jamaica. From Trinidad, an island contiguous to the continent, the new sugar-cane of the South Sea passed to the neighbouring coasts of Caracas. Here it has become of greater importance than the bread-fruit tree, which will probably never supersede so valuable and nutritious a plant as the banana. The Tahitian sugar-cane is more succulent than the common species, which is generally supposed to be a native of Eastern Asia. It likewise yields one-third more sugar on the same area than the *Caña criolla*, which is thinner in its stalk, and more crowded with joints. As, moreover, the West Indian Islands are beginning to suffer great scarcity of fuel (on the island of Cuba the sugar-pans are heated with orange-wood), the new plant acquires additional value from the fact of its yielding a thicker and more ligneous cane (*bagaso*). If the introduction of this new product had not been nearly simultaneous with the outbreak of the sanguinary Negro war in St. Domingo, the prices of sugar in Europe would have risen even higher than they did, owing to the interruption occasioned to agriculture and trade. The important question which here arises, whether the sugar-cane of Otaheiti, when removed from its indigenous soil, will not gradually degenerate and merge into the common sugar-cane, has been decided in the negative, from the experience hitherto obtained on this subject. In the island of Cuba a *caballeria*, that is to say, an area of 34,969 square toises (nearly 33 English acres), produces 870 cwt. of sugar, if it be planted with the Tahitian sugar-cane. It is remarkable enough that this important product of the South Sea Islands should be cultivated precisely in that portion of the Spanish colonies which is most remote from the South Sea. The voyage from the Peruvian shore to Otaheiti may be made in twenty-five days, and yet, at the period of my travels in Peru and Chili, the Tahitian sugar-cane was not yet known in those provinces. The natives of Easter Island, who suffer great distress from want of fresh water, drink the juice of the sugar-cane, and, what is very remarkable in a physiological point of view, likewise sea-water. On the Society, Friendly, and Sandwich Islands, the light green and thick stemmed sugar-cane is everywhere cultivated.

In addition to the *Caña de Otaheiti* and the *Caña criolla*, a reddish African sugar-cane is cultivated in the West Indies, which is known as the *Caña de Guinea*. It is less succulent than the common Asiatic variety, but its juice is esteemed especially well adapted for the preparation of rum.

In the province of Caracas the light green of the Tahitian sugar-cane forms a beautiful contrast with the dark shade of the cacao plantations. Few tropical trees have so thick a foliage as the *Theobroma Cacao*. This noble tree thrives best in hot and humid valleys. Extreme fertility of soil and insalubrity of atmosphere are as inseparably connected in South America as in Southern Asia. Nay, it has even been observed that in proportion as the cultivation of the land increases, and the woods are removed, the soil and the climate become less humid, and the cacao plantations thrive less luxuriantly. But while they diminish in numbers in the province of Caracas, they spread rapidly in the eastern provinces of New Barcelona and Cumana, more especially in the humid woody region lying between Cariaco and the Golfo Triste.

(2) p. 1—"*The natives term this phenomenon 'banks.'*"

The Llanos of Caracas are covered with a widely-extended formation of ancient conglomerate. On passing from the valleys of Aragua over the most southern range of the coast chain of Guigue and Villa de Cura, descending towards Parapara, the traveller meets successively with strata of gneiss and micaceous schist, a probably *Silurian* transition rock of argillaceous schist and black limestone; serpentine and greenstone in detached spheroidal masses; and lastly, on the margin of the great plain, small elevations of augitic amygdaloid and porphyritic schist. These hills between Parapara and Ortiz appear to me to be produced by volcanic eruptions on the old sea-shore of the Llanos. Further to the north, rise the far-famed cavernous and grotesquely-shaped elevations known as the Morros de San Juan, which form a species of devil's dyke, the grain of which is crystalline, like upheaved dolomite. They are, therefore, to be regarded rather as portions of the shore than as islands in the ancient gulf. I consider the Llanos to have been a gulf, for when their inconsiderable elevation above the present sea level, the adaptation of their form to the rotation current, running from east to west, and the lowness of the eastern shore between the mouth of the Orinoco and the Essequibo are taken into account, it can scarcely be doubted that the sea once overflowed the whole of this basin between the coast chain and the Sierra de la Parime, extending westward to the mountains of Merida and Pamplona (in the same manner as it probably passed through the plains of Lombardy to the Cottian and Pennine Alps). Moreover, the inclination or line of strike of these Llanos is directed from west to east. Their elevation at Calabozo, a distance of 100 geographical (400 English) miles from the sea, scarcely amounts to 30 toises, or 192 English feet; consequently 15 toises (96 English feet) less than the elevation of Pavia, and 45 toises (288 English feet) less than that of Milan in the plain of Lombardy between the Swiss Lepontine Alps and the Ligurian Apennines. This conformation of the land reminds us of Claudian's expression, "curvata tumore parvo planities." The surface of the Llanos is so perfectly horizontal that in many parts over an area of some 480 English square miles, not a single point appears elevated one foot above the surrounding level. When it is further borne in mind that there is a total absence of all shrubs, and that in some parts, as in the Mesa de Pavones, there is not even a solitary palm-tree to be seen, it may easily be supposed that this sea-like and dreary plain presents a most singular aspect. Far as the eye can range, it scarcely rests on any object elevated many inches above the general level. If the boundary of the horizon did not continually present an undefined flickering and undulating outline, owing to the condition of the lower strata of air, and the refraction of light, solar elevations might be determined by the sextant above the margin of the plain as above the horizon of the sea. This perfect flatness of the ancient sea-bottom renders the *banks* even more striking. They are composed of broken floetz-strata, which rise abruptly about two or three feet above the surrounding level, and extend uniformly over a length of from 10 to 12 geographical (*i.e.*, 40 to 48 English) miles. It is here that the small rivers of the Steppe take their origin.

On our return from the Rio Negro, we frequently met with traces of landslips in passing over the Llanos of Barcelona. We here found in the place of elevated banks, isolated strata of gypsum lying from 3 to 4 toises, or 19 to 25 English feet, below the contiguous rock. Further westward, near the confluence of the River Caura and the Orinoco, a large tract of thickly grown forest land to the east of the Mission of San Pedro de Alcantara, fell in after an earthquake in the year 1790. A lake was immediately formed in the plain, which measured upwards of 300 toises (1919 feet) in diameter. The lofty trees, as the Desmanthus, Hymenæa, and Malpighia, retained their verdure and foliage for a long time after their submersion.

(3) p. 2—"*A shoreless ocean seems spread before us.*"

The distant aspect of the Steppe is the more striking when the traveller emerges from dense forests, where his eye has been familiarised to a limited prospect and luxuriant natural scenery. I shall ever retain an indelible impression of the effect produced on my mind by the Llanos, when, on our return from the Upper Orinoco, they first broke on our view from a distant mountain, opposite the mouth of the Rio Apure, near the Hato del Capuchino. The last rays of the setting sun illumined the Steppe, which seemed to swell before us like some vast hemisphere, while the rising stars were refracted by the lower stratum of the atmosphere. When the plain has been excessively heated by the vertical rays of the sun, the evolution of the radiating heat, the ascent of currents of air, and the contact of atmospheric strata of unequal density, continue throughout the night.

(4) p. 2—"*The naked stony crust.*"

The deserts of Africa and Asia acquire a peculiar character from the frequent occurrence of immense tracts of land, covered by one flat uniform surface of naked rock. In the Schamo, which separates Mongolia and the mountain chain of Ulangom and Malakha-Oola from the north-west part of China, such rocky banks are termed *Tsy*. In the woody plains of the Orinoco they are found to be surrounded with the most luxuriant vegetation.* In the midst of these flat, tabular masses of granite and syenite, several thousands of feet in diameter, presenting merely a few scattered lichens, we find in the forests, or on their margins, little islands of light soil, covered with low and ever-flowering plants, having the appearance of small gardens. The monks settled on the Upper Orinoco, singularly enough regard the whole of these horizontal naked stony plains, when extending over a considerable area, as conducive to fevers and other diseases. Many of the villages belonging to the mission have been transferred to other spots on account of the general prevalence of this opinion. Do these stony flats (*laxas*) act chemically on the atmosphere or influence it only by means of a greater radiation of heat?

(5) p. 2—"*Compared with the Llanos and Pampas of South America, or even with the Prairies on the Missouri.*"

Our physical and geognostic knowledge of the western mountain region of North America has recently been enriched by the acquisition of many accurate data yielded by the admirable labours of the enterprising traveller Major Long, and his companion Edwin James, but more especially by the comprehensive investigations of Captain Frémont. The knowledge thus established clearly corroborates the accuracy of the different facts which in my work on New Spain I could merely advance as hypothetical conjectures regarding the northern plains and mountains of America. In natural history, as well as in historical research, facts remain isolated until by long-continued investigation they are brought into connection with each other.

The eastern shore of the United States of North America inclines from south-west to north-east, as does the Brazilian coast south of the equator from the Rio de la Plata to Olinda. on both these regions there rise, at a short distance from the coast line, two ranges of mountains more nearly parallel to each other than to the western Andes, (the Cordilleras If Chili and Peru), or to the North Mexican chain of the Rocky Mountains. The South American or Brazilian moun-

* *Relation Hist.*, t. ii., p. 279.

tain system, forms an isolated group, the highest points of which, Itacolumi and Itambe, do not rise above an elevation of 900 toises, or 5755 English feet. The eastern portion of the ridge most contiguous to the sea is the only part that follows a regular inclination from S.S.W. to N.N.E., increasing in breadth and diminishing in general elevation as it approaches further westward. The chain of the Parecis hills approximates to the rivers Itenes and Guaporé, in the same manner as the mountains of Aguapehi and San Fernando (south of Villabella) approach the lofty Andes of Cochabamba and Santa Cruz de la Sierra.

There is no direct connection between the two mountain systems of the Atlantic and South-sea coasts (the Brazilian and the Peruvian Cordilleras); Western Brazil being separated from Eastern or Upper Peru by the low lands of the province of Chiquitos, which is a longitudinal valley that inclines from north to south, and communicates both with the plains of the Amazon and of the Rio de la Plata. In these regions, as in Poland and Russia, a ridge of land, sometimes imperceptible (termed in Slavonic *Uwaly*), forms the line of separation between different rivers; as for instance, between the Pilcomayo and Madeira, between the Aguapehi and Guaporé, and between the Paraguay and the Rio Topayos. The ridge (*seuil*) extends from Chayanta and Pomabamba (19°—20° lat.,) in a south-easterly direction, and after intersecting the depressed tracts of the province of Chiquitos, (which has become almost unknown to geographers since the expulsion of the Jesuits,) forms to the north-east, where some scattered mountains are again to be met with, the *divortia aquarum* at the sources of the Baures and near Villabella (15°—17° lat.)

This water-line of separation which is so important to the general intercourse and growing civilization of different nations corresponds in the northern hemisphere of South America with a second line of demarcation (2°—3° lat.) which separates the district of the Orinoco from that of the Rio Negro and the Amazon. These elevations or risings in the midst of the plains (*terræ tumores*, according to Frontinus) may almost be regarded as undeveloped mountain-systems, designed to connect two apparently isolated groups, the Sierra Parime and the Brazilian highlands, to the Andes chain of

Timana and Cochabamba. These relations, to which very little attention has hitherto been directed, form the basis of my division of South America into three depressions or basins, viz., those of the Orinoco in its lower course, of the Amazon, and of the Rio de la Plata. Of these three basins, the exterior ones, as I have already observed, are Steppes or Prairies; but the central one between the Sierra Parime and the Brazilian chain of mountains must be regarded as a wooded plain or *Hylæa.*

In endeavouring by a few equally brief touches to give a sketch of the natural features of North America, we must first glance at the chain of the Andes, which, narrow at its origin, soon increases in height and breadth as it follows an inclination from south-east to north-west, passing through Panama, Veragua, Guatimala, and New Spain. This range of mountains, formerly the seat of an ancient civilization, presents a like barrier to the general current of the sea between the tropics, and to a more rapid intercommunication between Europe, Western Africa, and Eastern Asia. From the 17th degree of latitude at the celebrated Isthmus of Tehuantepec, the chain deflects from the shores of the Pacific, and inclining from south to north becomes an inland Cordillera. In Northern Mexico, the Crane Mountains (Sierra de las Grullas) constitute a portion of the Rocky Mountains. On their western declivity rise the Columbia and the Rio Colorado of California; on the eastern side the Rio Roxo of Natchitoches, the Canadian river, the Arkansas, and the shallow river Platte, which latter has recently been converted by some ignorant geographers, into a Rio *de la Plata*, or a river yielding silver. Between the sources of these rivers rise in the parallels of 37° 20′ and 40° 13′ lat., three huge peaks composed of granite, containing little mica, but a large proportion of hornblende. These have been respectively named Spanish Peak, James or Pike's Peak, and Big Horn or Long's Peak.* Their elevation exceeds that of the highest summits of the North Mexican Andes, which indeed nowhere attain the height of the line of perpetual snow from the parallels of 18° and 19° lat., or from the group of Orizaba, (2717 toises, or 17,374 English feet), and of Popocatepetl (2771 toises, or 17,720 English feet) to Santa Fé and Taos in New Mexico.

* See my *Essai Politique sur la Nouvelle Espagne.* 2me édit., t. i., pp. 82 and 109.

James' Peak (38° 48′ lat.) is said to have an elevation of 11,497 English feet. Of this only 8537 feet have been determined by trigonometrical measurement, the remainder being deduced in the absence of barometrical observations, from uncertain calculations of the declivity or fall of rivers. As it is scarcely ever possible, even at the level of the sea, to conduct a purely trigonometrical measurement, determinations of impracticable heights are always in part barometrical. Measurements of the fall of rivers, of their rapidity and of the length of their course, are so deceptive, that the plain at the foot of the Rocky Mountains, more especially near those summits mentioned in the text, was, before the important expedition of Captain Frémont, estimated sometimes at 8000 and sometimes at 3000 feet above the level of the sea.* From a similar deficiency of barometrical measurements, the true height of the Himalaya remained for a long time uncertain; now, however, science has made such advances in India, that when Captain Gerard had ascended on the Tarhigang, near the Sutledge, north of Shipke, to the height of 19,411 feet, he still had, after having broken three barometers, four equally correct ones remaining.†

Frémont, in the expedition which he made between the years 1842 and 1844, at the command of the United States Government, discovered and measured barometrically the highest peak of the whole chain of the Rocky Mountains to the north-north-west of Spanish, James', Long's, and Laramie's Peaks. This snow-covered summit, which belongs to the group of the Wind River Mountains, bears the name of Frémont's Peak on the great chart published under the direction of Colonel Abert, chief of the topographical department at Washington. This point is situated in the parallel of 43° 10′ north lat., and 110° 7′ west long., and therefore nearly 5° 30′ north of Spanish Peak. The elevation of Frémont's Peak, which according to direct measurement is 13,568 feet, must therefore exceed by 2072 feet that given by Long to James' Peak, which would appear from its position to be identical with Pike's Peak, as given in the map above referred to. The Wind River Mountains constitute the dividing ridge (*divortia aquarum*) between the two seas. "From the summit,"

* See *Long's Expeditions*, vol. ii., pp. 36, 362, 382. Ap. p. xxxvii.

† *Critical Researches on Philology and Geography*, 1824, p. 144.

says Captain Frémont in his official report,* "we saw on the one side numerous lakes and streams, the sources of the Rio Colorado, which carries its waters through the Californian Gulf to the South Sea; on the other, the deep valley of the Wind River, where lie the sources of the Yellowstone River, one of the main branches of the Missouri which unites with the Mississippi at St. Louis. Far to the north-west we could just discover the snowy heads of the Trois Tetons, which give rise to the true sources of the Missouri not far from the primitive stream of the Oregon or Columbia river, which is known under the name of Snake River, or Lewis Fork."

To the surprise of the adventurous travellers, the summit of Frémont's Peak was found to be visited by bees. It is probable that these insects, like the butterflies which I found at far higher elevations in the chain of the Andes, and also within the limits of perpetual snow, had been involuntarily drawn thither by ascending currents of air. I have even seen large winged lepidoptera, which had been carried far out to sea by land-winds, drop on the ship deck at a considerable distance from land in the South Sea.

Frémont's map and geographical researches embrace the immense tract of land extending from the confluence of Kanzas River with the Missouri, to the cataracts of the Columbia and the Missions of Santa Barbara and Pueblo de los Angeles in New California, presenting a space amounting to 28 degrees of longitude (about 1360 miles) between the 34th and 45th parallels of north latitude. Four hundred points have been hypsometrically determined by barometrical measurements, and for the most part, astronomically: so that it has been rendered possible to delineate the profile above the sea's level of a tract of land measuring 3,600 miles with all its inflections, extending from the north of Kanzas River to Fort Vancouver and to the coasts of the South Sea (almost 720 miles more than the distance from Madrid to Tobolsk). As I believe I was the first who attempted to represent, in geognostic profile, the configuration of entire countries, as the Spanish Peninsula, the highland of Mexico, and the Cordilleras of South America (for the half-perspec-

* *Report of the Exploring Expedition to the Rocky Mountains in the year* 1842, *and to Oregon and North California, in the years* 1843-1844, p. 78.

tive projections of the Siberian traveller, the Abbé Chappe,* were based on mere and for the most part on very inaccurate estimates of the falls of rivers); it has afforded me special satisfaction to find the graphical method of representing the earth's configuration in a vertical direction, that is, the elevation of solid over fluid parts, achieved on so vast a scale. In the mean latitudes of 37° to 43° the Rocky Mountains present, besides the great snow-crowned summits, whose height may be compared to that of the Peak of Teneriffe, elevated plateaux of an extent scarcely to be met with in any other part of the world, and whose breadth from east to west is almost twice that of the Mexican highlands. From the range of the mountains, which begin a little westward of Fort Laramie, to the further side of the Wahsatch Mountains, the elevation of the soil is uninterruptedly maintained from five to upwards of seven thousand feet above the sea's level; nay, this elevated portion occupies the whole space between the true Rocky Mountains and the Californian snowy coast range from 34° to 45° north latitude. This district, which is a kind of broad longitudinal valley, like that of the lake of Titicaca, has been named *The Great Basin* by Joseph Walker and Captain Frémont, travellers well acquainted with these western regions. It is a *terra incognita* of at least 8000 geographical (or 128,000 English) square miles, arid, almost uninhabited, and full of salt lakes, the largest of which is 3940 Parisian (or 4200 English) feet above the level of the sea, and is connected with the narrow Lake Utah,† into which the "Rock River" (*Timpan Ogo* in the Utah language) pours its copious stream. Father Escalante, in his wanderings from Santa Fé del Nuevo Mexico to Monterey in New California, discovered Frémont's "Great Salt Lake" in 1776, and confounding together the river and the lake, called it Laguna de Timpanogo. Under this name I inserted it in my map of Mexico, which gave rise to much uncritical discussion regarding the assumed non-existence of a large inland salt lake,‡—a

* Chappe d'Auteroche, *Voyage en Sibérie, fait en* 1761. 4 vols., 4to., Paris, 1768.

† Frémont, *Report of the Exploring Expedition*, pp. 154, and 273–276.

‡ Humboldt, *Atlas Mexicain*, plch. 2; *Essai politique sur la Nouv. Esp.*, t. i. p. 231; t. ii. pp. 243, 313, and 420. Frémont, *Upper California*, 1848, p. 9. See also Duflot de Mofras, *Exploration de l'Orégon*, 1844, t. ii. p. 140.

question previously mooted by the learned American traveller Tanner. Gallatin expressly says in his memoir on the aboriginal races*—"General Ashley and Mr. J. S. Smith have found the Lake Timpanogo in the same latitude and longitude nearly as had been assigned to it in Humboldt's Atlas of Mexico."

I have purposely dwelt at length on these considerations regarding the remarkable elevation of the soil in the region of the Rocky Mountains, since by its extension and height it undoubtedly exercises a great, although hitherto unappreciated influence on the climate of the northern half of the new continent, both in its southern and eastern portions. On this vast and uniformly elevated plateau Frémont found the water covered with ice every night in the month of August. Nor is the configuration of the land less important when considered in reference to the social condition and progress of the great North American United States. Although the mountain range which divides the waters attains a height nearly equal to that of the passes of Mount Simplon (6170 Parisian or 6576 English feet), Mount Gothard (6440 Parisian or 6863 English feet), and the great St. Bernard (7476 Parisian or 7957 English feet), the ascent is so prolonged and gradual that no impediments oppose a general intercourse by means of vehicles and carriages of every kind between the Missouri and Oregon territories, between the Atlantic States, and the new settlements on the Oregon (or Columbia) river, or between the coast-lands lying opposite to Europe on the one side of the continent, and to China on the other. The distance from Boston to the old settlement of Astoria on the Pacific at the mouth of the Oregon when measured in a direct line, and taking into account the difference of longitude, is 550 geographical, *i.e.*, 2200 English miles, or one-sixth less than the distance between Lisbon and Katherinenburg in the Ural district. On account of this gentle ascent of the elevated plains leading from the Missouri to California and the Oregon territory (all the resting-places measured between the Fort and River Lamarie on the northern branch of the Platte river to Fort Hall on the Lewis Fork of the Columbia, being situated at an elevation of from five to upwards of seven thousand feet, and that in Old Park even at the height of 9760 Parisian or 10,402 English feet!), consi-

* In the *Archæologia Americana,* vol. ii. p. 140.

derable difficulty has been experienced in determining the culminating point, or that of the *divortia aquarum*. It is south of the Wind River Mountains, about midway between the Mississippi and the coast line of the Southern Ocean, and is situated at an elevation of 7490 feet, or only 480 feet lower than the pass of the Great Bernard. The emigrants call this culminating point the South Pass.* It is situated in a pleasant region, embellished by a profusion of artemisiæ, especially A. tridentata (Nuttall), and varieties of asters and cactuses, which cover the micaceous slate and gneiss rocks. Astronomical determinations place its latitude in the parallel of 42° 24′, and its longitude in that of 109° 24′ W. Adolf Erman has already drawn attention to the fact, that the line of strike of the great east-Asiatic Aldanian mountain-chain, which separates the basin of the Lena from the rivers flowing towards the Great Southern Ocean, if extended in the form of a great circle on the surface of the globe, passes through many of the summits of the Rocky Mountains between 40° and 55° north lat. "An American and an Asiatic mountain-chain," he remarks, "appear therefore to be only portions of one and the same fissure erupted by the shortest channels."†

The western high mountain coast chain of the Californian maritime Alps, the *Sierra Nevada de California*, is wholly distinct from the Rocky Mountains, which sink towards the Mackenzie River (that remains covered with ice for a great portion of the year), and from the high table land on which rise individual snow-covered peaks. However injudicious the choice of the appellation of *Rocky Mountains* may be, when applied to the most northerly prolongation of the Mexican central chain, I do not deem it expedient to substitute for it the denomination of the Oregon Chain, as has frequently been attempted. These mountains do indeed give rise to the sources of three main branches constituting the great Oregon or Columbia river (viz., Lewis', Clarke's, and North Fork); but this mighty stream also intersects the chain of the ever snow-crowned maritime Alps of California. The name of Oregon Territory is also employed, politically and officially, to designate the lesser territory of land west of the

* Frémont's *Report*, pp. 3, 60, 70, 100, and 129.

† Compare Erman's *Reise um die Erde*, Abth. i. Bd. 3, s. 8, Abth. ii. Bd. 1. s. 386, with his *Archiv für Wissenschaftliche Kunde von Russland*. Bd. vi. s. 671.

coast chain, where Fort Vancouver and the Walahmutti settlements are situated; and it would therefore seem better to abstain from applying the name of Oregon either to the central or to the coast chain. This denomination, moreover, led the celebrated geographer Malte-Brun into a misconception of the most remarkable kind. He read in an old Spanish chart the following passage:—"And it is still unknown (*y aun se ignora*) where the source of this river" (now called the Columbia) "is situated," and he believed that the word *ignora* signified the name of the Oregon.*

The rocks which give rise to the cataracts of the Columbia at the point where the river breaks through the chain, mark the prolongation of the Sierra Nevada of California from the 44th to the 47th degree of latitude.† In this northern prolongation of the chain lie the three colossal elevations of Mount Jefferson, Mount Hood, and Mount St. Helen's, which rise 14,540 Parisian (or 15,500 English) feet above the sea-level. The height of this coast chain or range far exceeds therefore that of the Rocky Mountains. "During an eight months' journey along these maritime Alps," says Captain Frémont,‡ "we were constantly within sight of snow-covered summits; and while we were able to cross the Rocky Mountains through the South Pass at an elevation of 7027 feet, we found that the passes in the maritime range, which is divided into several parallel chains, were more than 2000 feet higher"—and therefore only 1170 (English) feet below the summit of Mount Etna. It is also a very remarkable fact, and one which reminds us of the relations of the eastern and western Cordilleras of Chili, that volcanoes still active are only found in the Californian chain which lies in the closest proximity to the sea. The conical mountains of Regnier and of St. Helen's are almost invariably observed to emit smoke; and on the 23rd of November, 1843, the latter of these volcanoes erupted a mass of ashes which covered the shores of the Columbia for a distance of forty miles, like a fall of snow. To the volcanic Californian chain belong also in the far north of Russian America, Mount Elias (according to La Pérouse 1980 toises, or 12,660 feet, and according to Malaspina 2792 toises, or 17,850 feet in height), and Mount Fair

* See my *Essai polit. sur la Nouv. Espagne*, t. ii. p. 314.

† Fremont, *Geographical Memoir upon Upper California*, 1848, p. 6.

‡ *Report*, p. 274 (or *Narrative*, p. 300).

Weather (Cerro de Buen Tiempo, 2304 toises, or 14,733 feet high). Both these conical mountains are regarded as still active volcanoes. Frémont's expedition, which has proved alike useful in reference to botany and geognosy, likewise collected volcanic products in the Rocky Mountains (as scoriaceous basalt, trachyte, and true obsidian), and discovered an old extinct crater somewhat to the east of Fort Hall (43° 2′ north lat., and 112° 28′ west long.), but no traces of any still active volcanoes emitting lava and ashes, were to be met with. We must not confound with these the hitherto unexplained phenomenon termed *smoking hills*, *côtes brûlées*, and *terrains ardens*, in the language of the English settlers and the natives who speak French. "Rows of low conical hills," says the accurate observer M. Nicollet, "are almost periodically, and sometimes for two or three years continually, covered with dense black smoke, unaccompanied by any visible flames. This phenomenon is more particularly noticed in the territory of the Upper Missouri, and still nearer to the eastern declivity of the Rocky Mountains, where there is a river named by the natives Mankizitah-watpa, or the river of smoking earth. Scorified pseudo-volcanic products, a kind of porcelain jasper, are found in the vicinity of the smoking hills."

Since the expedition of Lewis and Clarke an opinion has generally prevailed that the Missouri deposits a true pumice on its banks; but here white masses of a delicate cellular texture have been mistaken for that substance. Professor Ducatel was of opinion that the phenomenon which is chiefly observed in the chalk formation, was owing to "the decomposition of water by sulphur pyrites and to a re-action on the brown coal floetzes."*

If before we close these general remarks regarding the configuration of North America we once more cast a glance at those regions which separate the two diverging coast chains from the central chain, we shall find in strong contrast, on the West, between that central chain and the Californian Alps of the Pacific, an arid and uninhabited elevated plateau nearly six thousand feet above the sea; and in the East, between the Rocky Mountains and the Alleghanies, (whose highest points, Mount Washington and Mount Marcy, rise,

* Compare Frémont's *Report*, pp. 164, 184, 187, 193, and 299, with Nicollet's *Illustration of the Hydrographical Basin of the Upper Mississippi River*, 1843, pp. 39–41.

according to Lyell, to the respective heights, of 6652 and 5400 feet,) we see the richly watered, fruitful, and thickly-inhabited basin of the Mississippi, at an elevation of from four to six hundred feet, or more than twice that of the plains of Lombardy. The hypsometrical character of this eastern valley, or in other words, its relation to the sea's level, has only very recently been explained by the admirable labours of the talented French astronomer Nicollet, unhappily lost to science by a premature death. His great chart of the Upper Mississippi, executed between the years 1836 and 1840, was based on two hundred and forty astronomical determinations of latitude, and one hundred and seventy barometrical determinations of elevation. The plain which encloses the valley of the Mississippi is identical with that of northern Canada, and forms part of one and the same depressed basin, extending from the Gulf of Mexico to the Arctic Sea.* Wherever the low land falls in undulations, and slight elevations which still retain their un-English appellation of *côteaux des prairies*, *côteaux des bois*, occur in connected rows between the parallels of 47° and 48° north lat., these rows and gentle undulations of the ground separate the waters between Hudson's Bay and the Gulf of Mexico. Such a line of separation between the waters is formed, north of Lake Superior or Kichi Gummi, by the Missabay Heights, and further west by the elevations known as *Hauteurs des Terres*, in which are situated the true sources of the Mississippi, one of the largest rivers in the world, and which were not discovered till the year 1832. The highest of these chains of hills hardly attains an elevation of from 1500 to 1600 feet. From its mouth (the old French Balize) to St. Louis, somewhat to the south of its confluence with the Missouri, the Mississippi has a fall of only 380 feet, notwithstanding that the itinerary distance between these two points exceeds 1280 miles. The surface of Lake Superior lies at an elevation of 618 feet, and as its depth in the neighbourhood of the island of Magdalena is fully 790 feet, its bottom must be 172 feet below the surface of the ocean.†

Beltrami, who in 1825 separated himself from Major Long's

* Compare my *Relation Historique*, t. iii. p. 234, and Nicollet, *Report to the Senate of the United States*, 1843, pp. 7, 57.

† Nicollet, op. cit. pp. 99, 125, 128.

expedition, boasted that he had found the sources of the Mississippi in Lake Cass. The river passes, in its upper course, through four lakes, the second of which is the one referred to, while the outermost one, Lake Istaca (47° 13′ north lat., and 95° west long.), was first recognised as the true source of the Mississippi, in 1832, in the expedition of Schoolcraft and Lieutenant Allen. This stream, which subsequently becomes so mighty, is only 17 feet in width, and 15 inches deep, when it issues from the singular horse-shoe-shaped Lake Istaca. The local relations of this river were first fully established on a basis of astronomical observations of position by the scientific expedition of Nicollet, in the year 1836. The height of the sources, that is to say, of the last access of water received by Lake Istaca from the ridge of separation, called *Hauteur de Terre*, is 1680 feet above the level of the sea. Near this point, and at the southern declivity of the same separating ridge, lies Elbow Lake, the source of the small Red River of the north, which empties itself, after many windings, into Hudson's Bay. The Carpathian Mountains exhibit similar relations in reference to the origin of the rivers which empty themselves into the Baltic and the Black Sea. M. Nicollet gave the names of celebrated astronomers, opponents as well as friends, with whom he had become acquainted in Europe, to the twenty small lakes which combine together to form narrow groups in the southern and western regions of Lake Istaca. His atlas is thus converted into a geographical album, reminding one of the botanical album of the *Flora Peruviana* of Ruiz and Pavon, in which the names of new families of plants were made to accord with the Court Calendar, and the various alterations made in the *Oficiales de la Secretaria*.

The east of the Mississippi is still occupied by dense forests; the west by prairies only, on which the buffalo (*Bos Americanus*) and the musk ox (*Bos moschatus*) pasture. These two species of animals, the largest of the new world, furnish the nomadic tribes of the Apaches-Llaneros and Apaches-Lipanos with the means of nourishment. The Assiniboins occasionally slay from seven to eight hundred bisons in the course of a few days in the artificial enclosures constructed for the purpose of driving together the wild herds, and known as *bison parks*.* The American bison, called by the Mexicans *Cibolo*, is killed chiefly on account of the tongue, which is regarded as a special delicacy. This animal is not a mere variety of the aurochs of the old world; although, like other species of animals, as for instance the elk (*Cervus alces*) and the reindeer (*Cervus tarandus*), no less than the stunted inhabitants of the polar regions, it may be regarded as common to the northern portions of *all* continents, and as affording a proof of their former long existing connection. The Mexicans apply to the European ox the Aztec term *quaquahue*, or horned animal, from *quaquahuitl*, a horn. The huge ox-horns which have been found in ancient Mexican buildings near Cuernavaca, south-west of the capital of Mexico, appear to me to belong to the bison. The Canadian bison can be used for agricultural labour, and will breed with the European cattle, although it is uncertain whether the hybrid thus engendered is capable of propagating its species. Albert Gallatin, who, before his appearance in Europe as a distinguished diplomatist, had acquired by personal observation a considerable amount of information regarding the uncultivated parts of the United States, assures us that the fruitfulness of the mixed breed of the American buffalo and European cattle is an undoubted fact: " the mixed breed," he writes, " was quite common fifty years ago in some of the north-western counties of Virginia, and the cows, the issue of that mixture, propagated like all others." "I do not remember," he further adds, " that full-grown buffaloes were tamed; but dogs would at that time occasionally bring in the young bison-calves, which were reared and bred with European cows. At Monongahela all the cattle for a long time were of this mixed breed. It was said, however, that the cows yielded but little milk." The favourite food of the buffalo is the *Tripsacum dactyloides* (known as buffalo-grass in North Carolina) and a hitherto undescribed species of clover allied to the *Trifolium repens*, and designated by Barton as *Trifolium bisonicum*.

I have elsewhere† drawn attention to the fact, that according to a passage of the trustworthy Gomara‡, there

* Maximilian, Prinz zu Wied, *Reise in das innere Nord-Amerika*, bd. i., 1839, s. 443.

† See *Cosmos*, vol. ii. p. 674 (Bohn's edition).

‡ *Historia general de las Indias*, cap. 214.

lived, as late as the sixteenth century, an Indian tribe in the north-west of Mexico, in 40° north lat., whose greatest wealth consisted in hordes of tamed buffaloes (*bueyes con una giba*). Yet, notwithstanding the possibility of taming the buffalo, and the abundance of milk it yields, and notwithstanding the herds of Lamas in the Peruvian Cordilleras, no pastoral tribes were met with on the discovery of America. Nor does history afford any evidence of the existence, at any period, of this intermediate stage of national development. It is also a remarkable fact that the North American bison or buffalo has exerted an influence on geographical discoveries in pathless mountain districts. These animals advance in herds of many thousands in search of a milder climate, during winter, in the countries south of the Arkansas river. Their size and cumbrous forms render it difficult for them to cross high mountains on these migratory courses, and a well-trodden buffalo-path is therefore followed wherever it is met with, as it invariably indicates the most convenient passage across the mountains. Thus buffalo-paths have indicated the best tracks for passing over the Cumberland Mountains in the south-western parts of Virginia and Kentucky, and over the Rocky Mountains, between the sources of the Yellowstone and Plate rivers, and between the southern branch of the Columbia and the Californian Rio Colorado. European settlements have gradually driven the buffalo from the eastern portions of the United States. Formerly these migratory animals passed the banks of the Mississippi and the Ohio, advancing far beyond Pittsburgh.*

From the granitic rocks of Diego Ramirez and the deeply-intersected district of Terra del Fuego (which in the east contains silurian schist, and in the west, the same schist metamorphosed into granite by the action of subterranean fire,)† to the North Polar Sea, the Cordilleras extend over a distance of more than 8000 miles. Although not the loftiest, they are the longest mountain chain in the world, being upheaved from one fissure, which runs in the direction of a meridian from pole to pole, and exceeding in linear

* *Archæologia Americana*, vol. ii., 1836, p. 139.

† Darwin, *Journal of Researches into the Geology and Natural History of the Countries visited* 1832—1836 *by the Ships Adventure and Beagle*, p. 266.

extent the distance which, in the old continent, separates the Pillars of Hercules from the Icy Cape of the Tschuktches, in the north-east of Asia. Where the Andes are divided into several parallel chains, those lying nearest the sea are found to be the seat of the most active volcanoes; and it has moreover been repeatedly observed that when the phenomenon of an eruption of subterranean fire ceases in one mountain chain, it breaks forth in some other parallel range. The cones of eruption usually follow the direction of the axis of the chain; but in the Mexican table-land, the active volcanoes are situated on a transverse fissure, running from sea to sea, in a direction from east and west.* Wherever the upheaval of mountain masses in the folding of the ancient crust of the earth has opened a communication with the fused interior, volcanic activity continued to be exhibited on the murally upheaved mass by means of the ramification of fissures. That which we call a mountain chain has not been raised to its present elevation, or manifested as it now appears, at one definite period; for we find that rocks, varying considerably in age, have been superimposed on one another, and have penetrated towards the surface through early formed channels. The diversity observable in rocks is owing to the outpouring and upheaval of rocks of eruption, as well as to the complicated and slow process of metamorphism going on in fissures filled with vapour, and conducive to the conduction of heat.

The following have for a long time, viz., from 1830 to 1848, been regarded as the highest or culminating points of the Cordilleras of the new continent:—

The *Nevado de Sorata*, also called Ancohuma or Tusubaya (15° 52′ south lat.), somewhat to the south of the village of Sorata or Esquibel, in the eastern chain of Bolivia: elevation, 25,222 feet.

The *Nevado de Illimani*, west of the mission of Yrupana (16° 38 south lat.), also in the eastern chain of Bolivia: elevation, 24,000 feet.

The *Chimborazo* (1° 27′ south lat.), in the province of Quito: clevation, 21,422 feet.

The Sorata and Illimani were first measured by the distinguished geologist, Pentland, in the years 1827 and 1838;

* Humboldt, *Essai politique*, t. ii. p. 173.

and since the publication of his large map of the basin of the Laguna de Titicaca, in June, 1848, we learn that the above elevations given for the Sorata and Illimani are 3960 feet and 2851 feet too high. His map gives only 21,286 feet for the Sorata, and 21,149 feet for the Illimani. A more exact calculation of the trigonometrical operations of 1838 led Mr. Pentland to these new results. He ascribes an elevation of from 21,700 to 22,350 feet to four summits of the western Cordilleras; and, according to his data, the Peak of Sahama would thus be 926 feet higher than the Chimborazo, but 850 feet lower than the Peak of Aconcagua.

(6) p. 2—" *The desert near the basaltic mountains of Harudsch.*"

Near the Egyptian Natron Lakes, which in Strabo's time had not yet been divided into the six reservoirs by which they are now characterized, there rises abruptly to the north a chain of hills, running from east to west past Fezzan, where it at length appears to form one connected range with the Atlas chain. It divides in north-eastern, as Mount Atlas does in north-western Africa the Lybia, described by Herodotus as inhabited and situated near the sea, from the land of the Berbirs, or Biledulgerid, famed for the abundance of its wild animals. On the borders of Middle Egypt the whole region, south of the 30th degree of latitude, is an ocean of sand, studded here and there with islands or oases abounding in springs and rich in vegetation. Owing to the discoveries of recent travellers, a vast addition has been made to the number of the Oases formerly known, and which the ancients limited to three, compared by Strabo to spots upon a panther's skin. The third Oasis of the ancients, now called Siwah, was the *nomos* of Ammon, a hierarchical seat and a resting-place for the caravans, which inclosed within its precincts the temple of the horned Ammon and the spring of the Sun, whose waters were supposed to become cool at certain periods. The ruins of Ummibida (*Omm-Beydah*) incontestably belong to the fortified caravanserai at the Temple of Ammon, and therefore constitute one of the most ancient monuments which have come down to us from the dawn of human civilization.*

* Caillaud, *Voyage à Syouah,* p. 14; Ideler, *Fundgruben des Orients* bd. iv. s. 399—411.

The word Oasis is Egyptian, and is synonymous with Auasis and Hyasis.* Abulfeda calls the Oases *el-Wah*. In the latter time of the Cæsars, malefactors were sent to the Oases, being banished to these islands in the sandy ocean, as the Spaniards and English transported their malefactors to the Falkland islands and New Holland. The ocean affords almost a better chance of escape than the desert surrounding the Oases; which, moreover, diminish in fruitfulness in proportion to the greater quantity of sand incorporated in the soil.

The small mountain range of Harudsch (*Harudje*†) consists of grotesquely-shaped basaltic hills. It is the *Mons Ater* of Pliny, and its western extremity, known as the Soudah mountain, has been recently explored by my unfortunate friend, the enterprising traveller Ritchie. These basaltic eruptions in the tertiary limestone, and rows of hills rising abruptly from fissures, appear to be analogous to the basaltic eruptions in the Vicentine territory.

Nature repeats the same phenomena in the most distant regions of the earth. Hornemann found an immense quantity of petrified fishes' heads in the limestone formations of the White Harudsch (*Harudje el-Abiad*), belonging probably to the old chalk. Ritchie and Lyon remarked that the basalt of the Soudah mountain was in many places intimately mingled with carbonate of lime, as is the case in Monte Berico; a phenomenon that is probably connected with eruptions through limestone strata. Lyon's chart even indicates dolomite in the neighbourhood. Modern mineralogists have found syenite and greenstone, but not basalt, in Egypt. Is it possible that the true basalt, from which many of the ancient vases found in various parts of the country were made, can have been derived from a mountain lying so far to the west? Can the *obsidius lapis* have come from there, or are we to seek basalt and obsidian on the coast of the Red Sea? The strip of the volcanic eruptions of Harudsch, on the borders of the African desert, moreover reminds the geologist of augitic vesicular amygdaloid, phonolite, and greenstone porphyry, which are only found on the northern and western limits of the steppes of Venezuela

* Strabo, lib. ii. p. 130, lib. xvii. p. 813, Cas.; Herod. lib. iii. cap. 26. p. 207, Wessel.

† See Ritter's *Afrika*, 1822, s. 885, 988, 993, and 1003.

and of the plains of the Arkansas, and therefore, as it were, on the ancient coast chains.*

(7) p. 3—"*When suddenly deserted by the tropical east wind, and the sea is covered with weeds.*"

It is a remarkable phenomenon, although one generally known to mariners, that in the neighbourhood of the African coast, (between the Canaries and the Cape de Verde islands, and more especially between Cape Bojador and the mouth of the Senegal,) a westerly wind often prevails instead of the usual east or trade wind of the tropics. The cause of this phenomenon is to be ascribed to the far-extending desert of Zahara, and arises from the rarefaction, and consequent vertical ascent of the air over the heated sandy surface. To fill up the vacuum thus occasioned, the cool sea-air rushes in, producing a westerly breeze, adverse to vessels sailing to America; and the mariner, long before he perceives any continent, is made sensible of the effects of its heat-radiating sands. As is well known, a similar cause produces that alternation of sea and land breezes, which prevails at certain hours of the day and night on all sea-coasts.

The accumulation of sea-weed in the neighbourhood of the western coasts of Africa has been often referred to by ancient writers. The local position of this accumulation is a problem which is intimately connected with the conjectures regarding the extent of Phœnician navigation. The Periplus, which has been ascribed to Scylax of Caryanda, and which, according to the investigations of Niebuhr and Letronne, was very probably compiled in the time of Philip of Macedon, contains a description of a kind of fucus sea, *Mar de Sargasso*, beyond Cerne; but the locality indicated appears to me very different from that assigned to it in the work "*De Mirabilibus Auscultationibus*," which for a long time, but incorrectly, bore the great name of Aristotle.† "Driven by the east wind," says the pseudo-Aristotle, "Phœnician

* Humboldt, *Relat. hist.*, t. ii. p. 142, and Long's *Expedition to the Rocky Mountains*, v. ii. pp. 91 and 405.

† Compare *Scyl. Caryand. Peripl.*, in Hudson, vol. ii. p. 53, with Aristot. *de Mirab. Auscult.* in *Op. omnia*, ex rec. Bekkeri, p. 884, § 136.

mariners came in a four days' voyage from Gades to a place where the sea was found covered with rushes and sea-weed (θρύον καὶ φῦκος). The sea-weed is uncovered at ebb, and overflowed at flood tide." Does he not here refer to a shoal lying between the 34th and 36th degrees of latitude? Has a shoal disappeared there in consequence of volcanic revolution? Vobonne refers to rocks north of Madeira.* In Scylax it is stated that "the sea beyond Cerne ceases to be navigable in consequence of its great shallowness, its muddiness, and its sea-grass. The sea-grass lies a span thick, and it is pointed at its upper extremity, so that it pricks." The sea-weed which is found between Cerne (the Phœnician station for merchant vessels, Gaulea; or, according to Gosselin, the small estuary of Fedallah, on the north-west coast of Mauritania,) and Cape Verde, at the present time by no means forms a great meadow or connected group, "*mare herbidum*," such as exists on the other side of the Azores. Moreover, in the poetic description of the coast given by Festus Avienus,† in which, as Avienus himself very distinctly acknowledges, he availed himself of the journals of Phœnician ships, the impediments presented by the sea-weed are described with great minuteness; but Avienus places the site of this obstacle much further north, towards Ierne, the Holy Isle.

> Sic nulla late flabra propellunt ratem,
> Sic segnis humor æquoris pigri stupet.
> Adjicit et illud, plurimum inter gurgites
> Exstare fucum, et sæpe virgulti vice
> Retinere puppim . . . .
> Hæc inter undas multa cæspitem jacet,
> Eamque late gens Hibernorum colit.

When we consider that the sea-weed (*fucus*) the mud or slime (πηλὸς), the shallowness of the sea, and the perpetual calms, are always regarded by the ancients as characteristic of the Western Ocean beyond the Pillars of Hercules, we feel inclined, especially on account of the reference to the *calms*, to ascribe this to Punic cunning, to the tendency of a great trading people to hinder others, by terrific descriptions, from competing with them in maritime trading westwards. But even

* See also Edrisi, *Geogr. Nub.*, 1619, p. 157.

† *Ora Maritima*, v. 109, 122, 388, and 408.

in the genuine writings of the Stagyrite,* the same opinion is retained regarding the absence of wind, and Aristotle attempts to explain a false notion, or, as it seems to me, more correctly speaking, a fabulous mariner's story, by an hypothesis regarding the depth of the sea. The stormy sea between Gades and the Islands of the Blest (Cadiz and the Canaries) can in truth in no way be compared with the sea, which lies between the tropics, ruffled only by the gentle trade-winds (*vents alisés*), and which has been very characteristically named by the Spaniards† *El Golfo de las Damas.*

From very careful personal researches and from comparison of the logs of many English and French vessels, I am led to believe that the old and very indefinite expression *Mar de Sargasso*, refers to two fucus banks, the larger of which is of an elongated form, and is the easternmost one, lying between the parallels of 19° and 34°, in a meridian 7° westward of the Island of Corvo, one of the Azores; while the smaller and westernmost bank is of a roundish form, and is found between Bermuda and the Bahama Islands (lat. 25°—31°, long. 66°—74°). The principal diameter of the small bank, which is traversed by ships sailing from Baxo de Plata (Caye d'Argent,) northward of St. Domingo to the Bermudas, appears to me to have a N. 60° E. direction. A transverse band of *fucus natans*, extending in an east-westerly direction between the latitudes of 25° and 30°, connects the greater with the smaller bank. I have had the pleasure of seeing these views adopted by my lamented friend Major Rennell, and confirmed, in his great work on Currents, by many new observations.‡ The two groups of sea-weed, together with the transverse band uniting them, constitute the Sargasso Sea of the older writers, and collectively occupy an area equal to six or seven times that of Germany.

The vegetation of the ocean thus offers the most remarkable example of *social plants* of a single species. On the main land the Savannahs or grass plains of America, the heaths (*ericeta*), and the forests of Northern Europe and Asia,

* Aristot. *Meteorol.*, ii. 1, 14.

† Acosta, *Historia natural y moral de las Indias*, lib. iii. cap. 4.

‡ Compare Humboldt, *Relation historique*, t. i. p. 202, and *Examen Critique*, t. iii. pp. 68–69, with Rennell's *Investigation of the Currents of the Atlantic Ocean*, 1832, p. 184.

in which are associated coniferous trees, birches, and willows, produce a less striking uniformity than do these thalassophytes. Our heaths present in the north not only the predominating Calluna vulgaris, but also Erica tetralix, E. ciliaris, and E. cinerea; and in the south, Erica arborea, E. scoparia, and E. Mediterranea. The uniformity of the view presented by the Fucus natans is incomparably greater than that of any other assemblage of social plants. Oviedo calls the fucus banks "meadows," *praderias de yerva.* If we consider that Pedro Velasco, a native of the Spanish harbour of Palos, by following the flight of certain birds from Fayal, discovered the Island of Flores as early as 1452, it seems almost impossible, considering the proximity of the great fucus bank of Corvo and Flores, that no part of these oceanic meadows should have been seen before the time of Columbus by Portuguese ships driven westward by storms.

We learn, however, from the astonishment of the companions of the admiral, when they were continuously surrounded by sea-grass from the 16th of September to the 8th of October, 1492, that the magnitude of the phenomenon was at that period unknown to mariners. In the extracts from the ship's journal given by Las Casas, Columbus certainly does not mention the apprehensions which the accumulation of sea-weed excited, or the grumbling of his companions. He merely speaks of the complaints and murmurs regarding the danger of the very weak but constant east winds. It was only his son, Fernando Colon, who in the history of his father's life, endeavoured to give a somewhat dramatic delineation of the anxieties of the sailors.

According to my researches, Columbus made his way through the great fucus bank in the year 1492, in latitude 28½°, and in 1493, in latitude 37°, and both times in the longitude of 38°–41°. This can be established with tolerable certainty from the estimation of the velocity recorded by Columbus, and "the distance daily sailed over;" not indeed by dropping the log, but by the information afforded by the running out of half-hour sand-glasses (*ampolletas*). The first certain and distinct account of the log, (*catena della poppa,*) which I have found, is in the year 1521, in Pigafetta's Journal of Magellan's Circumnavigation of the World.* The deter-

* See *Cosmos,* vol. ii. p. 631, and note; Bohn's edition.

mination of the ship's place during the days in which Columbus was crossing the great bank is the more important, because it shews us that for three centuries and a half the total accumulation of these socially-living thalassophytes, (whether consequent on the local character of the sea's bottom or on the direction of the recurrent Gulf stream,) has remained at the same point. Such evidences of the persistence of great natural phenomena doubly arrest the attention of the natural philosopher, when they occur in the ever-moving oceanic element. Although the limits of the fucus banks oscillate considerably, in accordance with the strength and direction of long predominating winds, yet we may still, in the middle of the nineteenth century, take the meridian of 41° west of Paris (or 8° 38′ west of Greenwich) as the principal axis of the *great bank.* Columbus, with his vivid imaginative force, associated the idea of the position of this bank with the great physical line of demarcation, which according to him, "separated the globe into two parts, and was intimately connected with the changes of magnetic deviation and of climatic erlations." Columbus when he was uncertain regarding the longitude, attempted to determine his place (February, 1493,) by the appearance of the first floating masses of tangled weed (*de la primera yerva*) on the eastern border of the great Corvo bank. The physical line of demarcation was, by the powerful influence of the Admiral, converted on the 4th of May, 1493, into a political one, in the celebrated *line of demarcation* between the Spanish and Portuguese rights of possession*.

(8) p. 3—"*The Nomadic Tribes of Tibbos and Tuaryks.*"

These two nations, which inhabit the desert between Bornou, Fezzan, and Lower Egypt, were first made more accurately known to us by the travels of Hornemann and Lyon. The Tibbos or Tibbous occupy the eastern, and the Tuaryks (Tueregs) the western portion of the great sandy ocean. The former, from their habits of constant moving, were named by the other tribes "birds." The Tuaryks are subdivided into two tribes—the Aghadez and the Tagazi. These are often caravan leaders and merchants. They speak

* See my *Examen Critique,* t. iii. pp. 64—99; and *Cosmos,* vol. ii. p. 655. Bohn's edition.

the same language as the Berbers, and undoubtedly belong to the primitive Lybian races. They present the remarkable physiological phenomenon that, according to the character of the climate, the different tribes vary in complexion from a white to a yellow, or even almost black hue; but they never have woolly hair or negro features.*

(9) p. 3—"*The ship of the desert.*"

In the poetry of the East, the camel is designated as *the land-ship*, or *the ship of the desert* (*Sefynet-el-badyet*†).

The camel is, however, not only the carrier in the desert, and the medium for maintaining communication between different countries, but is also, as Carl Ritter has shown in his admirable treatise on the sphere of distribution of this animal, "the main requirement of a nomadic mode of life in the patriarchal stage of national development, in the torrid regions of our planet, where rain is either wholly or in a great degree absent. No animal's life is so closely associated by natural bonds with a certain primitive stage of the development of the life of man, as that of the camel among the Bedouin tribes, nor has any other been established in like manner by a continuous historical evidence of several thousand years."‡ "The camel was entirely unknown to the cultivated people of Carthage through all the centuries of their flourishing existence, until the destruction of the city. It was first brought into use for armies by the Marusians, in Western Lybia, in the times of the Cæsars; perhaps in consequence of its employment in commercial undertakings by the Ptolemies, in the valley of the Nile. The Guanches, inhabiting the Canary Islands, who were probably related to the Berber race, were not acquainted with the camel before the fifteenth century, when it was introduced by Norman conquerors and settlers. In the probably very limited communication of the Guanches with the coast of Africa, the smallness of their boats must necessarily have impeded the transport of large animals. The true Berber race, which was diffused throughout the interior of Northern Africa, and to which the Tibbos and Tuaryks, as already observed, belong,

* *Exploration scientifique de l'Algerie*, t. ii. p. 343.
† Chardin, *Voyages*, nouv. éd. par Langlès, 1811, t. iii. p. 376.
‡ *Asien*, Bd. viii., Abth. 1, 1847, s. 610, 758.

is probably indebted to the use of the camel throughout the Lybian desert and its oases, not only for the advantages of internal communication, but also for its escape from complete annihilation and for the maintenance of its national existence to the present day. The use of the camel continued, on the other hand, to be unknown to the negro races, and it was only in company with the conquering expeditions and proselyting missions of the Bedouins through the whole of Northern Africa, that the useful animal of the Nedschd, of the Nabatheans, and of all the districts occupied by Aramean races, spread here, as elsewhere, to the westward. The Goths brought camels as early as the fourth century to the Lower Istros (the Danube), and the Ghaznevides transported them in much larger numbers to India as far as the banks of the Ganges." We must distinguish two epochs in the distribution of the camel throughout the northern part of the African continent; the first under the Ptolemies, which operated through Cyrene on the whole of the north-west of Africa, and the second under the Mahommedan epoch of the conquering Arabs.

It has long been a matter of discussion, whether those domestic animals which were the earliest companions of mankind, as oxen, sheep, dogs, and camels, are still to be met with in a state of original wildness. The Hiongnu, in Eastern Asia, are among the nations who earliest trained wild camels as domestic animals. The compiler of the great Chinese work, *Si-yu-wen-kien-lo**, states that in the middle of the eighteenth century, wild camels, as well as wild horses and wild asses, still roamed over Eastern Turkestan. Hadji Chalfa, in his Turkish Geography, written in the seventeenth century, speaks of the very frequent hunting of the wild camel in the high plains of Kashgar, Turfan, and Khotan. Schott finds in the writings of a Chinese author, Ma-dschi, that wild camels exist in the countries north of China and west of the basin of the Hoang-ho, in Ho-si or Tangut. Cuvier† alone doubts the present existence of wild camels in the interior of Asia. He believes that they have merely "become wild;" since Calmucks, and others professing kindred Bud-

* *Historia Regionum Occidentalium, quæ Si-yu vocantur, visu et auditu cognitarum.*

† *Règne animal,* t. i. p. 257.

dhist doctrines, set camels and other animals at liberty, in order "to acquire to themselves merit for the other world." The Ailanitic Gulf of the Nabatheans was the home of the wild Arabian camel, according to Greek witnesses of the times of Artemidorus and Agatharchides of Cnidus.* The discovery of fossil camel-bones of the ancient world in the Sewalik hills (which are projecting spurs of the Himalaya range), by Captain Cautley and Dr. Falconer, in 1834, is especially worthy of notice. These remains were found with antediluvian bones of mastodons, true elephants, giraffes, and a gigantic land tortoise (*Colossochelys*), twelve feet in length and six feet in height.† This camel of the ancient world has been named *Camelus sivalensis*, although it does not show any great difference from the still living Egyptian and Bactrian camels with one and two humps. Forty camels have very recently been introduced into Java, from Teneriffe‡. The first experiment has been made in Samarang. In like manner, reindeer were only introduced into Iceland from Norway in the course of the last century. They were not found there when the island was first colonised, notwithstanding its proximity to East Greenland, and the existence of floating masses of ice.§

(10) p. 3—"*Between the Altai and the Kuen-lün.*"

The great highland, or, as it is commonly called, the mountain plateau of Asia, which comprises the lesser Bucharia, Songaria, Thibet, Tangut, and the Mogul country of the Chalcas and Olotes, is situated between the 36th and 48th degrees of north latitude and the meridians of 81° and 118° E. long. It is an erroneous idea to represent this part of the interior of Asia as a single, undivided mountainous swelling, continuous like the plateaux of Quito and Mexico, and situated from seven to upwards of nine thousand feet above the level of the sea. I have already shown in my "*Researches respecting the Mountains of Northern India,*‖" that there is not in this sense any continuous mountain plateau in the interior of Asia.

* Ritter, *Asien*, Bd. viii. s. 670, 672, and 746.

† Humboldt, *Cosmos*, Bohn's ed., vol. i. p. 281.

‡ *Singapore Journal of the Indian Archipelago*, 1847, p. 286.

§ Sartorius von Waltershausen, *Physisch-geographische Skizze von Island*, 1847, s. 41.

‖ Humboldt, *Premier Mémoire sur les Montagnes de l'Inde*, in the *Annales de Chimie et de Physique*, t. iii. 1816, p. 303; *Second Mémoire*, t. xiv. 1820, pp. 5—55.

My views concerning the geographical distribution of plants, and the mean degree of temperature requisite for certain kinds of cultivation, had early led me to entertain considerable doubts regarding the continuity of a great Tartarian plateau between the Himalaya and the chain of the Altai. This plateau continued to be characterized, as it had been described by Hippocrates, as "the high and naked plains of Scythia, which, without being crowned with mountains, rise and extend to beneath the constellation of the Bear."* Klaproth has the undeniable merit of having been the first to make us acquainted with the true position and prolongation of two great and entirely distinct chains of mountains,—the Kuen-lün and the Thian-schan, in a part of Asia which better deserves to be termed "central," than Kashmeer, Baltistan, and the Sacred Lakes of Thibet (the Manasa and the Ravanahrada). The importance of the Celestial Mountains (the Thian-schan) had indeed been already surmised by Pallas, without his being conscious of their volcanic character; but this highly-gifted investigator of nature, led astray by the hypotheses of the dogmatic and fantastic geology prevalent in his time, and firmly believing in "chains of mountains radiating from a centre," saw in the Bogdo Oola (the *Mons Augustus*, or culminating point of the Thian-schan.) such "a central node, whence all the other Asiatic mountain chains diverge in rays, and which dominates over all the rest of the continent!"

The erroneous idea of a single boundless and elevated plain, occupying the whole of Central Asia, the "*Plateau de la Tartarie*," originated in France, in the latter half of the eighteenth century. It was the result of historical combinations, and of a not sufficiently attentive study of the writings of the celebrated Venetian traveller, as well as of the naïve relations of those diplomatic monks who, in the thirteenth and fourteenth centuries (thanks to the unity and extent of the Mogul empire at that time), were able to traverse almost the whole of the interior of the continent, from the ports of Syria and of the Caspian Sea to the east coast of China, washed by the great ocean. If a more exact acquaintance with the language and ancient literature of India were of an older date among us than half a century, the hypothesis of this central plateau, occupying the wide space between the Himalaya and

* *De Aëre et Aquis*, § xcvi. p. 74.

the south of Siberia, would no doubt have sought support from some ancient and venerable authority. The poem of the Mahabharata appears, in the geographical fragment Bhischmakanda, to describe "Meru" not so much as a mountain as an enormous swelling of the land, which supplies with water the sources of the Ganges, those of the Bhadrasoma (Irtysch), and those of the forked Oxus. These physico-geographical views were intermingled in Europe with ideas of other kinds, and with mythical reveries on the origin of mankind. The lofty regions from which the waters were supposed to have first retreated (for geologists in general were long averse to the theories of elevation) must also have received the first germs of civilization. Hebraic systems of geology, based on ideas of a deluge, and supported by local traditions, favoured these assumptions. The intimate connexion between time and space, between the beginning of social order and the plastic condition of the surface of the earth, lent a peculiar importance and an almost moral interest to the Plateau of Tartary, which was supposed to be characterized by uninterrupted continuity. Acquisitions of positive knowledge,—the late matured fruit of scientific travels and direct measurements,—with a fundamental study of the languages and literature of Asia, and more especially of China, have gradually demonstrated the inaccuracy and exaggeration of those wild hypotheses. The mountain plains (ὀροπέδια) of Central Asia are no longer regarded as the cradle of human civilization, and the primitive seat of all arts and sciences. The ancient nation of Bailly's Atlantis, which d'Alembert has happily described as "having taught us everything but its own name and existence," has vanished. The inhabitants of the Oceanic Atlantis were already treated, in the time of Posidonius, as having a merely apocryphal existence.*

A plateau of considerable but very unequal elevation runs with little interruption, in a S.S.W.–N.N.E. direction, from Eastern Thibet towards the mountain node of Kentei, south of Lake Baikal, and is known by the names of Gobi, Scha-mo, (sand desert,) Scha-ho, (sand river,) and Hanhai. This swelling of the ground, which is probably more ancient than the elevation of the mountain-chains by which it is intersected, is situated, as we have already remarked, between 81° and 118°

* Strabo, lib. ii. p. 102; and lib. xiii. p. 598, Casaub.

east longitude from Greenwich. Measured at right angles to its longitudinal axis, its breadth in the south, between Ladak, Gertop, and H'lassa (the seat of the great Lama), is 720 miles; between Hami in the Celestial Mountains, and the great curve of the Hoang-ho, near the In-schan chain, it is scarcely 480; but in the north, between the Khanggai, where the great city of Karakhorum once stood, and the chain of Khin-gan-Petscha, which runs in a meridian line (in the part of Gobi traversed in going from Kiachta to Pekin by way of Urga), it is 760 miles. The whole extent of this elevated ground, which must be carefully distinguished from the more eastern and higher mountain-range, may be approximately estimated, including its deflections, at about three times the area of France. The map of the mountain-ranges and volcanoes of Central Asia, which I constructed in 1839, but did not publish until 1843, shows in the clearest manner the hypsometric relations between the mountain-ranges and the Gobi plateau. It was founded on the critical employment of all the astronomical determinations accessible to me, and on many of the very rich and copious orographic descriptions in which Chinese literature abounds, and which were examined at my request by Klaproth and Stanislaus Julien. My map marks in prominent characters the mean direction and the height of the mountain-chains, together with the chief features of the interior of the continent of Asia from 30 to 60 degrees of latitude, between the meridians of Pekin and Cherson. It differs essentially from any map hitherto published.

The Chinese enjoyed a triple advantage, by means of which they were enabled to enrich their earliest literature with so considerable an amount of orographic knowledge regarding Upper Asia, and more especially those regions situated between the In-schan, the alpine lake of Khuku-noor, and the shores of the Ili and Tarim, lying north and south of the Celestial Mountains, and which were so little known to Western Europe. These three advantages were, besides the peaceful conquests of the Buddhist pilgrims, the warlike expeditions towards the west (as early as the dynasties of Han and Thang, one hundred and twenty-two years before our era, and again in the ninth century, when conquerors advanced as far as Ferghana and the shores of the Caspian Sea); the religious interest attached to certain high mountain summits, on account of the periodical performance of sacrifices, in accordance with pre-existing enactments; and lastly, the early and generally known use of the compass for determining the direction of mountains and rivers. This use, and the knowledge of the south-pointing of the magnetic needle, twelve centuries before the Christian era, gave a great superiority to the orographic and hydrographic descriptions of the Chinese over those of Greek and Roman authors, who treated less frequently of subjects of this nature. The acute observer Strabo was alike ignorant of the direction of the Pyrenees and of that of the Alps and Apennines.*

To the lowlands belong almost the whole of Northern Asia to the north-west of the volcanic Celestial Mountains (Thian-schan); the steppes to the north of the Altai and the Sayanic chain; and the countries which extend from the mountains of Bolor, or Bulyt-tagh (Cloud Mountains in the Uigurian dialect), which run in a north and south direction, and from the upper Oxus, whose sources were discovered in the Pamershian Lake, Sir-i-kol (Lake Victoria), by the Buddhist pilgrims Hiuen-thsang and Song-yun in 518 and 629, by Marco Polo in 1277, and by Lieutenant Wood in 1838, towards the Caspian Sea; and from Lake Tenghiz or Balkasch, through the Kirghis Steppe, towards the Aral and the southern extremity of the Ural Mountains. In the vicinity of mountainous plains, whose elevation varies from 6000 to more than 10,000 feet above the sea's level, we may assuredly be allowed to apply the term lowlands to districts which are only elevated from 200 to 1200 feet. The first of these heights correspond with that of the city of Mannheim, and the second with that of Geneva and Tübingen. If we extend the application of the word *plateau*, which has so frequently been misused by modern geographers, to elevations of the soil which scarcely present any sensible difference in the character of the vegetation and climate, physical geography, owing to the indefiniteness of the merely relatively important terms of *high* and *low* land, will be unable to distinguish the connexion between elevation above the sea's level and climate, between the decrease of the temperature and the increase in elevation. When I was in Chinese Dzungarei,

* Compare Strabo, lib. ii. pp. 71, 128; lib. iii. p. 137; lib. iv. pp. 199, 202; lib. v. p. 211, Casaub.

between the boundaries of Siberia and Lake Saysan (Dsaisang), at an equal distance from the Icy Sea and the mouth of the Ganges, I might assuredly consider myself to be in *Central Asia.* The barometer, however, soon showed me that the elevation of the plains watered by the Upper Irtysch between Ustkamenogorsk and the Chinese Dzungarian post of Chonimailachu (the sheep-bleating) was scarcely as much as from 850 to 1170 feet. Pansner's earlier barometric determinations of height, which were first made known after my expedition, have been confirmed by my own observations. Both afford a refutation of the hypotheses of Chappe D'Auteroche (based on calculations of the fall of rivers) regarding the elevated position of the shores of the Irtysch, in Southern Siberia. Even further eastward, the Lake of Baikal is only 1420 feet above the level of the sea.

In order to associate the idea of the *relation* between *lowlands* and *highlands*, and of the successive gradations in the elevation of the soil, with actual data based on accurate measurements, I subjoin a table, in which the heights of the elevated plains of Europe, Africa, and America are given in an ascending scale. With these numbers we may then further compare all that has as yet been made known regarding the mean height of the Asiatic plains, or true *lowlands.*

| | Toises. | Feet. |
|---|---|---|
| Plateau of Auvergne | 170 | 1,087 |
| „ of Bavaria | 260 | 1,663 |
| „ of Castille | 350 | 2,238 |
| „ of Mysore | 460 | 2,942 |
| „ of Caracas | 480 | 3,070 |
| „ of Popayan | 900 | 5,755 |
| „ of the vicinity of the Lake of Tzana, in Abyssinia | 950 | 6,075 |
| „ of the Orange River (in South Africa) | 1000 | 6,395 |
| „ of Axum (in Abyssinia) | 1100 | 7,034 |
| „ of Mexico | 1170 | 7,482 |
| „ of Quito | 1490 | 9,528 |
| „ of the Province de los Pastos | 1600 | 10,231 |
| „ of the vicinity of the Lake of Titicaca | 2010 | 12,853 |

No portion of the so-called Desert of Gobi, which consists in part of fine pasture lands, has been so thoroughly investigated in relation to its differences of elevations as the zone which extends over an area of nearly 600 miles, between the sources of the Selenga and the Chinese wall. A very accurate barometrical levelling was executed, under the auspices of the Academy of St. Petersburgh, by two distinguished savans—the astronomer George Fuss, and the botanist Bunge. They accompanied a mission of Greek monks to Pekin, in the year 1832, in order to establish there one of those magnetic stations whose construction I had recommended. The mean height of this portion of the Desert of Gobi amounts hardly to 4263 feet, and not to 8000 or 8500 feet, as had been too hastily concluded from the measurements of contiguous mountain summits by the Jesuits Gerbillon and Verbiest. The surface of the Desert of Gobi is not more than 2558 feet above the level of the sea between Erghi, Durma, and Scharaburguna; and scarcely more than 320 feet higher than the plateau of Madrid. Erghi is situated midway, in 45° 31′ north lat., and 111° 26′ east long., in a depression of the land extending in a direction from south-west to north-east over a breadth of more than 240 miles. An ancient Mongolian saga designates this spot as the former site of a large inland sea. Reeds and saline plants, generally of the same species as those found on the low shores of the Caspian Sea, are here met with; while there are in this central part of the desert several small saline lakes, the salt of which is carried to China. According to a singular opinion prevalent among the Mongols, the ocean will at some period return, and again establish its dominion in Gobi. Such geological reveries remind us of the Chinese traditions of the *bitter lake*, in the interior of Siberia, of which I have elsewhere spoken.*

The basin of Kashmir, which has been so enthusiastically praised by Bernier, and too moderately estimated by Victor Jacquemont, has also given occasion to great hypsometric exaggerations. Jacquemont found by an accurate barometric measurement that the height of the Wulur Lake, in the valley of Kashmir, near the capital Sirinagur, was 5346 feet. Uncertain determinations by the boiling point of water gave Baron Carl von Hügel 5819 feet, and Lieutenant Cunningham only 5052 feet.† The mountainous

* Humboldt, *Asie centrale*, t. ii. p. 141; Klaproth, *Asie polyglotta*, p. 232.

† Compare my *Asie centrale*, t. iii. p. 310, with the *Journal of the Asiatic Soc. of Bengal*, vol. x. 1841, p. 114.

districts of Kashmir, which has excited so great an interest in Germany, and whose climatic advantages have lost somewhat of their reputation since Carl von Hügel's account of the four months of winter snow in the streets of Sirinagur,* does not lie on the high crests of the Himalaya, as has commonly been supposed, but constitutes a true cauldron-like valley on their southern declivity. On the south-west, where the rampart-like Pir Panjal separates it from the Indian Punjaub, the snow-crowned summits are covered, according to Vigne, by basaltic and amygdaloid formations. The latter are very characteristically termed by the natives *schischak deyu*, or devil's pock-marks.† The charms of the vegetation have also been very differently described, according as travellers passed into Kashmir from the south, and left behind them the luxuriant and varied vegetation of India; or from the northern regions of Turkestan, Samarkand, and Ferghana.

Moreover, it is only very recently that we have obtained a clearer view regarding the elevation of Thibet, the level of the plateau having long been uncritically confounded with the mountain tops rising from it. Thibet occupies the space between the two great chains of the Himalaya and the Kuen-lün, and forms the elevated ground of the valley between them. The land is divided from east to west, both by the inhabitants and by Chinese geographers, into three parts. We distinguish Upper Thibet, with its capital, H'lassa (probably 9592 feet high); Middle Thibet, with the town of Leh or Ladak (9995 feet); and Little Thibet, or Baltistan, called the Thibet of Apricots (Sari-Butan), in which lie Iskardo (6300 feet), Gilgit, and south of Iskardo, but on the left bank of the Indus, the plateau Deotsuh, whose elevation was determined by Vigne (11,977 feet). On carefully examining all the notices we have hitherto possessed regarding the three Thibets, and which will have been abundantly augmented during the present year by the brilliant boundary surveying expedition under the auspices of the Governor-general, Lord Dalhousie, we soon become convinced that the region between the Himalaya and the Kuen-lün is no unbroken table-land, but that it is intersected by mountain groups, which undoubtedly belong to perfectly distinct systems of elevation.

* See his *Kashmir*, Bd. ii. s. 196.

† Vigne, *Travels in Kashmir*, 1842, vol. i. pp. 237—293.

Actual plains are very few in number: the most considerable are those between Gertop, Daba, Schang-thung (the Shepherd's Plain), the native country of the shawl-goat, and Schipke (10,449 feet); those round Ladak, which attain an elevation of 13,429 feet, and must not be confounded with the depressed land in which the town lies; and finally, the plateau of the Sacred Lakes, Manasa and Ravanahrada (probably 14,965 feet), which was visited by Father Antonio de Andrada as early as the year 1625. Other parts are entirely filled with compressed mountain masses, "rising," as a recent traveller observes, "like the waves of a vast ocean." Along the rivers, the Indus, the Sutledge, and the Yaru-dzangbo-tschu, which was formerly regarded as identical with the Buramputer (or correctly the Brahma-putra), points have been measured which are only between 6714 and 8952 feet above the sea; and the same is the case with the Thibetian villages Pangi, Kunawur, Kelu, and Murung.* From many carefully collected determinations of heights, I think that we are justified in assuming that the plateau of Thibet between 73° and 85° east long. does not attain a mean elevation of 11,510 feet: this is hardly the elevation of the fruitful plain of Caxamarca in Peru, and is 1349 and 2155 feet less than the plateau of Titicaca, and of the street pavement of the Upper Town of Potosi (13,665 feet).

That beyond the Thibetian highlands and the Gobi, whose outline has been already defined, Asia presents considerable depressions, and indeed true lowlands, between the parallels of 37° and 48°, where once an immeasurable continuous plateau was fabulously supposed to exist, is proved by the cultivation of plants which cannot flourish without a certain degree of temperature. An attentive study of the travels of Marco Polo, in which mention is made of the cultivation of the vine, and of the production of cotton in northern latitudes, had long ago directed the attention of the acute Klaproth to this point. In a Chinese work, bearing the title *Information respecting the recently conquered Barbarians* (Sin-kiang-wai-tan-ki-lio), it is stated that "the country of Aksu, somewhat to the south of the Celestial Mountains, near the rivers which form the great Tarim-gol, produces grapes, pomegranates, and numberless other fruits of singular excel-

* Humboldt, *Asie Centrale*, t. iii. pp. 281—325.

lence; also cotton (Gossypium religiosum), which covers the fields like yellow clouds. In summer the heat is extremely great, and in winter there is here, as at Turfan, neither intense cold nor heavy snow." The neighbourhood of Khotan, Kaschgar, and Yarkand still, as in the time of Marco Polo,* pays its tribute in home-grown cotton. In the oasis of Hami (Khamil), above 200 miles east of Aksu, orange trees, pomegranates, and the finer vines are found to flourish.

The products of cultivation which are here noticed lead to the belief that over extensive districts the elevation of the soil is very slight. At so great a distance from the sea side, and in the easterly situation which so much increases the degree of winter cold, a plateau, as high as Madrid or Munich, might indeed have a very hot summer, but would hardly have, in 43° and 44° latitude, an extremely mild and almost snowless winter. I have seen a high summer heat favour the cultivation of the vine, as at the Caspian Sea, 83 feet below the level of the Black Sea (at Astrachan, latitude 46° 21′); but the winter cold is there from − 4° to − 13°. Moreover, the vine is sunk to a greater depth in the ground after the month of November. We can understand that cultivated plants, which, as it were, live only in the summer, as the vine, the cotton plant, rice, and melons, may be cultivated with success between the latitudes of 40° and 44°, on plateaux at an elevation of more than 3000† feet, and may be favoured by the action of radiant heat; but how could the pomegranate trees of Aksu, and the orange trees of Hami, whose fruit Father Grosier extolled as excellent, endure a long and severe winter (the necessary consequence of a great elevation‡)? Carl Zimmerman§ has shown it to be extremely probable that the Tarim depression, or the desert between the mountain chain of Thian-schan and Kuen-lün, where the steppe river Tarimgol discharges itself into the Lake of Lop, formerly described as an alpine lake, is hardly 1280 feet above the level of the sea, or only twice the elevation of Prague. Sir Alexander Burnes also ascribes to Bokhara only

* *Il Milione di Marco Polo,* pubbl. dal Conte Baldelli, t. i. pp. 32 and 87.

† 500 toises in the German, accurately 3197 feet. Tr.

‡ *Asie centrale,* t. ii. pp. 48—52 and 429.

§ In the learned Analysis of his *Karte von Inner Asien,* 1841, s. 99.

an elevation of 1188 feet. It is most earnestly to be desired that all doubt regarding the elevation of the plateaux of Central Asia, south of 45° north latitude, should finally be removed by direct barometrical measurements, or by determinations of the boiling point of water, conducted with greater care than is usual in these cases. All our calculations of the difference between the limits of perpetual snow and the maximum elevation of vine cultivation in different climates, rest at present on too complex and uncertain elements.

In order as briefly as possible to rectify that which has been advanced in the former edition of the present work, regarding the great mountain systems which intersect the interior of Asia, I subjoin the following general review:—We begin with the four *parallel chains*, which run, with tolerable regularity, from east to west, and are connected together by means of a few detached transverse lines. Differences of direction indicate, as in the Alps of Western Europe, a difference in the epoch of elevation. After the four parallel chains (the *Altai*, the *Thian-schan*, the *Kuen-lün*, and the *Himalaya*) we must consider as following the direction of meridian, the Ural, the Bolor, the Khingan, and the Chinese chains, which, with the great inflection of the Thibetian and Assam-Birmese Dzangbo-tschu incline from north to south. The Ural divides a depressed portion of Europe from a similarly low portion of Asia. The latter was called by Herodotus,* and even earlier by Pherecydes of Syros, Scythian or Siberian Europe, and comprised all the countries to the north of the Caspian and of the Iaxartes, which flows from east to west, and may therefore be regarded as a continuation of our Europe, "as it now exists, extending lengthwise across the continent of Asia."

1. The great mountain system of the Altai (the "gold mountains" of Menander of Byzantium, an historical writer of the seventh century; the Altaï-alin of the Moguls, and the Kin-schan of the Chinese) forms the southern boundary of the great Siberian lowlands, and running between 50° and 52½° north latitude, extends from the rich silver mines of the Snake Mountains, and the confluence of the Uba and the Irtysch, to the meridian of Lake Baikal. The divisions and names of the "Great" and the "Little Altai," taken from an obscure passage of Abulghasi, should be wholly avoided.†

* Ed. Schweighaüser, t. v. p. 204. † *Asie centrale*, t. i. p. 247.

The mountain system of the Altai comprehends—(*a*) the Altai proper, or Kolywanski Altai, which is entirely under the Russian sceptre: it lies to the west of the intersecting fissures of the Telezki Lake, which follow the direction of the meridian; and in ante-historic times probably constituted the eastern shore of the great arm of the sea, by which, in the direction of the still existing lakes, Aksakal-Barbi and Sary-Kupa,* the Aralo-Caspian basin was connected with the Icy sea;—(*b*) East of the Telezki chains, which follow the direction of the meridian, the Sayani, Tangnu, and Ulangom, or Malakha ranges, all tolerably parallel with each other, and following an east and west direction. The Tangnu, which merges in the basin of the Selenga, has, from very remote times, constituted the national boundary between the Turkish race, to the south, and the Kïrghis (Hakas, identical with Σάκαι), to the north.† It is the original seat of the Samoieds or Soyotes, who wandered as far as the Icy Sea, and were long regarded in Europe as a race inhabiting exclusively the coasts of the Polar Sea. The highest snow-covered summits of the Kolywan Altai are the Bielucha and the Katunia Pillars. The latter attain only a height of about 11,000 feet, or about the height of Etna. The Daurian highland, to which the mountain node of Kemtei belongs, and on whose eastern margin lies the Jablonoi Chrebet, divides the depressions of the Baikal and the Amur.

2. The mountain system of the Thian-schan, or the chair of the Celestial Mountains, the Tengri-tagh of the Turks (Tukiu), and of the kindred race of the Hiongnu, is eight times as long, in an east and west direction, as the Pyrenees. Beyond, that is to say, to the west of its intersection with the meridian chain of the Bolor and Kosuyrt, the Thian-schan bears the names of Asferah and Aktagh, is rich in metals, and is intersected with open fissures, which emit hot vapours luminous at night, and which are used for obtaining sal-ammoniac.‡ East of the transverse Bolor and Kosyurt chain, there follow successively in the Thian-schan, the Kashgar Pass (Kaschgar-dawan), the Glacier Pass of Djeparle, which leads to Kutch and Aksu in the Tarim basin; the volcano of Pe-schan, which

* *Asie centrale,* t. ii. p. 138.

† Jacob Grimm, *Gesch. der deutschen Sprache,* 1848, Th. i. s. 227.

‡ *Asie centrale,* t. ii. pp. 18—20.

erupted fire and streams of lava at least as late as the middle of the seventh century; the great snow-covered massive elevation of Bogdo-Oola; the Solfatara of Urumtsi, which furnishes sulphur and sal-ammoniac (nao-scha), and lies in a coal district; the volcano of Turfan (or volcano of Ho-tscheu or Bischbalik), almost midway between the meridians of Turfan (Kune Turpan), and of Pidjan, and which is still in a state of activity. The volcanic eruptions of the Thian-schan chain reach, according to Chinese historians, as far back as the year 89, A.D., when the Hiongnu were pursued by the Chinese from the sources of the Irtysch as far as Kutch and Kharaschar*. The Chinese General, Teu-hian, crossed the Thian-schan, and saw "the Fire Mountains, which sent out masses of molten rock that flow to the distance of many *Li*."

The great distance of the volcanoes of the interior of Asia from the sea coast is a remarkable and isolated phenomenon. Abel Rémusat, in a letter to Cordier†, first directed the attention of geologists to this fact. This distance, for instance, in the case of the volcano of Pe-schan, from the north or the Icy Sea at the mouth of the Obi, is 1528 miles; and from the south or the mouths of the Indus and the Ganges, 1512 miles; so central is the position of fire-emitting volcanoes in the Asiatic continent. To the west its distance from the Caspian at the Gulf of Karuboghaz, is 1360 miles, and from the east shores of the Lake of Aral, 1020 miles. The active volcanoes of the New World had hitherto offered the most remarkable examples of great distance from the sea coast, but in the case of the volcano of Popocateptl, in Mexico, this distance is only one hundred and thirty-two miles, and only ninety-two, one hundred and four, and one hundred and fifty-six, respectively in the South American volcanoes Sangai, Tolima, and de la Fragua. All extinct volcanoes, and all trachytic mountains, which have no permanent connexion with the interior of the earth, have been excluded from these statements‡. East of the volcano of Turfat, and of the fruitful Oasis of Hami, the chain of the Thian-schan merges into the great elevated tract of Gobi, which runs in a S.W. and N.E. direction. This interruption

* Klaproth, *Tableau hist. de l'Asie*, p. 108.

† *Annales des Mines*, t. v. 1820, p. 137.

‡ *Asie centrale*, t. ii. pp. 16—55, 69—77, 341, 356.

of the mountain chain continues for more than 9½ degrees of longitude; it is caused by the transversal intersection of the Gobi, but beyond the latter, the more southern chain of In-schan (Silver Mountains), proceeding from west to east, to the shores of the Pacific near Pekin (north of the Pe-tscheli), forms a continuation of the Thian-schan. As we may regard the In-schan as an eastern prolongation of the fissure from which the Thian-schan is upheaved, so we may also be inclined to consider the Caucasus as a western prolongation of the same range, beyond the Great Aralo-Caspian basin or of the low-lands of Turan. The mean parallel or axis of elevation of the Thian-schan oscillates between 40° 40′ and 43° north latitude; that of the Caucasus (inclining, according to the map of the Russian Staff, from E.S.E. to W.N.W.) between 41° and 44°.* Of the four parallel chains that traverse Asia, the Thian-schan is the only one of which no summit has as yet been measured.

3. The mountain system of the Kuen-lün (Kurkun or Kul-kun), including the Hindoo-Coosh, with its western prolongation in the Persian Elburz and Demavend, and the American chain of the Andes, constitute the longest lines of elevation on our planet. At the point where the meridian chain of the Bolor intersects the Kuen-lün at right angles, the latter receives the name of Onion Mountains (Tchsung-ling), a term also applied to a portion of the Bolor at the inner eastern angle of intersection. Bounding Thibet in the north, the Kuen-lün runs in a regular direction from east to west, in the parallel of 36° north latitude; until the chain is broken in the meridian of H'lassa, by the vast mountain node which surrounds the *Sea of Stars*, *Sing-so-hai* (so celebrated in the mythical geography of the Chinese), and the Alpine lake of Khuku-noor. The chains of Nan-schan and Kilian-schan, lying somewhat further north, and extending to the Chinese wall near Liang-tsheu, may almost be regarded as the eastern prolongation of the Kuen-lün. To the west of the intersection of the Bolor and the Kuen-lün (Tchsung-ling), the regular direction of the axes of elevation (inclining from east to west in the Kuen-lün and Hindoo-Coosh, and from south-

* Baron von Meyendorff in the *Bulletin de la Société Géologique de France*, t. ix. 1837—1838, p, 230.

east to north-west in the Himalaya) proves, as I have elsewhere attempted to show, that the Hindoo-Coosh is a prolongation of the Kuen-lün and not of the Himalaya.* From the Taurus in Lycia to the Kafiristan, the chain follows the parallel of Rhodes (the diaphragm of Dicæarchus) over a distance of 45 degrees of longitude. The grand geological views of Eratosthenes,† which were further developed by Marinus of Tyre, and by Ptolemy, and according to which "the prolongation of the Taurus in Lycia was continued, in the same direction, through all Asia as far as India," appear in part to be based on representations derived by the Persians and Indians from the Punjaub.

"The Brahmins maintain," says Cosmas Indicopleustes, in his Christian Topography‡, "that a line drawn from Tzinitza (Thinæ) across Persia and Romania, would exactly pass over the centre of the inhabited earth." It is remarkable, as Eratosthenes observes, that this greatest axis of elevation in the old world passes directly through the basin (the depression) of the Mediterranean, in the parallels of 35½° and 36° north latitude, to the Pillars of Hercules.§ The most eastern portion of Hindoo-Coosh is the Paropanisus of the ancients, the Indian Caucasus of the companions of the great Macedonian. The name of *Hindoo-Coosh*, which is so frequently used by geographers, does not in reality apply to more than one single mountain pass, where the climate is so severe, as we learn from the travels of the Arabian writer, Ibn Batuta, that many Indian slaves frequently perish from the cold.‖ The Kuen-lün still exhibits active fire-emitting eruptions at the distance of several hundred miles from the sea-coast. Flames, visible at a great distance, burst from the cavern of the mountain of Schinkhieu, as I learn from a translation of the Yuen-thong-ki, made by my friend Stanislaus Julien.¶ The loftiest summit in the Hindoo-Coosh, north-west of Jellalabad, is 20,232 feet above the level of the sea; to the west, towards Herat, the

* *Asie centrale,* t. i. pp. xxiii et 118—159; t. ii. pp. 431—434, 465.

† Strabo, lib. ii. p. 68; lib. xi. pp. 490, 511; lib. xv. p. 689.

‡ Montfaucon, *Collectio nova Patrum,* t. ii. p. 137.

§ Compare *Asie centrale,* t. i. pp. xxiii et 122—138; t. ii. pp. 430—434, with *Cosmos,* vol. ii. p. 543, Bohn's ed.

‖ *Travels,* p. 97.

¶ *Asie centrale,* t. ii. pp. 427, 483.

chain sinks to 2558 feet, rising again north of Teheran, in the volcano of Demavend, to the height of 14,675 feet.

4. The mountain system of the Himalaya has a normal direction from east to west, running more than 15 degrees of longitude (from 81 to 97°), or from the colossal mountain Dhawalagiri (28,072 feet) to the intersection of the Dzangbo-tscheu (the Irawaddy of Dalrymple and Klaproth), whose existence was long regarded as problematical, and to the meridian chains, which cover the whole of Western China, and form the great mountain group, from which spring the sources of the Kiang, in the provinces of Sse-tschuan, Hu-kuang, and Kuang-si. Next to the Dhawalagiri, the Kinchinjinga, and not the more eastern peak of Schamalari, as has hitherto been supposed, is the highest point of this portion of the Himalaya, which inclines from east to west. The Kinchinjinga, in the meridian of Sikhim, between Butan and Nepal, between the Schamalari (23,980 feet) and the Dhawalagiri, is 28,174 feet in height.

It is only within the present year that it has been trigonometrically measured with exactness, and as I learn from India through the same channel, "that a new measurement of the Dhawalagiri still leaves it the first place among all the snow-crowned summits of the Himalaya," this mountain must necessarily have a greater elevation than the 28,072 feet hitherto ascribed to it.* The point of deflection in the direction of the chain is, near the Dhawalagiri, in 81° 22', east longitude. From thence the Himalaya no longer follows a due west direction, but runs from S.E. to N.W., as a vast connecting system of veins between Mozufer-abad and Gilgit, merging into a part of the Hindoo-Coosh chain in the south of Kafiristan. Such a turn and alteration in the line of the axis of elevation of the Himalaya (from E.—W. to S.E —N.W.) certainly indicates, as in the western region of our European Alpine mountains, a different age or period of elevation. The course of the Upper Indus, from the sacred lakes of Manasa and Ravana-hrada, (at an elevation of 14,965 feet,) in the vicinity of which this great river takes its origin, to Iskardo, and to the plateau of Deotsuh (at an elevation of 12,994 feet), measured by Vigne, follows in the Thi-

* From a letter of Dr. Joseph Hooker, the learned botanist to the last Antarctic expedition, dated Darjeeling, 25th of July, 1848.

betian highlands the same north-westerly direction as the Himalaya.

Here are situated the Djawahir, whose height was long since accurately determined at 26,902 feet, and the Alpine valley of Caschmere (never visited by winds or storms), where, at an elevation of only 5346 feet, lies the lake of Wulur, which freezes every winter, and whose surface is never broken by a single ripple.

After considering the four great mountain systems of Asia, which, in their normal geognostic character, are true parallel chains, we must turn to the long series of *alternating* elevations following a direction from north to south, and which extend from Cape Comorin, opposite to the island of Ceylon, to the Icy Sea, alternating between the parallels of 66° and 77° east longitude, from S.S.E. to N.N.W. To this system of meridian chains, whose alternations remind us of faults in veins, belong the Ghauts, the Soliman chain, the Paralasa, the Bolor, and the Ural range. This interruption of the profile of the elevation is so constituted, that each new chain begins in a degree of latitude beyond that to which the preceding one had attained, all alternating successively in an opposite direction. The importance which the Greeks (probably not earlier than the second century of our era) attached to these chains running from north to south, induced Agathodæmon and Ptolemy (*Tab.* vii. et viii.) to regard the Bolor under the name of Imaus as an axis of elevation, which extended as far as 62° north latitude into the basin of the lower Irtysch and Obi.*

As the vertical height of mountain summits above the sea's level (however unimportant the phenomenon of the more or less extensive folding of the crust of a planetary sphere may be in the eyes of geognosists) will always continue, like all that is difficult of attainment, to be an object of general curiosity, the present would appear to furnish a fitting place for the introduction of an historical notice relative to the gradual advance of hypsometric knowledge. When I returned to Europe in 1804, after an absence of four years, not one of the high snow-crowned summits of Asia (in the Himalaya, the Hindoo-Coosh, or the Caucasus) had been yet measured with any degree of accuracy. I was unable, therefore, to

* *Asie centrale,* t. i. pp. 138, 154, 198; t. ii. p. 367.

compare my determinations of the heights of perpetual snow in the Cordilleras of Quito or the mountains of Mexico, with any results obtained in India. The important travels of Turner, Davis, and Saunders to the highlands of Thibet, were indeed accomplished in the year 1783; but the intelligent Colebrooke justly observed that the height of the Schamalari (28° 5′ north latitude, 89° 30′ east longitude, somewhat north of Tassisudan), as given by Turner, rested on a foundation quite as slight as the assumed measurements of the heights seen from Patna and Kafiristan by Colonel Crawford and Lieutenant Macartney.* The admirable labours of Webb, Hodgson. Herbert, and the brothers Gerard, have indeed thrown considerable light on the question concerning the heights of the colossal summits of the Himalaya; but yet, in 1808, the hypsometric knowledge of the East Indian mountain chains was still so uncertain, that Webb wrote to Colebrooke, "The height of the Himalaya still remains undetermined. It is true that I have ascertained that the summits visible from the elevated plains of Rohilkand are 21,000 feet higher than that plateau, but we are ignorant of their absolute height above the sea."

In the year 1820 it first began to be currently reported in Europe that there were not only much higher summits in the Himalaya than in the Cordilleras, but that Webb had seen in the pass of Niti, and Moorcroft in the Thibetian plateau of Daba, and the sacred lakes, fine corn-fields and fertile pasture-lands at elevations far exceeding the height of Mont Blanc. This announcement was received in England with great incredulity, and opposed by doubts regarding the influence of the refraction of light. I have shown the unsoundness of such doubts in two printed treatises on the mountains of India, in the *Annales de Chimie et de Physique*. The Tyrolese Jesuit, Father Tiefenthaler, who in 1766 penetrated as far as the provinces of Kemaun and Nepal, had already divined the importance of the Dhawalagiri. We read on his map: "*Montes Albi, qui Indis Dolaghir, nive obsiti.*" Captain Webb always employs the same name. Until the measurements of the Djawahir (30° 22′ north latitude, and 79° 58′

* Compare Turner in the *Asiatic Researches*, vol. xii. p. 234, with Elphinstone, *Account of the Kingdom of Caubul*, 1815, p. 95, and Francis Hamilton, *Account of Nepal*, 1819, p. 92.

east longitude, 26,902 feet in elevation), and of the Dhawalagiri (28° 40 north latitude, and 83° 21′ east longitude, 28,072 feet in elevation), were made known in Europe, the Chimborazo, which, according to my trigonometrical measurement, was 21,422 feet, in height,* was still everywhere regarded as the loftiest summit on the earth. The Himalaya appeared, therefore, at that time, to be 4323 feet or 6620 feet higher than the Cordilleras, according as the comparison was made with the Djawahir or the Dhawalagiri. Pentland's South American travels, in the years 1827 and 1838, directed attention to two snow-crowned summits of Upper Peru, east of the lake of Titicaca, which were conjectured to be respectively 3824 and 2578 feet higher than the Chimborazo.† It has been already observed,‡ that the most recent computations in the measurements of the Sorata and Illimani have shown the error of this hypsometric assertion. The Dhawalagiri, therefore, on whose declivity in the river-valley of Ghandaki, the Salagrana Ammonites, so celebrated in the Brahminical ritual as symbols of the testaceous incarnation of Vishnu, are collected, still indicates a difference of elevation between both continents of more than 6600 feet.

The question has been asked, whether there may not be still greater heights in the rear of the southernmost chain, which has been as yet measured with more or less exactitude. Colonel George Lloyd, who in 1840 edited the important observations of Captain Alexander Gerard and his brother, entertains the opinion, that in that part of the Himalaya, which he somewhat indefinitely names the "Tartaric Chain" (and consequently in Northern Thibet, in the direction of the Kuen-lün, perhaps in the Kailasa of the sacred lakes or beyond Leh) there are mountain-summits which attain an elevation of from 29,000 to 30,000 feet, one or two thousand feet higher, therefore, than the Dhawalagiri.§ No definite opinion can be formed on the subject until we are in the possession

* *Recueil d'Observations astronomiques,* t. i. p. 73.

† *Annuaire du Bureau des Longitudes pour* 1830, pp. 320, 323.

‡ See Illustration (5), p. 44.

§ See Lloyd and Gerard, *Tour in the Himalaya,* 1840, vol. i., pp. 143, 312, and *Asie centrale,* t. iii., p. 324.

of actual measurements, since the indication which led the natives of Quito, long before the arrival of Bouguer and La Condamine, to regard the summit of the Chimborazo as the culminating point—or the highest point within the region of perpetual snow—is rendered very deceptive in the temperate zone of Thibet, where the radiation of the table-land is so effective, and where the lower limit of perpetual snow does not constitute a regular line of equal level as in the tropics. The greatest elevation above the level of the sea that has been reached by man on the sides of the Himalaya is 19,488 feet. This elevation was gained by Captain Gerard, with seven barometers, as we have already observed, on the mountain of Tarhigang, somewhat to the north-west of Schipke.* This happens to be almost the same height as that to which I myself ascended up on the Chimborazo (on the 23rd of June, 1802), and which was reached thirty years later (16th of December, 1831) by my friend Boussingault. The unattained summit of the Tarhigang is, moreover, 1255 feet higher than the Chimborazo.

The passes across the Himalaya from Hindostan to Chinese Tartary, or rather to Western Thibet, especially between the rivers Buspa and Schipke, or Langzing Khampa, are from 15,347 to 18,544 feet in height. In the chain of the Andes I found that the pass of Assuay, between Quito and Cuenca, at the Ladera de Cadlud, was also fully 15,566 feet above the level of the sea. A great part of the Alpine plains of the interior of Asia would lie buried throughout the whole year in snow and ice, if the limits of perpetual snow were not singularly elevated, probably to about 16,626 feet, by the force of the heat radiated from the Thibetian plain, the constant serenity of the sky, the rarity of the formation of snow in the dry atmosphere, and by the powerful solar heat peculiar to the eastern continental climate, which characterizes the northern declivity of the Himalaya. Fields of barley (of *Hordeum hexastichon*) have been seen in Kunawur at an elevation of 14,700 feet and another varitey of barley, called Ooa, and allied to *Hordeum cœleste*, even

* Colebrooke, in the *Transactions of the Geological Society*, vol. vi. p. 411.

much higher. Wheat thrives admirably well in the Thibetian highlands, up to an elevation of 12,000 feet. On the northern declivity of the Himalaya, Captain Gerard found that the upper limits of the birch woods ascend to 14,069 feet; and small brushwood used by the natives for fuel in their huts is even found within the parallels of 30° 45′ and 31° north latitude, at an elevation of 16,946 feet, and therefore nearly 1280 feet higher than the lower snow-limit in the equatorial regions. It follows from the data hitherto collected that on the northern declivity of the Himalaya the mean of the lower snow-line is at least 16,626 feet, whilst on the southern declivity it falls to 12,980 feet. But for this remarkable distribution of heat in the upper strata of the atmosphere, the mountain plain of Western Thibet would be rendered uninhabitable for the millions of men who now occupy it.*

In a letter which I have lately received from India from Dr. Joseph Hooker, who is engaged in meteorological and geological observations, as well as in the study of the geography of plants, he says, "Mr. Hodgson, whom we here consider more thoroughly conversant than any other geographer with the hypsometric relations of the snow ranges, recognises the correctness of the opinions you have advanced in the third part of your *Asie centrale*, regarding the cause of the unequal height of the limit of perpetual snow on the northern and the southern declivity of the Himalaya range. In the trans-Sutledge region (in 36° north latitude) we often observed the snow limit as high as 20,000 feet, whilst in the passes south of Brahmaputra, between Assam and Birmah (in 27° north latitude), where the most southern snow-capped mountains of Asia are situated, the snow limit sinks to 15,000 feet." I believe we ought to distinguish between the extreme and the mean elevations, but in both we find the formerly disputed difference between the Thibetian and the Indian declivities manifested in the clearest manner.

* Compare my investigation regarding the snow-limit on both declivities of the Himalaya in my *Asie centrale*, t. ii., pp. 435—437; t. iii, pp. 281—326; and in *Cosmos*, vol. i., p. 337, Bohn's ed.

| My results for the mean height of the snow line as given in *Asie centrale*, t. iii., p. 326. | Feet. | Extremes according to Dr. Hooker's Letter. | Feet. |
|---|---|---|---|
| Northern declivity ......... | 16,626 | Northern declivity ...... .. | 20,000 |
| Southern „ ......... | 12,981 | Southern „ ......... | 15,000 |
| Difference ......... | 3,645 | Difference ......... | 5,000 |

The local differences vary still more, as may be seen from the series of extremes given in *Asie centrale*, t. iii., p. 295. Alexander Gerard saw the snow-limit ascend to 20,463 feet on the Thibetian declivity of the Himalaya; and Jacquemont found it as low as 11,500 feet on the south-Indian declivity, north of Cursali on the Jumnautri.

[The recent investigations of Lieutenant Strachey show that M. Humboldt has been led astray, when treating of the Himalaya, by the very authorities on whom he placed the most reliance. The results of his inquiries on this point are given in the first volume of the Cosmos (Bohn's Ed.), pp. 9 and 338. As the subject is one of considerable interest we give a brief sketch of Lieutenant Strachey's* recent labours, confining ourselves to his own views, and omitting (for want of space) his somewhat lengthy exposition of the errors committed by the authorities quoted by Humboldt. The following are his personal observations regarding the *southern limit of the belt of perpetual snow.*

"In this part of the Himalaya it is not, on an average of years, till the beginning of December, that the snow line appears decidedly to descend for the winter. After the end of September, indeed, when the rains are quite over, light falls of snow are not of very uncommon occurrence on the higher mountains, even down to 12,000 feet; but their effects usually disappear very quickly, often in a few hours. The latter part of October, the whole of November, and the beginning of December, are here generally characterised by the beautiful serenity of the sky; and it is at this season, on the southern edge of the belt, that the line of perpetual snow is seen to attain its greatest elevation.

"The following are the results of trigonometrical measure-

* *Journal of the Asiatic Society of Bengal.* New Series. No. xxviii. p. 287.

ments of the elevation of the inferior edge of snow on spurs of the Treslú and Nandádevi groups of peaks, made, before the winter snow had begun, in November, 1848.

| Point observed. | Height as observed on face exposed to the East. | | | Height on face exposed to West. Observed from Almorah. |
|---|---|---|---|---|
| | From Almorah, (height, 5586 ft.) | From Binsar, (height, 7969 ft.) | Mean. | |
| No. | Feet. | Feet. | et. | Feet. |
| 1 | 16,599 | 16,767 | 16,683 | 15,872 |
| 2 | 16,969 | 17,005 | 16,987 | .. .. |
| 3 | 17,186 | 17,185 | 17,185 | 14,878 |
| 4 | 15,293 | 15,361 | 15,327 | .. .. |

The points 1, 2 and 3 are in ridges that run in a south-westerly direction. The dip of the strata being to the north-east, the faces exposed to view from the south are for the most part very abrupt, and snow never accumulates on them to any great extent. This in some measure will account for the height to which the snow is seen to have receded on the eastern exposures, that is, upwards of 17,000 feet. On the western exposures the ground is less steep, and the snow is seen to have been observed at a considerable less elevation; but it was in very small quantities, and had probably fallen lately, so that I am inclined to think that its height, viz., about 15,000 feet, rather indicates the elevation below which the light autumnal falls of snow were incapable of lying, than that of the inferior edge of the perpetual snow. It is further to be understood, that below this level of 15,000 feet the mountains were absolutely without snow, excepting those small isolated patches that are seen in ravines, or at the head of glaciers, which, of course, do not affect such calculations as these. On the whole, therefore, I consider that the height of the snow-line on the more prominent points of the southern edge of the belt may be fairly reckoned at 16,000 feet at the very least.

"The point No. 4 was selected as being in a much more retired position than the others. It is situate not far from the head of the Pindur river. It was quite free from snow at 15,300 feet, and I shall therefore consider 15,000 feet as the elevation of the snow-line in the re-entering angles of the chain.

"I conclude, then, that 15,500 feet, the mean of the heights at the most and least prominent points, should be assigned as the mean elevation of the snow-line at the southern limit of the belt of perpetual snow in Kumaon; and I conceive that whatever error there may be in this estimate will be found to lie on the side of diminution rather than of exaggeration.

"This result appears to accord well with what has been observed in the Bissehir range. The account given by Dr. Gerard of his visit to the Shátúl Pass on this range, which he undertook expressly for the purpose of determining the height of the snow-line, contains the only definite information as to the limit of the perpetual snow at the southern edge of the belt that is to be found in the whole of the published writings of the Gerards; and the following is a short abstract of his observations. Dr. Gerard reached the summit of the Shátúl Pass, the elevation of which is 15,500 feet, on the 9th of August, 1822, and remained there till the 15th of the same month. He found the southern slope of the range generally free from snow, and he states that it is sometimes left without any whatever. On the top of the pass itself there was no snow; but on the northern slope of the mountain it lay as far down as about 14,000 feet. On his arrival rain was falling, and out of the four days of his stay on this pass it either rained or snowed for the greater part of three. The fresh snow that fell during this time did not lie below 16,000 feet, and some of the more precipitous rocks remained clear even up to 17,000 feet.

"The conclusion to which Dr. Gerard comes from these facts is, that the snow-line on the southern face of the Bissehir range is at 15,000 feet above the sea. But I should myself be more inclined, from his account, to consider that 15,500 feet was nearer the truth; and in this view I am confirmed by verbal accounts of the state of the passes on this range, which I have obtained from persons of my acquaintance, who have crossed them somewhat later in the year. The difference, however, is after all trifling.

"Such is the direct evidence that can be offered on the height of the snow-line at the southern limit of the belt of perpetual snow: some additional light, may, however, be thrown on the subject generally by my shortly explaining the state in which I have found the higher parts of the mountains at the different seasons during which I have visited them.

"In the beginning of May, on the mountains to the east of the Rámganga river, near Námik, I found the ground on the summit of the ridge, called Champwá, not only perfectly free from snow at an elevation of 12,000 feet, but covered with flowers, in some places golden with calsha and ranunculus polypetalus, in others purple with primulus. The snow had in fact already receded to upwards of 12,500 feet, behind which even a few little gentians proclaimed the advent of spring.

"Towards the end of the same month, at the end of the Pindur, near the glacier from which that river rises, an open spot on which I could pitch my tent could not be found above 12,000 feet. But here the accumulation of snow, which was considerable in all ravines even below 11,000 feet, is manifestly the result of avalanches and drift. The surface of the glacier, clear ice as well as moraines, was quite free from snow up to nearly 13,000 feet; but the effect of the more retired position of the place in retarding the melting of the snow, was manifest from the less advanced state of the vegetation. During my stay at Pinduri the weather was very bad, and several inches of snow fell; but, excepting where it had fallen on the old snow, it all melted off again in a few hours, even without the assistance of the sun's direct rays. On the glacier, at 13,000 feet, it had all disappeared twelve hours after it fell.

"On revisiting Pinduri about the middle of October, the change that had taken place was very striking. Now not a sign of snow was to be seen on any part of the road up to the very head of the glacier; a luxuriant vegetation had sprung up, but had already almost entirely perished, and its remains covered the ground as far as I went. From this elevation, about 13,000 feet, evident signs of vegetation could be seen to extend far up the less precipitous mountains. The place is not one at which the height of the perpetual snow can be easily estimated, for on all sides are glaciers, and the vast accumulations of snow from which they are supplied, and these cannot always be readily distinguished from snow *in situ*; but as far as I could judge, those places which might be considered as offering a fair criterion were free from snow up to 15,000, or even 16,000 feet.

"Towards the end of August I crossed the Barjikang Pass, between Rálam and Juhár, the elevation of which is about 15,300 feet. There was here no vestige of snow on the ascent to the pass from the south-east, and only a very small patch remained on the north-western face. The view of the continuation of the ridge in a southerly direction was cut off by a prominent point, but no snow lay on that side within 500 feet of the pass, while to the north I estimated that there was no snow in considerable quantity within 1500 feet or more, that is, nearly up to 17,000 feet. The vegetation on the very summit of the pass was far from scanty, though it had already begun to break up into tufts, and had lost that character of continuity which it had maintained to within a height of 500 or 600 feet. Species of Potentilla, Sedum, Saxifraga, Corydalis, Aconitum, Delphinium, Thalictrum, Ranunculus Saussurea, Gentiana, Pedicularis, Primula, Rheum, and Polygonum, all evidently flourishing in a congenial climate, showed that the limits of vegetation and region of perpetual snow were still far distant.

"In addition to these facts, it may not be out of place to mention that there are two mountains visible from Almorah, Rigoli-gúdri, in Garhwal between the Kailganga and Nandákni, and Chipula, in Kumaon, between the Gori and Dauli (of Darma), both upwards of 13,000 feet in elevation, from the summits of which the snow disappears long before the end of the summer months, and which do not usually again become covered for the winter till late in December."

These remarks are followed by an exposition of the errors into which Webb, Colebrooke, Hodgson, A. Gerard, and Jacquemont, have fallen. The heights assigned by these travellers "must all be rejected; nor can it be considered at all surprising that any amount of mistake, as to the height of the snow-line, should be made, so long as travellers cannot distinguish snow from glacier ice, or look for the boundary of perpetual snow at the beginning of the spring."

With regard to *the northern limit of the belt of perpetual snow*, Lieutenant Strachey's observations were made in September, 1848, on his way from Milam into Hundes, *viá* Untadhúra, Kyungar-ghát, and Balch-dhúra, at the beginning of the month; and on his road back again, *viâ* Lakhur-ghât, at the end of the month.

"Of the three passes that we crossed on our way from Milam, all of them being about 17,000 feet in elevation, the first is Wata-dhára, and we saw no snow on any part of the way up to its top, which was reached in a very disagreeable drizzle of rain and snow. The final ascent to the pass from the south is about 1000 feet. The path leads up the side of a ravine, down which a small stream trickles, the ground having a generally even and rounded surface. Neither on any part of this nor on the summit of the pass itself, which is tolerably level, were there any remains of snow whatever. On the ridge to the right and left there were patches of snow a few hundred feet above; and on the northern face of the pass an accumulation remained that extended about 200 feet down, apparently the effect of the drift through the gap in which the pass lies. Below this again the ground was everywhere quite free from snow. On the ascent to Wata-dhára, at perhaps 17,000 feet, a few blades of grass were seen, but on the whole it may be said to have been utterly devoid of vegetation. On the north side of the pass, 300 or 400 feet below the summit, a cruciferous plant was the first met with.

"The Kyungar pass, which is four or six miles north of Wata-dhára, was found equally free from snow on its southern face and summit, which latter is particularly open and level. The mountains on either side were also free from snow to some height; but on the north a large bed lay a little way down the slope, and extended to about 500 feet from the top. On this pass a boragineous plant in flower was found above 17,000 feet; a species of *Urtica* was also got about the same altitude, and we afterwards saw it again nearly as high up on the Lákhur pass.

"In our ascent to the Balch pass no snow was observed on any of the southern spires of the range, and only one or two very small patches could be seen from the summit on the north side. The average height of the top of this range can hardly be more than 500 feet greater than that of the pass; and as a whole it certainly does not enter the region of perpetual snow. As viewed from the plains of Handes, it cannot be said to appear snowy, a few only of the peaks being tipped.

"We returned to Milam *viâ* Chirchun. The whole of the ascent to the Lákhur pass was perfectly free from snow to the very top, *i.e.* 18,300 feet, and many of the neighbouring mountains were bare still higher. The next ridge on this route is Jainti-dhára, which is passed at an elevation of 18,500 feet, but still without crossing the least portion of snow. The line of perpetual snow is however evidently near; for though the Jainti ridge was quite free, and some of the peaks near us were clear probably to upwards of 19,000 feet, yet in more sheltered situations unbroken snow could be seen considerably below us; and on the whole I think that 18,500 feet must be near the average height of the snow-line at this place."

A brief recapitulation of the principal results of Lieutenant Strachey's inquiries shows us that "the snow-line or the southern edge of the belt of perpetual snow in this portion of the Himalaya is at an elevation of 15,000 feet, while on the northern edge it reaches 18,500 feet; and that on the mountains to the north of the Sutlej, or still further, it recedes even beyond 19,000 feet. The greater elevation which the snow-line attains on the northern edge of the belt of perpetual snow is a phenomenon not confined to the Thibetan declivity alone, but extending far into the interior of the chain; and it appears to be caused by the quantity of snow that falls on the the northern portion of the mountains being much less than that which falls farther to the south along the line where the peaks, covered with perpetual snow, first rise above the less elevated ranges of the Himalaya."

The letters of Dr. Joseph Hooker published during the present year (1849) in the *Athenæum* (pp. 431 and 1039) may also be consulted with advantage.]

(11) p. 5—"*A tawny tribe of Herdsmen.*"

The Hiongnu (Hioung-nou), whom Deguignes and with him many other historians long believed to be identical with the Huns, inhabited the vast Tartarian tract of land which is bordered on the east by Uo-leang-ho, the present territory of the Mant-schu, on the south by the Chinese wall, on the west by the U-siün, and on the north by the land of the Eleuthes But the Hiongnu belong to the Turkish, and the Huns to the Finnish or Uralian race. The *northern* Huns, a rude people of herdsmen, unacquainted with agriculture, were of a blackish brown complexion. The *southern* Huns, or Hajatehah, called by the Byzantines Euthalites or Nephthalites, and inhabiting the eastern shore of the Caspian Sea, had fairer skins. These pursued agriculture, and dwelt in towns. They are frequently termed *White Huns*, and d'Herbelot even regards them as Indo-Scythians. In Deguignes* an account will be found of the Punu, the leader or Tanju of the Huns, and of the great drought and famine which led to the migration of a portion of the nation northwards about the year 46 A.D. All the details, given in his celebrated work regarding the Hiongnu, have been recently submitted by Klaproth to a rigid and learned scrutiny. From the result of his investigations it would appear, that the Hiongnu belong to the widely diffused Turkish races of the Altai and Tangnu mountain districts. The name of Hiongnu was a general name for the Ti, Thu-kiu or Turks, in the north and north-west of China, even in the third century before the Christian era. The southern Hiongnu submitted themselves to the Chinese, and in conjunction with the latter destroyed the empire of the northern Hiongnu, who were in consequence compelled to flee to the west, and thus appear to have given the first impulse to the migration of nations in Central Asia. The Huns, who were long confounded with the Hiongnu (as the Uigures were with the Ugures and Hungarians) belonged, according to Klaproth,† to the Finnish race of the Uralian mountains, which race has been variously intermixed with Germans, Turks, and Samoiedes.

The Huns (Οὖννοι) are first mentioned by Dionysius Periegetes, a writer who was able to obtain more accurate information than others regarding the interior of Asia, because, as a learned man and a native of Charax on the Arabian Gulf, he was sent back to the East by Augustus, to accompany thither his adopted son, Caius Agrippa. Ptolemy, a century later, writes the word Χοῦνοι with a strong aspiration, which, as St. Martin observes, is agan met with in the geographical name of Chunigard.

* *Hist. gén. des Huns, des Turcs, etc.*, 1756, t. i. P. 1, p. 217, P. 2, pp. 111, 125, 223, 447.

† See Klaproth, *Asia Polyglotta*, pp. 183, 211; *Tableaux Historiques de l'Asie*, pp. 102, 109.

(12) p. 6—"*No hewn stone.*"

Representations of the sun and figures of animals have certainly been found graven in rocks on the banks of the Orinoco, near Caicara, where the woody region borders on the plain, but in the Llanos themselves not a trace of these rough memorials of earlier inhabitants has ever been discovered. It is to be regretted that no accurate account has reached us of a monument which was sent to Count Maurepas, in France, and which, according to Kalm, was discovered in the prairies of Canada, 900 French leagues (about 2700 English miles) west of Montreal, by M. de Verandrier, while engaged on an expedition to the coast of the Pacific Ocean.* This traveller met in the plains with huge masses of stone erected by the hand of man, on one of which there was an inscription believed to be in the Tartar language†. How can so important a monument have remained uninvestigated? Can it actually have borne an alphabetical inscription, or are we not rather to believe that it must have been an historical picture, like the so-called Phœnician inscription, which has been discovered on the bank of the Taunton river, and whose authenticity has been questioned by Court de Gebelin? I indeed regard it as highly probable that these plains were once traversed by civilised nations, and it seems to me that this fact is proved by the existence of pyramidal grave-works or burrows and bulwarks of extraordinary length, between the Rocky Mountains and the Alleghanys, on which Squier and Davis have now thrown new light in their account of the ancient monuments of the Mississippi valley.‡ M. de Verandrier was despatched, about the year 1746, on this expedition by the Chevalier de Beauharnois, Governor-General of Canada; and several Jesuits in Quebec assured Kalm that they had actually had this so-called inscription in their hands, and that it was graven on a small tablet which was found inlaid in a hewn pillar. I have in vain requested several of my friends in France to make inquiries regarding this monument, in the event of its being in the Collection of Count Maurepas. I have also found equally uncertain ac-

* See Kalm's *Reise,* Th. iii. p. 416.

† *Archæologia, or Miscellaneous Tracts published by the Society of Antiquarians of London,* vol. viii. 1787, p. 304.

‡ *Relat. hist.* t. iii. p. 155.

counts of the alphabetical writing of the American aboriginal races, in a work of Pedro de Cieça de Leon,* in Garcia,† and in Columbus's‡ journal of his first voyage. M. de Verandrier maintained also that traces of the ploughshare were observed for days together in travelling over the grassy plains of Western Canada; a circumstance that other travellers, prior to him, likewise profess to have noticed. But the utter ignorance of the primitive nations of North America regarding this implement of agriculture, the want of beasts of draught, and the vast extent of surface over which these tracks extend through the prairie, tend rather to make me adopt the opinion that this singular appearance of furrows is owing to some movement of water over the earth's surface.

(13) p. 6—"*It spreads like an arm of the sea.*"

The great steppe, which extends from the mouth of the Orinoco to the snowy mountains of Merida, from east to west, deflects towards the south in the parallel of 8° north latitude, and occupies the whole space between the eastern declivity of the elevated mountains of New Granada and the Orinoco, which here flows in a northerly direction. That portion of the Llanos, which is watered by the Meta, Vichada, Zama, and Guaviare, connects as it were the valley of the Amazon with that of the Lower Orinoco. The word *Paramo*, which I have frequently employed in this work, signifies in the Spanish colonies all alpine regions which are situated from 11,000 to 14,000 feet above the level of the sea, and whose climate is rude, ungenial, and misty. In the higher Paramos hail and snow fall daily for many hours continuously, and yield a beneficial supply of humidity to the alpine plants, not from the absolute quantity of vapour in the higher strata of the air, but by the frequency of the aqueous deposits occasioned by the rapidly changing currents of air, and the variations of the electric tension. The trees found in these regions are low, and spread out in an umbrella-like form, have gnarled branches, which are constantly covered with fresh and evergreen foliage. They are

* *Chronica del Peru,* P. 1, cap. 87. (Losa con letras en los edificios de Vinaque.)

† *Origen de los Indios,* 1607, lib. iii. cap. 5, p. 258.

‡ Navarrete, *Viages de los Españoles,* t. i. p. 67.

mostly large-flowering laurel and myrtle-leaved alpine shrubs, *Escallonia tubar*, *Escallonia myrtilloides*, *Chuquiraga insignis*, *Araliæ*, *Weinmanniæ*, *Frezieræ*, *Gualtheriæ*, and *Andromeda reticulata*, may be regarded as the representatives of the physiognomy of this vegetation.* To the south of the town of Santa Fé de Bogota lies the celebrated *Paramo de la Suma Paz*, an isolated mountain group, in which, according to Indian legends, great treasures are concealed; and hence issues a small stream or brook, which pours its foaming waters through a remarkable natural bridge in the rocky ravine of Icononzo.

In my Latin treatise, *De Distributione geographica Plantarum secundum cœli temperiem et altitudinem montium*, 1817, p. 104, I have thus endeavoured to characterise these Alpine regions: "Altitudine 1700—1900 hexapod: asperrimæ solitudines, quæ a colonis hispanis uno nomine *Paramos* appellantur, tempestatum vicissitudinibus mire obnoxiæ, ad quas solutæ et emollitæ defluunt nives; ventorum flatibus ac nimborum grandinisque jactu tumultuosa regio, quæ æque per diem et per noctes riget, solis nubila et tristi luce fere nunquam calefacta. Habitantur in hac ipsa altitudine sat magnæ civitates, ut Micuipampa Peruvianorum, ubi thermometrum centes. meridie inter 5° et 8°, noctu — 0°.4 consistere vidi; Huancavelica, propter cinnabaris venas celebrata, ubi altitudine 1835 hexap. fere totum per annum temperies mensis Martii Parisiis."

(14) p. 6—"*The Cordilleras of Cochabamba and the Brazilian mountains approximate to one another by means of separate transverse chains.*"

The immense space between the eastern coasts of South America and the eastern declivity of the chain of the Andes is contracted by two mountain masses, which partially separate from one another the three valleys or plains of the Lower Orinoco, the Amazon, and the Rio de la Plata. The more northern mountain mass, called the group of the Parime, is opposite to the Andes of Cundinamarca, which, after extending far towards the east, assume the form of one elevated mountain, between the parallels of 66° and 68° west longitude. It is connected by the narrow mountain ridge of Pacaraima with the granitic hills of French Guiana, as I have clearly indicated in the map of Columbia which I drew up from my own

* Humboldt et Bonpland, *Plantæ æquinoctiales*, fasc. ii.

astronomical observations. The Caribs, in their long expeditions from the missions of Carony to the plains of Rio Branco, and even to the Brazilian frontier, are obliged to traverse the crests of Pacaraima and Quimiropaca. The second group of mountains, which separates the valley of the Amazon from that of La Plata, is the Brazilian, which approximates to the promontory of Santa Cruz de la Sierra, in the province of Chiquitos, west of the Parecis hills. As neither the group of the Parime, which gives rise to the cataracts of the Orinoco, nor the Brazilian group, is directly connected with the chain of the Andes, the plains of Venezuela and those of Patagonia are directly connected with one another.*

(15) p. 6—"*Herds of wild dogs.*"

In the Pampas of Buenos Ayres the traveller meets with European dogs, which have become wild. They live gregariously in holes and excavations, in which they conceal their young. When the horde becomes too numerous, several families go forth, and form new settlements elsewhere. The European dog barks as loudly after it has become wild, as does the indigenous American hairy species. Garcilaso asserts that, prior to the arrival of the Spaniards, the Peruvians had a race of dogs called *Perros gozques;* and he calls the indigenous dog *Allco.* In order to distinguish this animal from the European variety, it is called in the Qquichua language Runa-allco, Indian dog, or dog of the natives. The hairy Runa-allco appears to be a mere variety of the shepherd's dog. It is, however, smaller, has long yellow-ochry coloured hair, is marked with white and brown spots, and has erect and pointed ears. It barks continually, but seldom bites the natives, however it may attack the whites. When the Inca Pachacutec, in his religious wars, conquered the Indians of Xauxa and Huanca (the present valley of Huancaya and Jauja), and compelled them by force to submit to the worship of the sun, he found that dogs were made the objects of their adoration, and that the priests used the skulls of these animals as wind instruments. It would also appear that the flesh of this canine divinity was eaten by the believers.† The veneration of dogs in the valley of the

* See Humboldt's geognostic view of South America, in his *Relation historique,* t. iii. pp. 188—244.

† Garcilaso de la Vega, *Commentarios Reales,* P. i., p. 184.

Huancaya is probably the reason why the skulls, and even whole mummies, of these animals are sometimes found in the Huacas, or Peruvian graves of the most ancient period. Von Tschudi, the author of an admirable treatise on the *Fauna Peruana*, has examined these skulls, and believes them to belong to a peculiar species, which he calls *Canis ingæ*, and which is different from the European dog. The Huancas are still, in derision, called "dog-eaters" by the inhabitants of other provinces. Among the natives of the Rocky Mountains of North America, cooked dog's flesh is placed before the stranger guest, as a feast of honour. Captain Frémont was present at such a dog-feast in the neighbourhood of Fort Laramie, which is one of the stations of the Hudson's Bay Company for trading in skins and peltries with the Sioux Indians.*

The Peruvian dogs were made to play a singular part during eclipses of the moon, being beaten as long as the darkness continued. The Mexican *Techichi*, a variety of the common dog, which was called in Anahuac *Chichi*, was the only completely dumb dog. The literal signification of the word *Techichi* is "stone-dog," from the Aztec, *tetl*, a stone. This dog was eaten according to the ancient Chinese custom, and the Spaniards found this food so indispensable before the introduction of horned cattle, that the race was gradually almost entirely extirpated.† Buffon confounds the Techichi with the Koupara of Guiana,‡ which is, however, identical with the Procyon or *Ursus cancrivorus*, the *Raton crabier*, or the crab-eating Aguara-guaza of the coasts of Patagonia.§ Linnæus, on the other hand, confounds the dumb dog with the Mexican *Itzcuintepotzotli*, a canine species which has not hitherto been perfectly described, and which is said to be characterised by a short tail, a very small head, and a large hump on the back. The name signifies a hump-backed dog, and is derived from the Aztec *itzcuintli*, another word for dog, and *tepotzotli*, humped or a humpback. I was much struck in America, especially in Quito and Peru, with the great number of black hairless dogs. They are termed *Chiens turcs* by Buffon, and are the *Canis ægyptius* of Linnæus. This species is common amongst the Indians, who, however,

* Frémont's *Exploring Expedition*, 1845, p. 42.

† Clavigero, *Storia antica del Messico*, 1780, t. i. p. 73.

‡ Buffon, t. xv., p. 155.

§ Azara, *Sur les Quadrupèdes du Paraguay*, t. i. p. 315.

generally despise them, and treat them ill. All European dogs multiply rapidly in South America; and if no species are to be met with equal to those of Europe, it is partly owing to want of care, and partly to the circumstance that the finest varieties (as the elegant greyhound and the Danish tiger breed) have never been introduced.

Von Tschudi makes the singular remark, that on the Cordilleras, at elevations of more than 12,000 feet, delicate breeds of dogs and the European domestic cat are exposed to a particular kind of mortal disease. "Innumerable attempts have been made to keep cats as domestic animals in the town of Cerro de Pasco (lying at an elevation of 14,100 feet above the sea's level); but such endeavours have invariably been frustrated, as both cats and dogs have died in convulsions at the end of a few days. The cats, after being attacked by convulsive fits, attempt to climb the walls, but soon fall to the ground exhausted and motionless. I frequently observed instances in Yauli of this chorea-like disease; and it seems to arise from insufficient atmospheric pressure." In the Spanish colonies, the hairless dog, which is called *Perro chinesco*, or *chino*, is supposed to be of Chinese origin, and to have been brought from Canton, or from Manila. According to Klaproth, the race has been very common in the Chinese Empire from the earliest ages of its culture. Among the animals indigenous to Mexico, there was a very large, totally hairless, and dog-like wolf, named *Xoloitzcuintli*, from the Mexican *xolo* or *xolotl*, a servant or slave.*

The result of Tschudi's observations regarding the American indigenous races of dogs are as follows:—There are two varieties almost specifically different—1. The *Canis caraibicus* of Lesson, totally hairless, with the exception of a small tuft of white hair on the forehead and at the tip of the tail; of a slate-gray colour, and without voice. This variety was found by Columbus in the Antilles, by Cortes in Mexico, and by Pizarro in Peru (where it suffers from the cold of the Cordilleras); and it is still very frequently met with in the warmer districts of Peru, under the name of *Perros chinos*. 2. The *Canis ingæ*, which belongs to the barking species, and has a pointed nose and pointed ears; it is now used for watching sheep and cattle; it exhibits many variations of colour, in-

* On the dogs of America, see Smith Barton's *Fragments of the Natural History of Pennsylvania*, p. i., p. 34.

duced by being crossed with European breeds. The *Canis ingæ* follows man up the heights of the Cordilleras. In the old Peruvian graves, the skeleton of this dog is sometimes found resting at the feet of the human mummy, presenting an emblem of fidelity frequently employed by the mediæval sculptors.* European dogs, that had become wild, were found in the island of St. Domingo, and in Cuba, in the early periods of the Spanish conquest.† In the savannahs between the Meta, Arauca, and Apure, dumb dogs (*perros mudos*) were used as food as late as the sixteenth century. The natives called them *Majos* or *Auries*, says Alonzo de Herrera, who undertook an expedition to the Orinoco, in 1535. The highly intelligent traveller Gisecke found this variety of non-barking dogs in Greenland. The dogs of the Esquimaux live entirely in the open air, scraping for themselves at night holes in the snow, and howling like wolves, in concert with one of the troop, who sits in the middle, and takes the lead in the chorus. The Mexican dogs were castrated, in order that their flesh might become more fat and delicate. On the borders of the province of Durango, and further north, near the Slave Lake, the natives load the larger dogs with their buffalo-skin tents, (at all events they did so formerly,) when, on the change of seasons, they seek a different place of abode. These various details may all be regarded as characteristic of the mode of life led by the nations of Eastern Asia.‡

(16) p. 7—"*Like the greater part of the Desert of Sahara, the Llanos lie within the Torrid Zone.*"

Significant denominations, particularly such as refer to the form of the earth's surface, and which arose at a period when there was only very uncertain information respecting different regions and their hypsometric relations, have led to various and long-continued geographical errors. The ancient Ptolemaic denomination of the "Greater and Lesser Atlas"§ has exercised the injurious influence here indicated. There is no doubt that the snow-covered western summits of the Atlas of Morocco may

* J. J. von Tschudi, *Untersuchungen über die Fauna Peruana*, s. 247—251.

† Garcilaso, P. i. 1723, p. 326.

‡ Humboldt, *Essai polit.*, t. ii. p. 448, and *Relation hist.*, t. ii. p. 625.

§ *Geogr.*, lib. iii. cap 1.

be regarded as the Great Atlas of Ptolemy; but where is the limit of the Little Atlas? Are we still to maintain the division into two Atlas chains (which the conservative tendency of geographers has retained for 1700 years) in the territory of Algiers, and even between Tunis and Tlemse? Are we to seek a Greater and a Lesser Atlas between the coast and the parallel chains of the interior? All travellers familiar with geognostic views, who have visited Algeria since it has been in the possession of the French, contest the meaning conveyed by the generally adopted nomenclature. Among the parallel chains, that of Jurjura is generally supposed to be the highest of those which have been measured; but the well-informed Fournel (who was long *Ingénieur en chef des Mines de l'Algérie*) affirms that the mountain range of Aurès, near Batnah, which even at the end of March was found covered with snow, has a greater elevation. Fournel contests the existence of a Little and a Great Atlas, as I do that of a Little and a Great Altai*. There is but one Atlas, formerly called Dyris by the Mauritanians, "a name that must be applied to the foldings (*rides, suites de crêtes*), which form the division between the waters flowing to the Mediterranean and towards the lowland of the Sahara." The lofty Atlas chain of Morocco inclines from north-east to south-west, and not, like the Eastern Mauritanian portion of the Atlas, from east to west. It rises into summits which, according to Renou, attain an elevation of 11,400 feet, exceeding, therefore, the height of Etna†. A singularly formed highland, of an almost square shape (Sahab el-Marga), is situated in 33° north lat., and is bounded to the south by high elevations. From thence the Atlas declines in height in a westerly direction towards the sea, about a degree south of Mogador. This south-western portion bears the name of Idrar-N-Deren.

The northern boundaries of the extended low region of the Sahara in Mauritania, as well as its southern limits towards the fertile Sudan, have hitherto been but imperfectly investigated. If we take the parallels of 16½° and 32½° north lat. as the outer limits, we obtain for the Desert, including its oases, an area of more than 1,896,000 square miles;

* Humboldt, *Asie centrale*, t. i. pp. 247, 252.

† *Exploration scientifique de l'Algérie, de* 1840 *à* 1842, *publiée par ordre du Gouvernement; Sciences hist. et géogr.*, t. viii., 1846, pp. 364, 373.

or between nine and ten times the extent of Germany, and almost three times that of the Mediterranean, exclusive of the Black Sea. The best and most recent intelligence, for which we are indebted to the French observers, Colonel Daumas, and MM. Fournel, Renou, and Carette, shows us that the Desert of Sahara is composed of several detached basins, and that the number and the population of the fertile Oases is very much greater than had been imagined from the awfully desert character of the country between Insalah and Timbuctoo, and the road from Mourzouk, in Fezzan, to Bilma, Tirtuma, and Lake Tschad. It is now generally affirmed that the sand covers only the smaller portion of the lowlands. A similar opinion had been previously advanced by my Siberian travelling companion, the acute observer Ehrenberg, from what he had himself seen*. Of larger wild animals, only gazelles, wild asses, and ostriches are to be met with.

"That lions exist in the desert," says M. Carette, "is a myth popularised by the dreams of artists and poets, and has no foundation but in their imagination. This animal does not quit the mountains where it finds shelter, food, and drink. When the traveller questions the natives concerning these wild beasts, which Europeans suppose to be their companions in the desert, they reply, with imperturbable *sang froid*, 'Have you, then, lions in your country which can drink air and eat leaves? With us lions require running water and living flesh; and therefore they only appear where there are wooded hills and water. We fear only the viper (*lefa*), and, in humid spots, the innumerable swarms of mosquitoes which abound there†'"

While Dr. Oudney, in his long journey from Tripoli to Lake Tschad, estimated the elevation of the Southern Sahara at 1637 feet, and German geographers even ventured to add an additional thousand feet, Fournel, the engineer, has, by careful barometric measurements, based on corresponding observations, made it tolerably probable that a part of the northern desert is below the sea s level. The portion of the desert which is now called "Le Zahara d'Algérie," advances to the chains of hills of Metlili and el-Gaous, where lies the most northern of all the Oases, el-Kantara, fruitful in dates. This low basin, which reaches the parallel of 34° lat., receives

* *Exploration scientif. de l'Algérie, Hist. et géogr.,* t. ii. p. 332.
† Ibid. t. ii. pp. 126—129, and t. vii. pp. 94, 97.

the radiant heat of a stratum of chalk, inclined at an angle of 65° towards the south, and which is full of the shells of Inoceramus*. "Arrived at Biscara (Biskra)," says Fournel, "an indefinite horizon, like that of the sea, lay spread before us." Between Biscara and Sidi Ocba the land is only 243 feet above the sea's level. The inclination increases considerably towards the south. In another work†, where I have brought together all the points that refer to the depression of some portions of continents below the level of the sea, I have already noticed that, according toLe Père, the bitter lakes (*lacs amers*) on the isthmus of Suez, when they have but little water, and, according to General Andréossy, the Natron lakes of Fayoum, are also lower than the level of the Mediterranean.

Among other manuscript notices of M. Fournel, I possess a geognostic vertical profile, with all the inflexions and inclinations of the strata, representing the surface the whole way from the coast near Philippeville to a spot near the Oasis of Biscara in the Desert of Sahara. The direction of the line on which the barometric measurements were taken is south 20° west; but the points of elevation determined are projected, as in my Mexican profiles, on a different plane, one from N. to S. Ascending uninterruptedly from Constantine, whose elevation is 2123 feet, the highest point is found between Batnah and Tizur, at only 3581 feet. In the part of the desert which lies between Biscara and Tuggurt, Fournel has succeeded in digging a series of artesian wells‡. We learn from the old accounts of Shaw, that the inhabitants of the country were acquainted with a subterranean supply of water, and related fabulous tales of a "sea under the earth (bahr tôhl el-erd)." Fresh waters, which flow between clay and marl strata of the old chalk and other sedimentary formations, under the action of hydrostatic pressure, form gushing fountains when the strata are pierced§. The phenomenon of fresh water being often found near beds of rock salt, need not surprise the geognosist, acquainted with mining operations, since Europe offers many analogous phenomena.

* Fournel, *Sur les Gisemens de Muriate de Soude en Algérie*, p. 6, in the *Annales des Mines*, 4me serie, t. ix. 1846, p. 546.

† *Asie centrale*, t. ii. p. 320.

‡ *Comptes rendus de l'Académie des Sciences*, t. xx. 1845, pp. 170, 882, 1305.

§ See Shaw, *Voyages dans plusieurs parties de la Berbérie*, t. i. p. 169, and Rennel, *Africa*, Append. p. lxxxv.

The riches of the desert in rock-salt, and its employment for purposes of building, have been known since the time of Herodotus. The salt zone of the Sahara (zone salifère du désert) is the most southern of the three zones which pass through Northern Africa from south-west to north-east, and is believed to be connected with the beds of rock-salt in Sicily and Palestine described by Friedrich Hoffman, and by Robinson*.

The trade in salt with Sudan, and the possibility of cultivating the date-tree in the many Oasis-like depressions, caused probably by earth-slips in the beds of tertiary chalk or Keuper-gypsum, have equally contributed to animate the desert, at various parts, by human intercourse. The high temperature of the air, which renders the day's march so oppressive across the Sahara, makes the coolness of the night (of which Denham and Sir Alexander Burnes frequently complained in the African and Asiatic deserts) so much the more remarkable. Melloni† ascribes this coolness (which is probably produced by the radiation of heat from the ground), not to the great purity of the heavens (irraggiamento calorifico per la grande serenità di cielo nell' immensa e deserta pianura dell' Africa centrale), but to the extreme calm, and the absence of all movement in the air throughout the whole night‡.

The river Quad-Dra (Wadi Dra), which is almost dry the greater part of the year, and which, according to Renou§, is one-sixth longer than the Rhine, flows into the Sahara in 32° north latitude, from the southern declivity of the Atlas of Morocco. It runs at first from north to south, until in 29° north lat., and 5° 8′ west long., it deflects at right angles to the west, and traversing the great fresh-water lake of Debaid, flows into the sea at Cape Nun, in lat. 28° 46′, and long. 11° 8′. This region, which was first rendered celebrated by the Portuguese discoveries of the fifteenth century, and whose geography has subsequently been shrouded in the deepest obscurity, is now known on the coast as the country of the Scheik of Beirouk

* Fournel, *Sur les Gisemens de Muriate de Soude en Algérie*, pp. 28–41; and Karsten, *Ueber das Vorkommen des Kochsalzes auf der Oberfläche der Erde*, 1846, s. 497, 648, 741.

† *Memoria sull' abbassamento di temperatura durante le notti placide e serene*, 1847, p. 55.

‡ Consult, also, on African Meteorology, Aimé, in the *Exp'or. de l'Algérie, Phys. Géner.* t. ii., 1846, p. 147.

§ *Explor. de l'Alg., Hist. et Géogr.* t. viii. pp. 65–78.

(whose dominions are independent of the Emperor of Morocco). It was explored, in the months of July and August, 1840, by the French Count, Captain de Bouet-Villaumez, under the orders of his government. From manuscript and official reports it would appear that the mouth of the Quad-Dra is at present so much blocked up by sand as to have an open channel of only about 190 feet. The Saguiel-el-Hamra,—still very little known,—which comes from the south, and is supposed to have a course of at least 600 miles, flows into the same mouth at a point somewhat farther eastward. The length of these deep, but generally dry, river-beds is astonishing. They are ancient furrows, similar to those which I observed in the Peruvian desert at the foot of the Cordilleras, between the latter and the shores of the Pacific. In Bouet's manuscript narrative*, the mountains which rise to the north of Cape Nun are estimated at the great height of 9,186 feet.

It is generally supposed that Cape Nun was discovered in 1433 by the Knight Gilianez, despatched under the order of the celebrated Infante, Henry, Duke of Viseo, and founder of the Academy of Sagres, which was presided over by the pilot and cosmographer, Mestre Jacomè, of Majorca; but the *Portulano Mediceo*,—the work of a Genoese navigator of the year 1351,—already contains the name of "Cavo di Non." The doubling of this Cape was as much dreaded as has been since then the passage round Cape Horn; although it is only 23′ north of the parallel of Teneriffe, and might be reached by a few days' sail from Cadiz. The Portuguese adage, "Quem passa o Cabo de Num, ou tornarà ou não," could not intimidate the Infante, whose heraldic French motto of "Talent de bien faire," well expressed his noble, enterprising, and vigorous character. The name of this Cape, which has long been supposed to originate in a play of words on the negative particle, does not appear to me to be of Portuguese origin. Ptolemy placed on the north-west coast of Africa a river *Nuius*, in the Latin version *Nunii ostia*. Edrisi refers to a town, Nul, or Wadi Nun, somewhat further south, and about three days' journey in the interior, named by Leo Africanus *Belad de Non*. Several European navigators had penetrated far to the south of Cape Nun before the Portuguese squadron under Gilianez. The Catalan, Don Jayme Ferrer, in 1346, as we learn from the *Atlas Catalan*, published at

* *Relation de l'Expédition de la Malouine.*

Paris by Buchon, had advanced as far as the Gold River (*Rio do Ouro*), in 23° 56′ north lat.; while the Normans, at the close of the fourteenth century, reached Sierra Leone in 8° 30′ north latitude. The merit of having been the first to cross the equator in the Western Ocean incontrovertibly belongs, like so many other great achievements, to the Portuguese.

(17) p. 7.—"*As a grassy plain, resembling many of the Steppes of Central Asia.*"

The Llanos of Caracas, of the Rio Apure and the Meta, which are the abode of numerous herds of cattle, are, in the strictest sense of the word, grassy plains. The two families of the Cyperaceæ and the Gramineæ, which are the principal representatives of the vegetation, yield numerous forms of Paspalum (*Paspalum leptostachyum*, *P. lenticulare*), of Kyllingia (*Kyllingia monocephala* (*Rottb.*) *K. odorata*), of Panicum (*Panicum granuliferum*, *P. micranthum*), of Antephora, Aristida, Vilfa, and Anthisteria (*Anthisteria reflexa*, *A. foliosa*). It is only here and there that any herbaceous dicotyledon, as the low-growing species of Mimosa intermedia and M. dormiens, which are so grateful to the wild horses and cattle, are found interspersed among the Gramineæ. The natives very characteristically apply to this group the name of "Dormideras," or sleepy plants, because the delicate and feathery leaves close on being touched. For many square miles not a tree is to be seen; but where a few solitary trees are found, they are, in humid districts, the Mauritia Palm, and, in arid spots, a Proteacea described by Bonpland and myself, the Rhopala complicata (*Chaparro bobo*), which Willdenow regarded as an Embothrium; also the useful Palma de Covija or de Sombrero; and our Corypha inermis, an umbrella palm allied to Chamærops, and used by the natives for the covering of their huts. How much more varied and rich is the aspect of the Asiatic plains! In a great portion of the Kirghis and Kalmuck Steppes which I have traversed (extending over a space of 40 degrees of longitude), from the Don, the Caspian Sea and the Orenburg Ural river Jaik, to the Obi and the Upper Irtysch, near the Lake Dsaisang, the extreme range of view is never bounded by a horizon in which the vault of heaven appears to rest on an unbroken sea-like plain, as is so frequently the case in the Llanos, Pampas, and Prairies of America. I have, indeed, never observed anything approaching to this phenomenon, excepting, perhaps, where I have looked only towards one quarter of the heavens, for the Asiatic plains are frequently intersected by chains of hills, or clothed with coniferous woods. The Asiatic vegetation, too, in the most fruitful pasture lands, is by no means limited to the family of the Cyperaceæ, but is enriched by a great variety of herbaceous plants and shrubs. In the season of spring, small snowy white and red flowering Rosaceæ and Amygdaleæ (*Spiræa*, *Cratægus*, *Prunus spinosa*, *Amygdalus nana*), present a pleasing appearance. I have elsewhere spoken of the tall and luxuriant Synanthereæ (*Saussurea amara*, *S. salsa*, *Artemisiæ*, and *Centaureæ*), and of leguminous plants, (species of the Astragalus, Cytisus and Caragana). Crown Imperials (*Fritillaria ruthenica* and *F. meleagroides*), Cypripediæ and tulips gladden the eye with their varied and bright hues.

A contrast is presented to this charming vegetation of the Asiatic plains by the dreary Salt Steppes, especially by that portion of the Barabinski Steppe which lies at the base of the Altai Mountains, between Barnaul and the Serpent Mountain, and by the country to the east of the Caspian. Here the social Chenopodiæ, species of Salsola, Atriplex, Salicorniæ, and Halimocnemis crassifolia*, cover the clayey soil with patches of verdure. Among the five hundred phanerogamic species which Claus and Göbel collected on the Steppes, Synanthereæ, Chenopodiæ, and Cruciferæ were more numerous than the grasses; the latter constituting only $\frac{1}{11}$th of the whole, and the two former $\frac{1}{7}$th and $\frac{1}{9}$th. In Germany, owing to the alternation of hills and plains, the Glumaceæ (comprising the Gramineæ, Cyperaceæ, and Juncaceæ) constitute $\frac{1}{7}$th, the Synanthereæ (Compositæ) $\frac{1}{8}$th, and the Cruciferæ $\frac{1}{18}$th of all the German Phanerogamic species. In the most northern part of the flat land of Siberia, the extreme limit of tree and shrub vegetation (*Coniferæ* and *Amentaceæ*) is, according to Admiral Wrangell's fine map, 67° 15′ north lat., in the districts contiguous to Behring's Straits, while more to the west, towards the banks of the Lena, it is 71°, which is the parallel of the North Cape of Lapland. The plains bordering on the Polar Sea are the domain of Cryptogamic plants. They are called Tundra (Tuntur in Finnish), and are vast swampy districts, covered

* Göbel, *Reise in die Steppe des südlichen Russlands*, 1838, th. ii. s. 244, 301.

partly with a thick mantle of Sphagnum palustre and other Liverworts, and partly with a dry snowy-white carpet of Cenomyce rangiferina (Reindeer-moss), Stereocaulon paschale, and other lichens. "These *Tundra*," says Admiral Wrangell, in his perilous expedition to the Islands of New Siberia, so rich in fossil wood, "accompanied me to the extremest Arctic coast. Their soil is composed of earth that has been frozen for thousands of years. In the dreary uniformity of the landscape, and surrounded by reindeer, the eye of the traveller rests with pleasure on the smallest patch of green turf that shows itself on a moist spot."

(18) p. 7.—*A diversity of causes diminishes the dryness and heat of the New Continent.*

I have endeavoured to compress the various causes of the humidity and lesser heat of America into one general category. It will of course be understood, that I can only have reference here to the *general* hygroscopic condition of the atmosphere, and the temperature of the *whole* continent; for in considering individual regions, as for instance, the island of Margarita, or the coasts of Cumana and Coro, it will be found that these exhibit an equal degree of dryness and heat with any portion of Africa.

The maximum of heat, at certain hours of a summer's day, considered with reference to a long series of years, has been found to be almost the same in all regions of the earth, whether on the Neva, the Senegal, the Ganges, or the Orinoco, namely, between 93° and 104° Fahr., and on the whole not higher; provided that the observation be made in the shade, far from solid radiating bodies, and not in an atmosphere filled with heated dust or granules of sand, and not with spirit-thermometers, which absorb light. The fine grains of sand (forming centres of radiant heat) which float in the air, were probably the cause of the fearful heat (122° to 133° Fahr. in the shade) in the Oasis of Mourzouk to which my unhappy friend Ritchie, who perished there, and Captain Lyon, were exposed for weeks. The most remarkable instance of a high temperature, in an air probably free from dust, is mentioned by an observer who well knew how to arrange and correct all his instruments with the greatest accuracy. Rüppel found the temperature 110°.6 Fahr. at Ambukol, in Abyssinia, with a cloudy sky, a strong south-west wind, and an approaching thunder-storm.

The *mean* annual temperature of the tropics, or the actual climate of the region of palms, is on the main land between 78°.2 and 85°.5 Fahr., without any sensible difference between the observations made in Senegal, Pondichery, and Surinam*.

The great coolness, one might almost say *coldness*, which prevails during a great portion of the year in the tropics, on the coast of Peru, and which causes the mercury to fall to 59° Fahr., is, as I hope to show in another place, not to be attributed to the effect of neighbouring mountains covered with snow, but rather to the mist (*garua*) which obscures the sun's disk, and to a current of *cold sea-water* commencing in the antarctic regions, and which coming from the south-west, strikes the coast of Chili near Valdivia and Concepcion, and is thence propelled with violence, in a northerly direction, to Cape Pariña. On the coast of Lima, the temperature of the Pacific is 60°.2 Fahr., whilst it is 79°.2 Fahr. under the same parallel of latitude when outside the current. It is singular, that so remarkable a fact should have remained unnoticed, until my residence on the coast of the Pacific, in October, 1802.

The variations of temperature, of many parts of the earth, depend principally on the character of the bottom of the aërial ocean, or in other words, on the nature of the solid or fluid (continental or oceanic) base on which the atmosphere rests. Seas, traversed in various directions by currents of warm and cold water (oceanic rivers), exert a different action from articulated or inarticulated continental masses or islands, which may be regarded as the shoals in the aërial ocean, and which, notwithstanding their small dimensions, exercise, even to great distances, a remarkable degree of influence on the climate of the sea. In continental masses, we must distinguish between barren sandy deserts, savannahs, (grassy plains,) and forest districts. In Upper Egypt and in South America, Nouet and myself found, at noon, the temperature of the ground, which was composed of granitic sand, 154° and 141° Fahr. Numerous careful observations instituted at Paris, have given, according to Arago, 122° and 126°.5 Fahr.† The Savannahs, which, between the Missouri and the Mississippi, are called *Prairies*, and which appear in the south at

* Humboldt, *Mémoire sur les Lignes Isothermes*, 1817, p. 54. *Asie centrale*, t. iii. Mahlmann, Table IV.

† *Asie centrale*, t. iii. p. 176.

the Llanos of Venezuela and the Pampas of Buenos Ayres, are covered with small monocotyledons, belonging to the family of the Cyperaceæ, and with grasses, whose dry pointed stalks, and whose delicate, lanceolate leaves radiate towards the unclouded sky, and possess an extraordinary power of emission. Wells and Daniell* have even seen in our latitude, where the atmosphere has a much less considerable degree of transparency, the thermometer fall to 14°.5, or 18° Fahr. on being placed on the grass. Melloni has most ably shown† that in a calm, which is a necessary condition of a powerful radiation, and of the formation of dew, the cooling of the stratum of grass is promoted by the falling to the ground of the cooler particles of air, as being the heavier.

In the vicinity of the equator, under the cloudy sky of the Upper Orinoco, the Rio Negro and the Amazon, the plains are covered with dense primeval forests; but to the north and south of this woody district, there extend, from the zone of palms and of tall dicotyledonous trees in the northern hemisphere, the Llanos of the Lower Orinoco, the Meta, and Guaviare; and in the south, the Pampas of the Rio de la Plata and of Patagonia. The area thus covered by grassy plains, or Savannahs, in South America, is at least nine times greater than that of France.

The forest region acts in a threefold manner, by the coolness induced by its shade, by evaporation, and by the cooling process of radiation. Forests uniformly composed in our temperate zone of "social" plants, belonging to the families of the Coniferæ or Amentaceæ (the oak, beech, and birch), and under the tropics composed of plants not living socially, protect the ground from direct insolation, evaporate the fluids they have themselves produced, and cool the contiguous strata of air by the radiation of heat from their leafy appendicular organs. The leaves are by no means all parallel to one another, and present different inclinations towards the horizon; and according to the laws established by Leslie and Fourier, the influence of this inclination on the quantity of heat emitted by radiation is such, that the radiating power of a given measured surface *a*, having a given oblique direction, is equal to the radiating power of a leaf of the size of *a* projected on a horizontal

* *Meteor. Essays,* 1827, pp. 230, 278.

† *Sull' Abbassamento di Temperatura durante le Notti placide e serene,* 1847, pp. 47, 53.

plane. In the initial condition of radiation of all the leaves which form the summit of a tree, and which partially cover each other, those which are directly presented towards the unclouded sky, will be first cooled.

This production of cold (or the exhaustion of heat by emission) will be the more considerable in proportion to the thinness of the leaves. A second stratum of leaves has its upper surface turned to the under surface of the former, and will give out more heat by radiation towards that stratum than it can receive from it. The result of this unequal exchange will then be a diminution of temperature for the second stratum also. A similar action will extend from stratum to stratum, till all the leaves of the tree, by their greater or less radiation, as modified by their difference of position, have passed into a condition of stable equilibrium, of which the law may be deduced by mathematical analysis. In this manner, in the serene and long nights of the equinoctial zone, the forest air, which is contained in the interstices between the strata of leaves, becomes cooled by the process of radiation; for a tree, a horizontal section of whose summit would hardly measure 2000 square feet, would, in consequence of the great number of its appendicular organs (the leaves), produce as great a diminution in the temperature of the air as a space of bare land or turf many thousand times greater than 2000 square feet.* I have thus sought to develope somewhat fully the complicated relations which the action of great forest regions exerts on the atmosphere, because they have so often been touched upon in connection with the important question of the climate of ancient Germany and Gaul.

As in the old continent, European civilization has had its principal seat on the western coast, it could not fail to be early remarked that under equal degrees of latitude the opposite eastern littoral region of the United States of North America was several degrees colder, in mean annual temperature, than Europe, which is, as it were, a western peninsula of Asia, and bears much the same relation to it as Brittany does to the rest of France. The fact, however, escaped notice that these differences decrease from the higher to the lower latitudes, and that they are hardly perceptible below 30°. For the west coast of the New Continent exact observations

* *Asie centrale*, t. iii. pp. 195—205.

of the temperature are still almost entirely wanting; but the mildness of the winter in New California shows that in reference to their mean annual temperature, the west coasts of America and Europe under the same parallels, scarcely present any differences. The annexed table gives the mean annual temperatures, which correspond to the same geographical latitudes, on the eastern coast of the New Continent and the western coast of Europe:—

| Similar Degrees of Latitude. | Eastern Coast of America. | Western Coast of Europe. | Mean Temperature of the Year, of Winter, and Summer. | | Difference between the annual Temperature of Eastern America and Western Europe. |
|---|---|---|---|---|---|
| 57° 10′ | Nain . . . | ... ... | 25°.7 | — 0°.4 / 45°.7 | 20°.7 |
| 57° 41′ | ... ... | Gottenburg | 46°.4 | 31°.5 / 62°.4 | |
| 47° 34′ | St. John's . . | ... ... | 38°.1 | 23° / 54° | 13°.6 |
| 47° 30′ | .. ... | Buda. . . | 50°.5 | 31°.1 / 69°.8 | |
| 48° 50′ | ... ... | Paris . . | 51°.7 | 37°.8 / 64°.6 | |
| 44° 39′ | Halifax . . | ... ... | 43°.5 | 24°.1 / 63°.0 | 13°.7 |
| 44° 50′ | ... ... | Bordeaux . | 57°.2 | 42°.8 / 71°.1 | |

(*continued*)

| Similar Degrees of Latitude. | Eastern Coast of America. | Western Coast of Europe. | Mean Temperature of the Year, of Winter, and Summer. | Difference between the annual Temperature of Eastern America and Western Europe. |
|---|---|---|---|---|
| 40° 43′ | New York. . | ... ... | 52°.5 $\frac{32°.2}{72°.9}$ | 9°.3 |
| 39° 57′ | Philadelphia . | ... ... | 52°.2 $\frac{32°.2}{72°.7}$ | |
| 38° 53′ | Washington . | ... ... | 54°.9 $\frac{36°.0}{71°.1}$ | |
| 40° 51′ | ... ... | Naples . . | 61°.0 $\frac{49°.5}{74°.9}$ | |
| 38° 52′ | .. .. | Lisbon . . | 61°.5 $\frac{52°.2}{71°.1}$ | |
| 29° 48′ | St. Augustin . | .. .. | 72°.3 $\frac{59°.5}{81°.5}$ | 0°.5 |
| 30° 2′ | .. .. | Cairo . . | 71°.8 $\frac{58°.5}{84°.6}$ | |

In the preceding table the number placed before the fraction represents the mean annual temperature, the numerator of the fraction, the mean winter temperature, and the denominator the mean summer temperature. Besides the more marked difference between the mean annual temperatures, there is also a very striking contrast between the opposite coasts in respect to the distribution of heat over the different seasons of the year; and it is indeed this distribution which exerts the greatest influence on our bodily feelings and on the process of vegetation. Dove* makes the general remark, that

* *Temperatur-tafeln nebst Bemerkungen über die Verbreitung der Wärme auf der Oberfläche der Erde*, 1848, s. 95.

the summer temperature of America is lower under equal degrees of latitude than that of Europe. The climate of St. Petersburgh (lat. 59° 56′), or to speak more correctly, the mean annual temperature of that city, is found on the eastern coast of America, in lat. 47° 30′, or 12° 30′ more to the south; and in like manner we find the climate of Königsberg (lat. 54° 43′) at Halifax in lat. 44° 39′. Toulouse (lat. 43° 36′) corresponds in its thermic relations to Washington.

It is very hazardous to attempt to obtain any general results respecting the distribution of heat in the United States of North America, since there are three regions to be distinguished—1, the region of the Atlantic States, east of the Alleghanys; 2, the Western States, in the wide basin between the Alleghanys and the Rocky Mountains, watered by the Mississippi, the Ohio, the Arkansas, and the Missouri; and 3, the elevated plains between the Rocky Mountains and the Coast Range of New California, through which the Oregon or Columbia river wends its course. Since the commendable establishment by John Calhoun of uninterrupted observations of temperature, made on a uniform plan, at thirty-five military stations, and reduced to diurnal, mensal, and annual means, we have attained more correct climatic views than were generally held in the time of Jefferson, Barton, and Volney. These meteorological stations extend from the point of Florida and Thompson's Island (West Key), lat. 24° 33′, to the Council Bluffs on the Missouri; and if we reckon Fort Vancouver (lat. 45° 37′), among them, they include a space extending over forty degrees of longitude.

It cannot be affirmed that on the whole the second region has a higher mean annual temperature than the first, or Atlantic. The further advance towards the north of certain plants on the western side of the Alleghanys, depends partly on the nature of those plants and partly on the different distribution through the seasons of the year of the same annual amount of heat. The broad valley of the Mississippi enjoys, at its northern extremity, the warming influence of the Canadian lakes, and at the south, that of the Mexican Gulf-Stream. These five lakes (Lakes Superior, Michigan, Huron, Erie, and Ontario,) cover an area of 92,000 square miles. The climate is so much milder and more uniform in the vicinity of the lakes, that at Niagara, for instance (in 43° 15 north lat.), the mean annual winter temperature is only half a degree below the freezing-point, whilst, at a distance from the lakes, in 44° 53′ north lat. at Fort Snelling, near the confluence of the river St. Peter with the Mississippi, the mean winter temperature is 15°.8 Fahr.* At this distance from the Canadian lakes, whose surface is from five to upwards of six hundred feet *above* the sea's level, whilst the bottom of Lakes Michigan and Huron is five hundred feet *below* it, recent observations have shown that the climate of the country possesses the actual continental character of hotter summers and colder winters. "It is proved," says Forry, "by our thermometrical data, that the climate west of the Alleghany chain is more *excessive* than that on the Atlantic side." At Fort Gibson, on the Arkansas river, which falls into the Mississippi, in lat. 35° 47′, where the mean annual temperature hardly equals that of Gibraltar, the thermometer was observed, in August, 1834, to rise to 117° Fahr. when in the shade, and without any reflected heat from the ground.

The statements so frequently advanced, although unsupported by measurements, that since the first European settlements in New England, Pennsylvania, and Virginia, the destruction of many forests on both sides of the Alleghanys, has rendered the climate more equable,—making the winters milder and the summers cooler,—are now generally discredited. No series of thermometric observations worthy of confidence extend further back in the United States than seventy-eight years. We find from the Philadelphia observations that from 1771 to 1824, the mean annual heat has hardly risen 2°.7 Fahr.;—an increase that may fairly be ascribed to the extension of the town, its greater population, and to the numerous steam-engines. This annual increase of temperature may also be owing to accident, for in the same period I find that there was an increase of the mean winter temperature of 2° Fahr.; but with this exception the seasons had all become somewhat warmer. Thirty-three years' observations at Salem in Massachusetts show scarcely any difference, the mean of each one oscillating within 1° of Fahrenheit, about the mean of the whole number; and the winters of Salem, instead of having been rendered more mild, as conjectured, from the eradication

* See the admirable treatise by Samuel Forry, on *The Climate of the United States*, 1842, pp. 37, 39, 102.

of the forests, have become colder by 4° Fahr. during the last thirty-three years.*

As the east coast of the United States may be compared, in equal latitudes, with the Siberian and Chinese eastern coasts of Europe, in respect to mean annual temperature, so the western coasts of Europe and America have also very justly been compared together. I will here only adduce a few instances from the western region of the Pacific, for two of which, viz., Sitka, (New Archangel,) in Russian America, and Fort George, (having the same latitudes respectively as Gottenburg and Geneva,) we are indebted to Admiral Lütke's voyage of circumnavigation. Iluluk and Danzig are situated in about the same parallel of latitude, and although the mean temperature of Iluluk, owing to its insular climate and the cold sea current contiguous to it, is lower than that of Danzig, the winters of the former are milder than those of the Baltic city.

| | | | | | |
|---|---|---|---|---|---|
| Sitka . . . | Lat. 57° 3′ | Long. | 135° 16′ W. | 44°.6 | 33°.3 / 55°.0 |
| Gottenburg . | Lat. 57° 41′ | Long. | 11° 59′ E. | 46°.4 | 31°.6 / 62°.4 |
| Fort George. | Lat. 46° 18′ | Long. | 123° 58′ W. | 50°.2 | 37°.8 / 60.°0 |
| Geneva . . | Lat. 46° 12′ | Altitude | 1298 feet | 49°.8 | 33°.6 / 63°.5 |
| Cherson . . | Lat. 46° 38′ | Long. | 32° 39′ E. | 53°.1 | 25°.0 / 71°.0 |

Snow is hardly ever seen on the banks of the Oregon or Columbia river, and ice on the river lasts only a few days. The lowest temperature which Mr. Ball ever observed there (in 1838) was 18°.4 Fahr.† A cursory glance at the summer and

* Forry, *Op. Cit.*, pp. 97, 101, 107.

† *Message from the President of the United States to Congress*, 1844, p. 160, and Forry, *Op. Cit.*, pp. 49, 67, 73.

winter temperatures given above, suffices to show that a true insular climate prevails on and near the western coasts; whilst the winter cold is less considerable than in the western part of the old continent, the summers are much cooler. This contrast is made most apparent when we compare the mouth of the Oregon with Forts Snelling and Howard, and the Council Bluffs in the interior of the Mississippi and Missouri basin, (44°—46° north lat.,) where, to speak with Buffon, we find an *excessive* or true *continental* climate,—a winter cold, which on some days is $-32°$ or even $-37°$ Fahr., followed by a mean summer's heat, which rises to 69° and 71°.4 Fahr.

(19) p. 8.—"*As if America had emerged later from the chaotic covering of waters.*"

The acute natural inquirer Benjamin Smith Barton, expresses himself thus accurately:*—"I cannot but deem it a puerile supposition, unsupported by the evidence of nature, that a great part of America has probably later emerged from the bosom of the ocean than the other continents." I have already elsewhere treated of this subject in a memoir on the primitive nations of America:†—"The remark has been too frequently made by authors of general and well-attested merit that America was in every sense of the word a *new* continent. The luxuriance of vegetation, the vast mass of waters in the rivers, and the continued activity of great volcanoes, confirm the fact (say these writers,) that the still agitated and humid earth is in a condition approximating more closely to the chaotic primordial state of our planet than the old continent. Such ideas appeared to me, long before my travels in those regions, no less unphilosophical than at variance with generally acknowledged physical laws. These imaginary representations of an earlier age and a want of repose, and of the increase of dryness and inertia with the increased age of our globe, could only have been framed by those who seek to discover striking contrasts between the two hemispheres, and who do not endeavour to consider the construction of our terrestrial planet from one grand and general point of view. Are we to regard the southern as more recent than the northern part of Italy, simply because the former is almost constantly disturbed by earthquakes and volcanic eruptions? How

* *Fragments of the Nat. Hist. of Pennsylvania*, P. I., p. 4.
† See *Neue Berlinische Monatschrift*, Bd. xv., 1806, § 190.

inconsiderable, moreover, are the phenomena presented by our volcanoes and earthquakes, when compared with the convulsions of nature which the geognosist must conjecture to have occurred in the chaotic condition of our globe, when mountain masses were upheaved, solidified, or cleft asunder? Different causes must also occasion a diversity of effects in the forces of nature in parts of the earth remote from one another. The volcanoes in the new continent," (of which I still count about twenty-eight,) "may probably have continued longer active, because the high mountain ridges on which they are erupted in rows upon long fissures are nearer to the sea, and because this vicinity appears to modify the energy of the subterranean fire, in a manner which, with few exceptions, has not yet been explained. Besides, both earthquakes and fire-erupting mountains act periodically. At present" (this I wrote forty-two years ago,) "physical disquietude and political repose prevail in the new continent, whilst in the old continent the calm repose of nature is contrasted with the dissensions of different nations. The time may however come, when this strange contrast between physical and moral forces may change its theatre of action from one quarter of the world to another. Volcanoes enjoy centuries of repose between their manifestations of activity; and the idea that in the older countries nature must be characterized by a certain repose and quietude, has no other foundation than in the mere caprice of the imagination. There exists no reason for assuming that one side of our planet is older or more recent than the other. Islands, as the Azores and many flat islands of the Pacific, which have been upheaved by volcanoes, or been gradually formed by coral animals, are indeed more recent than many plutonic formations of the European central chain. Small tracts of land, as Bohemia and Kashmeer, and many of the valleys in the moon, inclosed by a ring of mountains, may continue for a long time under the form of a sea, owing to partial inundations, and after the flowing off of these inland waters, the bottom, on which plants would gradually manifest themselves, might indeed be figuratively regarded as of more recent origin. Islands have been connected together into continental masses by upheaval, whilst other parts of the previously existing land have disappeared in consequence of the subsidence of the oscillating ground; but general submersions can, from hydrostatic laws, only be imagined as embracing simultaneously all parts of the earth. The sea cannot permanently submerge the vast lowlands of the Orinoco and the Amazon, without at the same time destroying our Baltic lands. Moreover the succession and identity of the floetz strata, and of the organic remains of plants and animals belonging to the primitive world, inclosed in those strata, show that several great depositions have occurred almost simultaneously over the whole earth."*

(20) p. 8.—"*The Southern Hemisphere is cooler and more humid than the Northern.*"

Chili, Buenos Ayres, the southern part of Brazil, and Peru, enjoy the cool summers and mild winters of a true *insular climate,* owing to the narrowness and contraction of the continent towards the south. This advantage of the Southern Hemisphere is manifested as far as 48° or 50° south lat., but beyond that point, and nearer the Antarctic Pole, South America is an inhospitable waste. The different degrees of latitude at which the *southern* extremities of Australia, including Van Diemen's Island, of Africa, and America, terminate, give to each of these continents its peculiar character. The Straits of Magellan lie between the parallels of 53° and 54° south lat.; and notwithstanding this, the thermometer falls to 41° Fahr. in the months of December and January, when the the sun is eighteen hours above the horizon. Snow falls almost daily in the lowlands, and the maximum of atmospheric heat observed by Churruca in 1788, during the month of December, and consequently in the summer of that region, did not exceed 52°.2 Fahr. The Cabo Pilar, whose turret-like rock is only 1394 feet in height, and which forms the southern extremity of the chain of the Andes, is situated in nearly the same latitude as Berlin.†

Whilst in the Northern Hemisphere all continents fall, in their prolongation towards the Pole, within a mean limit, which corresponds tolerably accurately with 70°, the southern extremities of America, (in Tierra del Fuego, which is so

* On the vegetable remains found in the lignite formations of the north of America and of Europe, compare Adolph Brongniart, *Prodrome d'une Hist. des Végétaux Fossiles,* p. 179, and Charles Lyell's *Travels in North America,* vol. ii., p. 20.

† *Relacion del Viage al Estrecho de Magallanes* (Apendice, 1793), p. 76.

deeply indented by intersecting arms of the sea,) of Australia, and of Africa, are respectively 34°, 46°30′ and 56° distant from the South Pole. The temperature of the unequal extents of ocean which separate these southern extremities from the icy Pole contributes essentially towards the modification of the climate. The areas of the dry land of the two hemispheres separated by the equator are as 3 to 1. But this deficiency of continental masses in the Southern Hemisphere is greater in the temperate than in the torrid zone, the ratio being in the former at 13 to 1, and in the latter as 5 to 4. This great inequality in the distribution of dry land exerts a perceptible influence on the strength of the ascending atmospheric current, which turns towards the South Pole, and on the temperature of the Southern Hemisphere generally. Some of the noblest forms of tropical vegetation, as for instance tree-ferns, advance south of the equator to the parallels of from 46° to 53°, whilst to the north of the equator they do not occur beyond the tropic of Cancer.* Tree-ferns thrive admirably well at Hobart Town in Van Diemen's Land (42° 53′ lat.), with a mean annual temperature of 52°.2 Fahr., and therefore on an isothermal line less by 3°.6 Fahr. than that of Toulon. Rome, which is almost one degree of latitude further from the equator than Hobart Town, has an annual temperature of 59°.7 Fahr.; a winter temperature of 46°.6 Fahr., and a summer temperature of 86° Fahr.; whilst in Hobart Town these three means are respectively 52°, 42°.1, and 63° Fahr. In Dusky Bay, New Zealand, tree-ferns thrive in 46° 8′ lat., and in the Auckland and Campbell Islands in 53° lat.†

In the Archipelago of Tierra del Fuego, having a mean winter temperature of 33° Fahr., and a mean summer temperature of only 50° Fahr., in the same latitude as Dublin, Captain King found "vegetation thriving most luxuriantly in large woody-stemmed trees of Fuchsia and Veronica;" whilst this vigorous vegetation, which, especially on the western coast of America (in 38° and 40° south lat.), has been so picturesquely described by Charles Darwin, suddenly disappears south of Cape Horn, on the rocks of the Southern Orkney and Shetland Islands, and of the Sandwich Archipelago. These islands, but scantily covered with grass, moss, and

* See Robert Brown, *Appendix to Flinders' Voyage*, pp. 575, 584; and Humboldt, *De Distributione Geographica Plantarum*, pp. 81—85.
† Jos. Hooker, *Flora Antarct.*, 1844, p. 107.

lichens, *Terres de Désolation*, as they have been called by French navigators, lie far to the north of the Antarctic Polar Circle; whilst in the Northern Hemisphere, in 70° lat., on the extremest verge of Scandinavia, fir-trees reach a height of more than 60 feet.* If we compare Tierra del Fuego, and more particularly Port Famine, in the Straits of Magellan, 53° 38′ lat., with Berlin, which is situated one degree nearer the equator, we shall find for Berlin, $47^\circ.3\ \frac{38^\circ.9}{62^\circ.3}$; and for Port Famine, $42^\circ.6\ \frac{34^\circ.7}{50^\circ.0}$ Fahr. I subjoin the few certain data of temperature which we at present possess of the temperate zones of the Southern Hemisphere, and which may be compared with the temperatures of northern regions in which the distribution of summer heat and winter cold is so unequal. I make use of the convenient mode of notation already explained in which the number standing before the fraction indicates the mean annual temperature, the numerator the winter, and the denominator the summer temperature.

| Places. | South Latitude. | Mean Annual, Winter, and Summer Temperatures. |
|---|---|---|
| Sydney and Paramatta (New Holland) . . | 33° 50′ | $64^\circ.6\ \frac{54^\circ.5}{77^\circ.5}$ |
| Cape Town (Africa) . | 33° 55′ | $65^\circ.7\ \frac{58^\circ.5}{73^\circ.2}$ |
| Buenos Ayres . . . | 34° 17′ | $62^\circ.4\ \frac{52^\circ.5}{73^\circ.0}$ |
| Monte Video . . . | 34° 54′ | $67^\circ\ \frac{57^\circ.4}{77^\circ.5}$ |
| Hobart Town (Van Diemen's Land) . | 42° 45′ | $52^\circ.5\ \frac{42^\circ.1}{63^\circ.0}$ |
| Port Famine (Straits of Magellan) . . . . | 53° 38′ | $42^\circ.6\ \frac{34^\circ.7}{50^\circ.0}$ |

* Compare Darwin in the *Journal of Researches*, 1845, p. 244, with King in vol. i. of the *Narr. of the Voyages of the Adventure and the Beagle*, p. 577.

(21) p. 9.—"*One connected sea of sand.*"

As we may regard the social *Erica* as furnishing one continuous vegetable covering spread over the earth's surface, from the mouth of the Scheldt to the Elbe, and from the extremity of Jutland to the Harz mountains, so may we likewise trace the sea of sand continuously through Africa and Asia, from Cape Blanco to the further side of the Indus, over an extent of 5,600 miles. The sandy region mentioned by Herodotus, which the Arabs call the Desert of Sahara, and which is interrupted by oases, traverses the whole of Africa like a dried arm of the sea. The valley of the Nile is the eastern boundary of the Lybian desert. Beyond the Isthmus of Suez and the porphyritic, syenitic, and greenstone rocks of Sinai begins the Desert mountain plateau of Nedschd, which occupies the whole interior of the Arabian Peninsula, and is bounded to the west and south by the fruitful and more highly favoured coast-lands of Hedschaz and Hadhramaut. The Euphrates forms the eastern boundary of the Arabian and Syrian desert. The whole of Persia, from the Caspian Sea to the Indian Ocean, is intersected by immense tracts of sand (*bejaban*), among which we may reckon the soda and potash Deserts of Kerman, Seistan, Beludschistan, and Mekran. The last of these barren wastes is separated by the Indus from the Desert of Moultan.

(22) p. 9.—"*The western portion of Mount Atlas.*"

The question of the position of the Atlas of the ancients has often been agitated in our own day. In making this inquiry, ancient Phœnician traditions are confounded with the statements of the Greeks and Romans regarding Mount Atlas at a less remote period. The elder Professor Ideler, who combined a profound knowledge of languages with that of astronomy and mathematics, was the first to throw light on this obscure subject; and I trust I may be pardoned if I insert the communications with which I have been favoured by this enlightened observer.

"The Phœnicians ventured at a very early period in the world's history to penetrate beyond the Straits of Gibraltar. They founded Gades and Tartessus on the Spanish, and Lixus, together with many other cities on the Mauritanian coasts of the Atlantic Ocean. They sailed northward along these shores to the Cassiterides, from whence they obtained tin, and to the Prussian coasts where they procured amber found there: whilst southward they penetrated as far as Madeira and the Cape de Verd Islands. Amongst other regions they visited the Archipelago of the Canary Isles, where their attention was arrested by the Peak of Teneriffe, whose great height appears to be even more considerable than it actually is from the circumstance of the mountain projecting directly from the sea. Through their colonies established in Greece, especially under Cadmus in Bœotia, the Greeks were made acquainted with the existence of this mountain which soared high above the region of clouds, and with the 'Fortunate Islands' on which this mountain was situated, and which were adorned with fruits of all kinds, and particularly with the golden orange. By the transmission of this tradition through the songs of the bards, Homer became acquainted with these remote regions, and he speaks of an Atlas to whom all the depths of ocean are known, and who bears upon his shoulders the great columns which separate from one another the heavens and the earth,* and of the *Elysian Plains*, described as a wondrously beautiful land in the west."† Hesiod expresses himself in a similar manner regarding Atlas, whom he represents as the neighbour of the Hesperides.‡ The Elysian Plains, which he places at the western limits of the earth, he terms the 'Islands of the Blessed.'§ Later poets have still further embellished these myths of Atlas, the Hesperides, their golden apples, and the Islands of the Blessed, which are destined to be the abode of good men after death, and have connected them with the expeditions of the Tyrian God of Commerce, Melicertes, the Hercules of the Greeks.

"The Greeks did not enter into rivalship with the Phœnicians and Carthaginians in the art of navigation until a comparatively late period. They indeed visited the shores of the Atlantic, but they never appear to have advanced very far. It is doubtful whether they had penetrated as far as the Canary Isles and the Peak of Teneriffe; but be this as it may, they were aware that Mount Atlas, which their poets had described

* *Od.*, i. 52,
† *Il.*, iv. 561.
‡ *Theog.*, v. 517.
§ *Op. et Dies*, v. 167.

as a very high mountain situated on the western limits of the earth, must be sought on the western coast of Africa. This too was the locality assigned to it by their later geographers Strabo, Ptolemy, and others. As however no mountain of any great elevation was to be met with in the north-west of Africa, much perplexity was entertained regarding the actual position of Mount Atlas, which was sought sometimes on the coast, sometimes in the interior of the country, and sometimes in the vicinity of the Mediterranean, or further southward. In the first century of the Christian era, when the armies of Rome had penetrated to the interior of Mauritania and Numidia, it was usual to give the name of Atlas to the mountain chain which traverses Africa from west to east in a parallel direction with the Mediterranean. Pliny and Solinus were both, however, fully aware that the description of Atlas given by the Greek and Roman poets did not apply to this mountain range, and they therefore deemed it expedient to transfer the site of Mount Atlas, which they described in picturesque terms, in accordance with poetic legends, to the *terra incognita* of Central Africa. The Atlas of Homer and Hesiod can, therefore, be none other than the Peak of Teneriffe, while the Atlas of Greek and Roman geographers must be sought in the north of Africa."

I will only venture to add the following remarks to the learned explanations of Professor Ideler. According to Pliny and Solinus, Atlas rises from the midst of a sandy plain (*e medio arenarum*), and its declivity affords pasture to elephants, which have undoubtedly never been known in Teneriffe. That which we now term Atlas is a long mountain ridge. How could the Romans have recognised one isolated conical elevation in this mountain range of Herodotus? May the cause not be ascribed to the optical illusion by which every mountain chain, when seen laterally from an oblique point of view, appears to be of a narrow and conical form? I have often, when at sea, mistaken long mountain ranges for isolated mountains. According to Höst, Mount Atlas is covered with perpetual snow near Morocco. Its elevation must therefore be upwards of 11,500 feet at that particular spot. It seems to me very remarkable that the barbarians, the ancient Mauritanians, if we are to believe the testimony of Pliny, called Mount Atlas *Dyris*. This mountain chain is still called by the Arabs *Daran*, a word that is almost identical in its consonants with *Dyris*. Hornius,* on the other hand, thinks that he recognises the term Dyris in the word *Ayadyrma*, the name applied by the Guanches to the Peak of Teneriffe.†

As our present geological knowledge of the mountainous parts of North Africa, which, however, must be admitted to be very limited, does not make us acquainted with any traces of volcanic eruptions within historic times, it seems the more remarkable that so many indications should be found in the writings of the Ancients of a belief in the existence of such phenomena in the Western Atlas and the contiguous west coast of the continent. The streams of fire so often mentioned in Hanno's Ship's Journal might indeed have been tracks of burning grass, or beacon fires lighted by the wild inhabitants of the coasts as a signal to warn each other of threatening danger on the first appearance of hostile vessels. The high summit of the "Chariot of the Gods," of which Hanno speaks (the θεῶν ὄχημα), may also have had some faint reference to the Peak of Teneriffe; but farther on he describes a singular configuration of the land. He finds in the gulf, near the Western Horn, a large island, in which there is a salt lake, which again contains a smaller island. South of the Bay of the Gorilla Apes the same conformation is repeated. Does he refer to coral structures, lagoon islands (Atolls), and to volcanic crater lakes, in the middle of which a conical mountain has been upheaved? The Triton Lake was not in the neighbourhood of the lesser Syrtis, but on the western shores of the Atlantic.‡ The lake disappeared in an earthquake, which was attended with great fire-eruptions. Diodorus§ says expressly πυρὸς ἐκφυτήματα μεγάλα. But the most wonderful configuration is ascribed to the hollow Atlas, in a passage hitherto but little noticed in one of the philosophical

* *De Originibus Americanorum*, p. 195.

† On the connexion of purely mythical ideas and geographical traditions, and on the manner in which the Titan Atlas gave occasion to the image of a mountain beyond the Pillars of Hercules supporting the heavens, see Letronne, *Essai sur les Idées cosmographiques qui se rattachent au nom d'Atlas*, in Férussac's *Bulletin universel des Sciences*, Mars 1831, p. 10.

‡ *Asie centrale*, t. i., p. 179.

§ Lib. iii., 53, 55.

Dialexes of Maximus Tyrius, a Platonic philosopher who lived in Rome under Commodus. *His* Atlas is situated "on the continent where the Western Lybians inhabit a projecting peninsula." The mountain has a deep semi-circular abyss on the side nearest the sea; and its declivities are so steep that they cannot be descended. The abyss is filled with trees, and "one looks down upon their summits and the fruits they bear as if one were looking into a well."* The description is so minute and graphic that it no doubt sprung from the recollection of some actual view.

(23) p. 9.—"*The Mountains of the Moon, Djebel-al-Komr.*"

The Mountains of the Moon described by Ptolemy,† σελήνης ὄρος, form on our older maps a vast uninterrupted mountain chain, traversing the whole of Africa from east to west. The existence of these mountains seems certain; but their extent, their distance from the equator, and their mean direction, still remain problematical. I have indicated in another work‡ the manner in which a more intimate acquaintance with Indian idioms and the ancient Persian or Zend teaches us that a part of the geographical nomenclature of Ptolemy constitutes an historical memorial of the commercial relations that existed between the West and the remotest regions of Southern Asia and Eastern Africa. The same direction of ideas is apparent in relation to a subject that has very recently become a matter of investigation. It is asked, whether the great geographer and astronomer of Pelusium merely meant in the denomination of Mountains of the Moon (as in that of "Island of Barley," (Jabadiu, Java) to give the Greek translation of the native name of those mountains; whether, as is most probable, El-Istachri, Edrisi, Ibn-al-Vardi, and other early Arabian geographers, simply transferred the Ptolemaic nomenclature into their own language; or whether similarity in the sound of the word and the manner in which it was written misled them? In the notes to the translation of Abd-Allatif's celebrated description of Egypt, my great teacher, Silvestre de Sacy,§ expressly says, "The

* Maximus Tyrius, viii., 7, ed. Markland.

† Lib. iv., cap. 9.

‡ *Cosmos,* vol. ii., p. 559. Bohn's ed.

§ Edition de 1810, pp. 7, 353.

name of the mountains regarded by Leo Africanus as furnishing the sources of the Nile, has generally been rendered 'Mountains of the Moon,' and I have adhered to the same practice. I do not know whether the Arabs originally borrowed this denomination from Ptolemy. It may indeed be inferred that at the present day they understand the word قمر in the sense of *moon*, pronouncing it *kamar;* I do not think however, that such was the practice of the older Arabs, who pronounced it *komr*, as has been proved by Makrizi. Aboulfeda positively rejects the opinion of those who would adopt the pronunciation *kamar*, and derive the word from the name of the moon. As, according to the author of *Kamous*, the word *komr*, considered as the plural of اقمر, signifies an object of a greenish or dirty white colour, it would appear that some authors have supposed that this mountain derived its name from its colour."

The learned Reinaud, in his recent excellent translation of Abulfeda (t. ii., p. i., pp. 81, 82), regards it as probable that the Ptolemaic interpretation of the name of Mountains of the Moon (ὄρη σεληναῖα) was that originally adopted by the Arabs. He observes that in the Moschtarek of Yakut, and in Ibn-Said, the mountain is written al-Komr, and that Yakut writes in a similar manner the name of the Island of Zendj (Zanguebar). The Abyssinian traveller Beke, in his learned and critical treatise on the Nile and its tributaries,* endeavours to prove that Ptolemy, in his σελήνης ὄρος, merely followed the native name, for the knowledge of which he was indebted to the extensive commercial intercourse which then existed. He says, "Ptolemy knew that the Nile rises in the mountainous district of Moezi, and in the languages which are spoken over a great part of Southern Africa (as, for instance, in Congo, Monjou, and Mozambique), the word *moezi* signifies the moon. A large tract of country situated in the south-west was called Mono-Muezi, or Mani-Moezi, *i.e.*, the land of the King of Moezi (or Moon-land); for in the same family of languages in which *moezi* or *muezi* signifies the moon, *mono* or *mani* signifies a king. Alvarez† speaks

* See *Journal of the Royal Geographical Society of London*, vol. xvii., 1847, pp. 74—76.

† *Viaggio nella Ethiopia* (Ramusio, vol. i., p. 249).

of the 'regno di Manicongo,' or territory of the king of Congo." Beke's opponent, Ayrton, seeks the sources of the White Nile (Bahr el-Abiad), not as do Arnaud, Werne. and Beke, near the equator, or south of it (in 31° 22′ E. long. from Greenwich), but far to the north-east, as does Antoine d'Abbadie, in the Godjeb and Gibbe of Eneara (Iniara), therefore in the high mountains of Habesch, in 7° 20′ north lat., and 35° 22′ east long. from Greenwich. He is of opinion that the Arabs, from a similarity of sound, may have interpreted the native name Gamaro, which was applied to the Abyssinian mountains lying south-west of Gaka, and in which the Godjeb (or White Nile) takes its rise, to signify a mountain of the moon (Djebel al-Kamar); so that Ptolemy himself, who was familiar with the intercourse existing between Abyssinia and the Indian Ocean, may have adopted the Semitic interpretation, as given by the descendants of the early Arab immigrants.*

The lively interest which has recently been felt in England for the discovery of the most southern sources of the Nile induced the Abyssinian traveller above referred to, (Charles Beke) at a recent meeting of the "British Association for the advancement of Science," held at Swansea, more fully to develope his ideas respecting the connection between the Mountains of the Moon and those of Habesch. "The Abyssinian elevated plain," he says, "generally above 8000 feet high, extends towards the south to nearly 9° or 10° north latitude. The eastern declivity of the highlands has, to the inhabitants of the coast, the appearance of a mountain chain. The plateau, which diminishes considerably in height towards its southern extremity, passes into the Mountains of the Moon, which run not east and west, but parallel to the coast, or from N.N.E. to S.S.W., extending from 10° north to 5° south latitude. The sources of the White Nile are situated in the Mono-Moezi country, probably in 2° 30′ south latitude, not far from where the river Sabaki, on the eastern side of the Mountains of the Moon, falls into the Indian Ocean, near Melindeh, north of Mombaza. Last autumn (1847), the two Abyssinian missionaries Rebmann and Dr. Krapf

* Compare Ayrton, in the *Journal of the Royal Geog. Soc.*, vol. xviii., 1848, pp. 53, 55, 59—63, with Ferd. Werne's instructive *Exped. zur Entd. der Nil-Quellen*, 1848, s. 534—536.

were still on the coast of Mombaza. They have established in the vicinity, among the Wakamba tribe, a missionary station, called Rabbay Empie, which seems likely to be very useful for geographical discoveries. Families of the Wakamba tribe have advanced westward five or six hundred miles into the interior of the country, as far as the upper course of the river Lusidji, the great lake Nyassi or Zambeze (5° south lat.?), and the vicinal sources of the Nile. The expedition to these sources, which Friedrich Bialloblotzky, of Hanover, is preparing to undertake" (by the advice of Beke), "is to start from Mombaza. The Nile coming from the west referred to by the ancients is probably the Bahr-el-Ghazal, or Keilah, which falls into the Nile in 9° north lat., above the mouth of the Godjeb or Sobat."

Russegger's scientific expedition—undertaken in 1837 and 1838, in consequence of Mehemed Ali's eager desire to participate in the gold washings of Fazokl on the Blue (Green) Nile, Bahr el-Azrek—has rendered the existence of a Mountain of the Moon very doubtful. The Blue Nile, the Astapus of Ptolemy, rising from Lake Coloe (now called Lake Tzana), winds through the colossal Abyssinian range of mountains; while to the south-west there appears a far extended tract of low land. The three exploring expeditions which the Egyptian Government sent from Chartum to the confluence of the Blue and the White Nile (the first under the command of Selim Bimbaschi, in November, 1839; the next, which was attended by the French engineers Arnaud, Sabatier, and Thibaut, in the autumn of 1840; and the third, in the month of August, 1841), first removed some of the obscurity which had hitherto shrouded our knowledge of the high mountains, which between the parallels of 6°—4°, and probably still further southward, extend first from west to east, and subsequently from north-west to south-east, towards the left bank of the Bahr-el-Abiad. The second of Mehemet Ali's expeditions first saw the mountain chain, according to Werne's account, in 11° 20′ north lat., where Gebel Abul and Gebel Kutak rise to the height of 3623 feet. The high land continued to approach the river more to the south from 4° 45′ north lat. to the parallel of the Island of Tchenker in 4° 4′, near the point at which terminated the expedition commanded by Selim and Feizulla Effendi. The shallow river breaks its way through the rocks, and separate mountains again rise in the land of Bari to the height of more than 3200 feet. These are probably a part of the Mountains of the Moon, as they are given in our most recent maps, although they are not covered with perpetual snow, as asserted by Ptolemy.* The line of perpetual snow would assuredly not be found in these parallels of latitude below an elevation of nearly 15,500 feet above the sea's level. It is not improbable that Ptolemy extended the knowledge he may have possessed of the high mountains of Habesch, near Upper Egypt and the Red Sea, to the country of the sources of the White Nile. In Godjam, Kaffa, Miecha, and Sami, the Abyssinian mountains rise from 10,000 to nearly 15,000 feet, as we learn from exact measurements; (not according to those of Bruce, who gives to Chartum an elevation of 5041 feet, instead of the true height, 1524 feet!) Rüppell, who ranks amongst the most accurate observers of the present day, found Abba Jarat (in 13° 10′ north lat.) only 70 feet below the elevation of Mont Blanc.† The same observer states that a plain, elevated 13,940 feet above the Red Sea, was barely covered with a thin layer of freshly fallen snow.‡ The celebrated inscription of Adulis, which, according to Niebuhr, is of somewhat later date than the age of Juba and Augustus, speaks of "Abyssinian snow that reaches to the knee," and affords, I believe, the most ancient record in antiquity of snow within the tropics,§ as the Paropanisus is 12° lat. north of that limit.

Zimmermann's map of the district of the Upper Nile shows the dividing line where the basin of the great river terminates in the south-east, and which separates it from the domain of the rivers belonging to the Indian Ocean, viz.; from the Doara which empties itself north of Magadoxo; from the Teb on the amber coast of Ogda; from the Goschop whose abundant waters are derived from the confluence of the Gibu and the Zebi, and which must be distinguished from the Godjeb, rendered celebrated since 1839 by Antoine d'Abbadie, Beke, and the Missionary Krapf. In a letter to Carl Ritter I hailed with the most lively joy the appearance of the

* Lib. iv., cap. 9.

† See Rüppell, *Reise in Abyssinien*, bd. i., s. 414; bd. ii., s. 443.

‡ Humboldt, *Asie centrale*, t. iii., p. 272.

§ *Op. cit.*, t. iii., p. 235.

combined results of the recent travels of Beke, Krapf, Isenberg, Russegger, Rüppel, Abbadie, and Werne, as ably and comprehensively brought together in 1843 by Zimmermann. "If a prolonged span of life," I wrote to him, "bring with it many inconveniences to the individual himself, and some to those about him, it yields a compensation in the mental enjoyment, afforded by comparing the earlier state of our knowledge with its more recent condition, and of seeing the growth and development of many branches of science that had long continued torpid, or whose actual fruits hypercriticism may even have attempted to set aside. This genial enjoyment has from time to time fallen to our lot in our geographical studies, and more especially in reference to those portions of which we could hitherto only speak with a certain timid hesitation. The internal configuration and articulation of a continent depends in its leading characters on several plastic relations which are usually among the latest to be elucidated. A new and excellent work of our friend, Carl Zimmermann, on the district of the Upper Nile and of the eastern portions of Central Africa, has made me more vividly sensible of these considerations. This new map indicates, in the clearest manner, by means of a special mode of shading, all that still remains unknown, and all that by the courage and perseverance of travellers of all nations (among which our own countrymen happily play an important part), has already been disclosed to us. We may regard it as alike important and useful that the actual condition of our knowledge, should, at different periods, be graphically represented by men well acquainted with the existing and often widely scattered materials of knowledge, and who not merely delineate and compile, but who know how to compare, select, and, where it is practicable, test the routes of travellers by astronomical determinations of place. Those who have contributed as much to the general stock of knowledge as you have done, have indeed an especial right to expect much, since their combinations have greatly increased the number of connecting points; yet I scarcely think that when, in the year 1822, you executed your great work on Africa, you could have anticipated so many additions as we have received." It must be admitted that, in some cases, we have only acquired a knowledge of rivers, their direction, their branches, and their numerous synonymes according to various languages and dialects; but the courses of rivers indicate the configuration of the surface of the earth, and exert a threefold influence; they promote vegetation, facilitate general intercourse, and are pregnant with the future destiny of man.

The northern course of the White Nile, and the south-eastern course of the great Goschop, show that both rivers are separated by an elevation of the surface of the earth; although we are as yet but imperfectly acquainted with the manner in which such an elevation is connected with the highlands of Habesch, or how it may be prolonged in a southerly direction beyond the equator. Probably, and this is also the opinion of my friend Carl Ritter, the Lupata Mountains, which, according to the excellent Wilhelm Peters, extend to 26° south lat., are connected by means of the Mountains of the Moon with this northern swelling of the earth's surface (the Abyssinian Highlands). *Lupata*, according to the last-named African traveller, signifies, in the language of Tette, *closed*, when used as an adjective. This mountain-range which is only intersected by some few rivers would thus be the *closed* or *barred*. "The Lupata chain of the Portuguese writers," says Peters, "is situated about 90 leagues from the mouth of the Zambeze, and has an elevation of little more than 2000 feet. This mural chain has a direction due north and south, although it frequently deflects to the east or the west. It is sometimes interrupted by plains. Along the coast of Zanzibar the traders in the interior appear to be acquainted with this long, but not very high range, which extends between 6° and 26° south lat. to the Factory of Lourenzo-Marques on the Rio de Espirito Santo (in the Delagoa Bay of the English). The further the Lupata chain extends to the south, the nearer it approaches the coast, until at Lourenzo-Marques it is only 15 leagues distant from it."

(24) p. 10—"*The consequence of the great rotatory movement of the waters.*"

The waters of the northern part of the Atlantic between Europe, Northern Africa, and the New Continent, are agitated by a continually recurring gyratory movement. Under the tropics the general current to which the term *rotation-stream* might appropriately be given in consideration of the cause from which it arises, moves, as is well known, like the trade wind from east to west. It accelerates the navigation of vessels sailing from the Canary Isles to South America; while it is nearly impossible to pursue a straight course against the current from Carthagena de Indias to Cumana. This bend to the west, attributed to the trade winds, is accelerated in the Caribbean Sea by a much stronger movement, which originates in a very remote cause, discovered as early as 1560 by Sir Humphrey Gilbert,* and confirmed in 1832 by Rennell. The Mozambique current, flowing from north to south between Madagascar and the eastern coast of Africa, sets on the Lagullas Bank, and bends to the north of it round the southern point of Africa. After advancing with much violence along the western coast of Africa beyond the equator to the island of St. Thomas, it gives a north-westerly direction to a portion of the waters of the South Atlantic, causing them to strike Cape St. Augustin, and follow the shores of Guiana beyond the mouth of the Orinoco, the Boca del Drago, and the coast of Paria.† The New Continent from the Isthmus of Panama to the northern part of Mexico forms a dam or barrier against the movements of the sea. Owing to this obstruction the current is necessarily deflected in a northerly direction at Veragua, and made to follow the sinuosities of the coast-line from Costa Rica, Mosquitos, Campeche, and Tabasco. The waters which enter the Mexican Gulf between Cape Catoche of Yucatan, and Cape San Antonio de Cuba, force their way back into the open ocean north of the Straits of Bahama, after they have been agitated by a great rotatory movement between Vera Cruz, Tamiagna, the mouth of the Rio Bravo del Norte, and the Mississippi. Here they form a warm, rapid current, known to mariners as the *Gulf Stream*, which deflects in a diagonal direction further and further from the shores of North America. Ships bound for this coast from Europe, and uncertain of their geographical longitude, are enabled by this oblique direction of the current to regulate their course as soon as they reach the Gulf Stream by observations of latitude only. The bearings of this current were first accurately determined by Franklin, Williams, and Pownall.

* Hakluyt, *Voyages*, vol. iii. p. 14.

† Rennell, *Investigation of the Currents of the Atlantic Ocean*, 1832, pp. 96, 136.

From the parallel of 41° north lat. this stream of warm water follows an easterly direction, gradually diminishing in rapidity as it increases in breadth. It almost touches the southern edge of the Great Newfoundland Bank, where I found the greatest amount of difference between the temperature of the waters of the Gulf Stream and those exposed to the cooling action of the banks. Before the warm current reaches the Western Azores it separates into two branches, one of which turns at certain seasons of the year towards Ireland and Norway, while the other flows in the direction of the Canary Isles and the western coast of Northern Africa.

The course of this Atlantic current, which I have described more fully in the first volume of my travels in the regions of the tropics, affords an explanation of the manner in which, notwithstanding the action of the trade winds, stems of the South American and West Indian *dicotyledons* have been found on the coasts of the Canary Islands. I made many observations on the temperature of the Gulf Stream in the vicinity of the Newfoundland Bank. This current bears the warmer water of lower latitudes with great rapidity into more northern regions. The temperature of the stream is therefore from about 4°½ to 7° Fahr. higher than that of the contiguous and unmoved water which constitutes the shore as it were of the warm oceanic current.

The flying-fish of the equinoctial zone (*Exocetus volitans*), is borne by its predilection for the warmth of the water of the Gulf Stream far to the north of the temperate zone. Floating sea-weed (*Fucus natans*), chiefly taken up by the stream in the Mexican Gulf, makes it easy for the navigator to recognize when he has entered the Gulf Stream, whilst the position of the branches of the sea-weed indicate the direction of the current. The mainmast of the English ship of war, the Tilbury, which was destroyed by fire in the seven years' war on the coasts of Saint Domingo, was carried by the Gulf Stream to the northern coasts of Scotland: and casks filled with palm-oil, the remains of the cargo of an English ship wrecked on a rock off Cape Lopez in Africa, were in like manner carried to Scotland, after having twice traversed the Atlantic Ocean, once from east to west between 2° and 12° north lat., following the course of the equinoctial current, and once from west to east between 45° and 55° north lat. by help of the Gulf Stream. Rennell, in the work already referred to, p. 347, relates the voyage of a bottle inclosing a written paper which had been thrown from the English ship Newcastle in 38° 52′ north lat., and 63° 58′ west long., on the 20th of January, 1819, and which was first seen on the 2nd of June, 1820, at the Rosses in the north-west of Ireland, near the Island of Arran. Shortly before my arrival at Teneriffe a stem of South American cedar-wood (*Cedrela odorata*), thickly covered with lichens, was cast ashore near the harbour of Santa Cruz.

The effects of the Gulf Stream in stranding on the Azorean Islands of Fayal, Flores, and Corvo, bamboos, artificially cut pieces of wood, trunks of an unknown species of pine from Mexico or the West Indies, and corpses of men of a peculiar race, having very broad faces, have mainly contributed to the discovery of America, as they confirmed Columbus in his belief of the existence of Asiatic countries and islands situated in the west. The great discoverer even heard from a settler on the Cap de la Verga in the Azores "that persons in sailing westward had met with covered barks, which were managed by men of foreign appearance, and appeared to be constructed in such a manner that they could not sink, *almadias con casa movediza que nunca se hunden.*" There are well authenticated proofs, however much the facts may have been called in question, that natives of America (probably Esquimaux from Greenland or Labrador), were carried by currents or streams from the north-west to our own continent. James Wallace* relates that in the year 1682 a Greenlander in his canoe was seen on the southern extremity of the Island of Eda by many persons, who could not, however, succeed in reaching him. In 1684 a Greenland fisherman appeared near the Island of Westram. In the church at Burra there was suspended an Esquimaux boat, which had been driven on shore by currents and storms. The inhabitants of the Orkneys call the Greenlanders who have appeared amongst them *Finnmen.*

In Cardinal Bembo's *History of Venice* I find it stated, that in the year 1508 a small boat, manned by seven persons of a foreign aspect, was captured near the English coast by a French ship. The description given of them applies perfectly to the form of the Esquimaux (*homines erant septem mediocri statura, colore subobscuro, lato et patente vultu, cicatriceque una*

* *Account of the Islands of Orkney* (1700), p. 60.

*violacea signato*). No one understood their language. Their clothing was made of fish skins sewn together. On their heads they wore *coronam e culmo pictam, septem quasi auriculis intextam*. They ate raw flesh, and drank blood as we would wine. Six of these men perished during the voyage, and the seventh, a youth, was presented to the King of France, who was then at Orleans.*

The appearance of men called *Indians* on the coasts of Germany under the Othos and Frederic Barbarossa in the tenth and twelfth centuries, and as Cornelius Nepos (in his *Fragments*),† Pomponius Mela,‡ and Pliny§ relate, when Quintus Metellus Celer was Proconsul in Gaul, may be explained by similar effects of oceanic currents and by the long continuance of north-westerly winds. A king of the Boii, or, as others say, of the Suevi, gave these stranded dark-coloured men to Metellus Celer. Gomara‖ regards these Indian subjects of the King of the Boii as natives of Labrador. He writes, *Si ya no fuesen de Tierra del Labrador, y los tuviesen los Romanos por Indianos engañados en el color.* It may be inferred that the appearance of Esquimaux on the northern shores of Europe was more frequent in earlier times, for we learn from the investigations of Rask and Finn Magnusen, that this race had spread in the eleventh and twelfth century in considerable numbers, under the name of Skrälingers, from Labrador as far south as the Good Vinland, *i.e.* the shore of Massachussets and Connecticut.¶

As the winter cold of the most northern part of Scandinavia is ameliorated by the action of the Gulf Stream, which carries American tropical fruits (as cocoa-nuts, seeds of *Mimosa scandens* and *Anacardium occidentale*) beyond 62° north lat.; so also Iceland enjoys from time to time the genial influence of the diffusion of the warm waters of the Gulf Stream far to the northward. The sea coasts of Iceland, like those of the Faroe Isles, receive a large number of trunks of

* Bembo, *Historiæ Venetæ*, ed. 1718, lib. vii. p. 257.
† Ed. Van. Staveren, cur. Bardili, t. ii. 1820, p. 356.
‡ Lib. iii, cap. 5, § 8.
§ *Hist. Nat.* ii. 67.
‖ *Historia Gen. de las Indias.* Saragossa, 1553, fol. vii.
¶ See *Cosmos*, vol. ii. p. 604 (Bohn's ed.) and *Examen critique de l'Hist. de la Géographie*, t. ii. pp. 247—278.

trees, driven thither from America; and this drift-wood, which formerly came in greater abundance, was used for the purposes of building, and cut into boards and laths. The fruits of tropical plants collected on the Icelandic shores, especially between Raufarhaven and Vapnafiord, show that the movement of the water is from a southerly direction.*

(25) p. 10.—"*Lecideæ and other Lichens.*"

In northern regions, the absence of plants is compensated for by the covering of *Bæomyces roseus*, *Cenomyce rangiferinus*, *Lecidea muscorum*, *Lecidea icmadophila*, and other cryptogamia which are spread over the earth, and which may be said to prepare the way for the growth of grasses and other herbaceous plants. In the tropical world, where mosses and lichens are only observed to abound in shady places, some few oily plants supply the place of the lowly lichen.

(26) p. 11.—"*The Care of Animals yielding milk.—Ruins of the Aztek fortress.*"

The two oxen already named, *Bos americanus* and *Bos moschatus*, are peculiar to the northern part of the American continent. But the natives—

*Queis neque mos, neque cultus erat, nec jungere tauros*
Virg. Æn. i. 316.

drank the fresh blood, and not the milk, of these animals. Some few exceptions have indeed been met with, but only among tribes who at the same time cultivated maize. I have already observed that Gomara speaks of a people in the north-west of Mexico who possessed herds of tame bisons, and derived their clothing, food, and drink from these animals. This drink was probably the blood,† for, as I have frequently remarked, a dislike of milk, or at least the absence of its use, appears before the arrival of Europeans to have been common to all the natives of the New Continent, as well as to the inhabitants of China and Cochin China, notwithstanding their great vicinity to true pastoral tribes. The herds of tame lamas which were found in the highlands of Quito, Peru, and Chili, belonged to a settled and agricultural

* Sartorius von Waltershausen, *Physisch-geographische Skizze von Island,* 1847, s. 22—35.

† Prescott, *Conquest of Mexico,* vol. iii. p. 416.

population. Pedro de Cieca de Leon* seems to imply, although assuredly as a very rare exception to the general mode of life, that lamas were employed on the Peruvian mountain plain of Callao for drawing the plough.† Ploughing was, however, generally conducted in Peru by men only.‡ Barton has made it appear probable that the American buffalo had from an early period been reared among some West Canada tribes on account of its flesh and hide.§ In Peru and Quito the lama is nowhere found in its original wild condition. According to the statements made to me by the natives, the lamas on the western declivity of the Chimborazo became wild at the time when Lican, the ancient residence of the rulers of Quito, was laid in ashes. In Central Peru, in the Ceja de la Montaña, cattle have in like manner become completely wild; a small but daring race that often attacks the Indians. The natives call them "Vacas del Monte" or "Vacas Cimarronas."‖ Cuvier's assertion that the lama had descended from the guanaco, still in a wild state, which had unfortunately been extensively propagated by the admirable observer, Meyen,¶ has now been completely refuted by Tschudi.

The Lama, the Paco or Alpaca, and the Guanaco are three originally distinct species of animals.** The Guanaco (Huanacu in the Qquichua language) is the largest of the three, and the Alpaca, measured from the ground to the crown of the head, the smallest. The Lama is next to the Guanaco in height. Herds of Llamas, when as numerous as I have seen them on the elevated plateaux between Quito and Riobamba, are a great ornament to the landscape. The Moromoro of Chili appears to be a mere variety of the lama. The different species of camel-like sheep found still wild at elevations of from 13,000 to upwards of 16,000 feet above the level of the sea, are the Vicuña, the Guanaco, and the Alpaca; of these the two latter species are also found tame, although this is but rarely the

* *Chronica del Peru*, Sevilla, 1553, cap. 110, p. 264.

† See Gay, *Zoologia de Chili, Mamiferos*, 1847, p. 154.

‡ See the Inca Garcilaso, *Commentarios reales*, P. 1, lib. v. cap. 2, p. 133; and Prescott, *Hist. of the Conquest of Peru*, 1847, vol. i. p. 136.

§ *Fragments of the Nat. Hist. of Pennsylvania*, P. 1, p. 4.

‖ Tschudi, *Fauna Peruana*, s. 256.

¶ *Reise um die Erde*, th. iii. s. 64.

** Tschudi, s. 228. 237.

case with the Guanaco. The alpaca does not bear a warm climate as well as the lama. Since the introduction of the more useful horse, mule, and ass (the latter of which exhibits great animation and beauty in tropical regions), the lama and alpaca have been less generally reared and employed as beasts of burden in the mining districts. But their wool, which varies so much in fineness, is still an important branch of industry among the inhabitants of the mountains. In Chili the wild and the tame guanaco are distinguished by special names, the former being called "Luan" and the latter "Chilihueque." The wide dissemination of the wild Guanacos from the Peruvian Cordilleras to Tierra del Fuego, sometimes in herds of 500 heads of cattle, has been facilitated by the circumstance that these animals can swim with great facility from island to island, and are not therefore impeded in their passage across the Patagonian channels or fiords.*

South of the river Gyla, which together with the Rio Colorado pours itself into the Californian Gulf (Mar de Cortes), lie in the midst of the dreary steppe the mysterious ruins of the Aztek Palace, called by the Spaniards "las Casas Grandes." When, about the year 1160, the Azteks first appeared in Anahuac, having migrated from the unknown land of Aztlan, they remained for a time on the borders of the Gyla river. The Franciscan monks, Garces and Font, who saw the "Casas Grandes" in 1773, are the last travellers who have visited these remains. According to their statement, the ruins extended over an area exceeding sixteen square miles. The whole plain was covered with the broken fragments of ingeniously painted earthenware vessels. The principal palace, if the word can be applied to a house formed of unburnt clay, is 447 feet in length and 277 feet in breadth.†

The Tayé of California, a delineation of which is given by the Padre Venegas, appears to differ but inconsiderably from the *Ovis musimon* of the Old Continent. The same animal has also been seen in the Stony Mountains near the source of the River of Peace, and differs entirely from

* See the pleasing descriptions in Darwin's *Journal,* 1845, p. 66.

† See a rare work printed at Mexico, in 1792, and entitled *Cronica seráfica y Apostólica del Colegio de Propaganda Fide de la Santa Cruz de Querétaro,* por Fray Juan Domingo Arricivita.

the small white and black spotted goat-like animal found on the Missouri and Arkansas. The synonyme of Antilope furcifer, A. tememazama, (Smith,) and Ovis montana is still very uncertain.

(27) p. 11.—"*The culture of farinaceous grasses.*"

The original habitat of the farinaceous grasses, like that of the domestic animals which have followed man since his earliest migrations, is shrouded in obscurity. Jacob Grimm has ingeniously derived the German name for corn, *Getraide*, from the old German "gitragidi," "getregede." "It is as it were the *tame* fruit (*fruges*, *frumentum*) that has fallen into the hands of man, as we speak of tame animals in opposition to those that are wild."*

"It is a most striking fact that on one half of our planet there should be nations who are wholly unacquainted with the use of milk and of the meal yielded by narrow-eared grasses, (*Hordeaceæ* and *Avenaceæ*) whilst in the other hemisphere nations may be found in almost every region who cultivate cereals and rear milch cattle. The culture of different cereals is common to both hemispheres; but while in the New Continent we meet with only one species, maize, which is cultivated from 52° north to 46° south lat., we find that in the Old World the fruits of Ceres, (wheat, barley, spelt, and oats,) have been everywhere cultivated from the earliest ages recorded in history. The belief that wheat grew *wild* in the Leontine plains as well as in other parts of Sicily was common to several ancient nations, and is mentioned as early as Diodorus Siculus.† Cereals were also found in the alpine meadow of Enna. Diodorus says expressly, "The inhabitants of the Atlantis were *unacquainted with the fruits of Ceres*, owing to their having separated from the rest of mankind before those fruits were made known to mortals." Sprengel has collected several interesting facts from which he is led to conjecture that the greater number of our European cereals originally grew wild in Northern Persia and India. He supposes for instance that summer wheat was indigenous in the land of the Musicani, a province of Northern India;‡ barley, *antiquissi-*

* Jacob Grimm, *Gesch. der Deutschen Sprache,* 1848, th. i. s. 62.
† lib. v. pp. 199, 232. Wessel.
‡ Strabo, xv. 1017.

*mum frumentum*, as Pliny terms it, and which was also the only cereal known to the Guansches of the Canaries, originated, according to Moses of Chorene,* on the banks of the Araxes or Kur in Georgia, and according to Marco Polo in Balascham, in Northern India;† and *Spelt* originated in Hamadan.

My intelligent friend and teacher, Link, has however shown in a comprehensive and critical treatise,‡ that these passages are open to much doubt. In a former essay of my own,§ I expressed doubts regarding the existence of wild cereals in Asia, and considered them to have become wild. Reinhold Forster, who before his voyage with Captain Cook made an expedition for purposes of natural history into the south of Russia by order of the Empress Catherine, reported that the two-lined summer barley (*Hordeum distichon*) grew wild near the confluence of the Samara and the Volga. At the end of September in the year 1829, Ehrenberg and myself also herborised on the Samara, during our journey from Orenburg and Uralsk to Saratow and the Caspian Sea. The quantity of wheat and rye plants growing wild on uncultivated ground in this district was certainly very remarkable; but the plants did not appear to us to differ from the ordinary kinds. Ehrenberg received from M. Carelin a species of rye, *Secale fragile*, that had been gathered on the Kirghis Steppe, and which Marshal Bieberstein for some time conjectured to be the mother plant of our cultivated rye, *Secale cereale*. Michaux's herbarium does not show (according to Achill Richard's testimony), that Spelt (*Triticum spelta*) grows wild at Hamadan in Persia, as Olivier and Michaux have been supposed to maintain. More confidence is due to the recent accounts obtained through the unwearied zeal of the intelligent traveller, Professor Carl Koch. He found a large quantity of rye (*Secale cereale var. β, pectinata*) in the Pontic Mountains, at heights of more than 5000 or 6000 feet above the level of the sea, on spots where this species of grain had not within the memory of the inhabitants been previously cultivated. "Its appearance here is the more important," he remarks, "because with us this grain never propagates

* *Geogr. Armen.*, ed. Whiston, 1736, p. 360.

† Ramusio, vol. ii. p. 10.

‡ *Abhandl. der Berl. Akad.* 1816, s. 123.

§ *Essai sur la Géographie des Plantes*, 1805, p. 28.

itself spontaneously." Koch collected in the Schirwan part of the Caucasus a kind of grain which he calls *Hordeum spontaneum*, and regards as the originally wild *Hordeum zeocriton*. (Linn.)*

A negro slave of the great Cortes was the first who cultivated wheat in New Spain, from three seeds which he found amongst some rice brought from Spain for the use of the troops. In the Franciscan convent at Quito I saw, preserved as a relic, the earthen vessel which had contained the first wheat sowed in Quito by the Franciscan monk, Fray Jodoco Rixi de Gante, a native of Ghent in Flanders. The first crop was raised in front of the convent, on the "Plazuela de S. Francisco," after the wood which then extended from the foot of the Volcano of Pichincha had been cleared. The monks, whom I frequently visited during my stay at Quito, begged me to explain the inscription on the cup, which according to their conjecture contained some hidden allusion to wheat. On examining the vessel, I read in old German the words "Let him who drinks from me, ne'er forget his God." This old German drinking cup excited in me feelings of veneration! Would that everywhere in the New Continent the names of those were preserved who, instead of devastating the soil by bloody conquests, confided to it the first fruits of Ceres! There are "fewer examples of a general affinity of names in terms relating to the different species of corn and objects of agriculture than to the rearing of cattle. Herdsmen when they migrated to other regions had still much in common, while the subsequent cultivators of the soil had to invent special words. But the fact that in comparison with the Sanscrit, Romans and Greeks seem to stand on the same footing with Germans and Slavonians, speaks in favour of the very early contemporaneous emigration of the two latter. Yet the Indian *java* (*frumentum hordeum*), when compared with the Lithuanian *jawai*, and the Finnish *jywa*, affords a striking exception."†

(28) p. 11.—"*Preferring to keep within a cooler climate.*"

Throughout the whole of Mexico and Peru we find the trace of human civilisation confined to the elevated table-

* Carl Koch, *Beiträge zur Flora des Orients*. Heft. 1, s. 139, 142.

† Jacob Grimm, *Gesch. der deutschen Sprache*, th. i. s. 69.

lands. We saw the ruins of palaces and baths on the sides of the Andes, at an elevation of from 10.230 to 11,510 feet. None but northern tribes migrating from the north towards the equato. could have remained from preference in such a climate.

(29) p. 12.—"*The history of the peopling of Japan.*"

I believe I have succeeded in showing, in my work on the monuments of the American primitive races,* by an examination of the Mexican and Thibetian-Japanese calendars, by a correct determination of the position of the Scansile Pyramids, and by the ancient myths which record four revolutions of the world and the dispersion of mankind after a great deluge, that the western nations of the New Continent maintained relations of intercourse with those of Eastern Asia, long before the arrival of the Spaniards. These observations have derived additional weight, since the appearance of my work, from the facts recently published in England, France, and the United States, regarding the remarkable pieces of sculpture carved in the Indian style, which have been discovered in the ruins of Guatimala and Yucatan.† The ancient architectural remains found in the peninsula of Yucatan testify more than those of Palenque, to an astonishing degree of civilization. They are situated between Valladolid, Merida, and Campeche, chiefly in the western portion of the country. But the monuments on the island of Cozumel, (properly Cuzamil,) east of Yucatan, were the first which were seen by the Spaniards in the expedition of Juan de Grijalva in 1518, and in that of Cortes in 1519. Their discovery tended to diffuse throughout Europe an exalted idea of the advanced condition of ancient

* *Vues des Cordillères et Monuments des peuples indigènes de l'Amérique*, 2 tomes.

† Compare the work of D. Antonio del Rio, entitled *Description of the Ruins of an Ancient City discovered near Palenque*, 1822, translated from the orig. manuscr. report by Cabrera, p. 9, tab. 12—14 (Rio's researches were made in the year 1787); with Stephens, *Incidents of Travel in Yucatan*, 1843, vol. i. pp. 391, 429—434, and vol. ii. pp. 21, 54, 56, 317, 323; with the magnificent work of Catherwood, *Views of Ancient Monuments in Central America, Chiapas, and Yucatan*, 1844; and lastly with Prescott, *The Conquest of Mexico*, vol. iii. Append. p. 360.

Mexican civilization. The most important ruins of the peninsula of Yucatan (unfortunately not yet thoroughly measured and drawn by architects) are those of the "Casa del Gobernador" of Uxmal, the Teocallis and vaulted constructions at Kabah, the ruins of Labnan with its domed pillars, those of Zayi which exhibit columns of an order of architecture nearly approaching the Doric, and those of Chiche with large ornamented pilasters. An old manuscript written in the Maya language by a Christian Indian, which is still in the hands of the "Gefe politico" of Peto, Don Juan Rio Perez, gives the different epochs (*Katunes* of 52 years) at which the Toltecs settled in different parts of the peninsula. Perez would infer from these data that the architectural remains of Chiche go back as far as the fourth century of our era, whilst those of Uxmal belong to the middle of the tenth century; but the accuracy of these historical deductions is open to great doubt.*

I regard the existence of a former intercourse between the people of Western America and those of Eastern Asia as more than probable, although it is impossible at the present time to say by what route and with which of the tribes of Asia this intercourse was established. A small number of individuals of the cultivated hierarchical castes may perhaps have sufficed to effect great changes in the social condition of Western America. The fabulous accounts formerly current regarding Chinese expeditions to the New Continent refer merely to expeditions to Fusang or Japan. It is, however, possible that Japanese and Sian-Pi may have been driven by storms from the Corea to the American coasts. We know as matters of history that Bonzes and other adventurers navigated the Eastern Chinese seas in search of a remedial agent capable of making man immortal. Thus under Tschin-chi-huang-ti three hundred young couples were dispatched to Japan in the year 209 before our era, who, instead of returning to China, settled on the Island of Nipon.† May not accident have led to similar expeditions to

* Stephens, *Incid. of Travel in Yucatan,* vol. i. p. 439, and vol. ii. p. 278.

† Klaproth, *Tableaux historiques de l'Asie,* 1824, p. 79; *Nouveau Journal asiatique,* t. x. 1832, p. 335; and Humboldt, *Examen critique,* t. ii. pp. 62—67.

the Fox Islands, to Alaschka, or New California? As the western coasts of the American continent incline from north-west to south-east, and the eastern coasts of Asia from north east to south-west, the distance between the two continents in the milder zone, which is most conducive to mental development (45° lat.), would appear too considerable to admit of an accidental settlement having been made in this latitude. We must therefore assume that the first landing took place in the ungenial climate of 55° and 65°, and that cultivation, like the general advance of population in America, progressed by gradual stations from north to south.* It was even believed in the beginning of the sixteenth century that the fragments of ships from Catayo, *i.e.* from Japan or China, had been found on the coasts of the Northern Dorado, called also Quivira and Cibora.†

We know as yet too little of the languages of America entirely to renounce the hope that, amid their many varieties, some idiom may be discovered, that has been spoken with certain modifications in the interior of South America and Central Asia, or that might at least indicate an ancient affinity. Such a discovery would undoubtedly be one of the most brilliant to which the history of the human race can hope to attain! But analogies of language are only deserving of confidence where mere resemblances of sound in the roots are not alone the object of research, but attention is also directed to the organic structure, the grammatical forms, and those elements of language which manifest themselves as the product of the intellectual power of man.

(30) p. 12—"*Many other forms of animal life.*"

The Steppes of Caracas abound in flocks of the so-called *Cervus mexicanus*. This stag when young is spotted, and resembles the roe. We have frequently met with perfectly white varieties, which is a very striking fact when the high temperature of this zone is taken into consideration. The *Cervus mexicanus* is not found on the declivities of the Andes in the equatorial region, at an elevation exceeding from 4476 to 5115 feet, but another white deer, which I could scarcely distinguish by any one specific characteristic from the

* *Rélat. hist.* t. iii. pp. 155—160.

† Gomara, *Hist. general de las Indias,* p. 117.

European species, ascends to an elevation of nearly 13,000 feet. The *Cavia capybara* is known in the province of Caracas by the name of *Chiguire*. This unfortunate animal is pursued in the water by the crocodile, and on land by the tiger or jaguar. It runs so badly that we were often able to catch it with our hands. The extremities are smoked and eaten as hams, but have a most unpleasant taste, owing to the flavour and smell of musk by which they are impregnated; and on the Orinoco we gladly ate monkey-hams in preference. These beautifully striped animals—the *Viverra mapurito*, *Viverra zorilla*, and *Viverra vittata*—exhale a fetid odour.

(31.) p. 12—" *The Guaranes and the fan-palm Mauritia.*"

The small coast tribe of the Guaranes (called in British Guiana, the Warraws, or Guaranos, and by the Caribs U-ara-u) inhabit not only the swampy delta and the river network of the Orinoco (more particularly the banks of the Manamo grande and the Caño Macareo), but also extend, with very slight differences in their mode of living, along the sea-shore, between the mouths of the Essequibo and the Boca de Navios of the Orinoco.* According to the testimony of Schomburgk, the admirable observer referred to in the note, there are still about 1700 Warraus or Guaranos living in the vicinity of Cumaca, and along the banks of the Barime river, which empties itself into the gulf of the Boca de Navios. The social habits of the tribes settled in the delta of the Orinoco were known to the great historian Cardinal Bembo, the cotemporary of Christopher Columbus, Amerigo Vespucci, and Alonzo de Hojeda. He says† *quibusdam in locis propter paludes incolæ domus in arboribus ædificant.* It is hardly probable that instead of the Guaranos at the mouth of the Orinoco, Bembo should here allude to the natives of the country near the mouth of the gulf of Maracaibo, where Alonzo de Hojeda, in August, 1499, (when accompanied by Vespucci and Juan de la Cosa) found a population having their dwellings *fondata sopra l'acqua come Venezia* ("built like Venice on the water').‡ Vespucci, in the account of his

* Compare my *Relation historique*, t. i. p. 492, t. ii. pp. 653, 703, 6ith Richard Schomburgk, *Reisen in Britisch Guiana*, th. i. 1847, s. 2, 120, 173, 194.

† *Historiæ Venetæ*, 1551, p. 88.

‡ See text of Riccardi in my *Examen crit.* t. iv. p. 496.

travels, in which we meet with the first traces of the etymology of the name of the province of *Venezuela* (Little Venice) as used for the province of Caracas, speaks only of houses built on a foundation of piles, and makes no mention of habitations in trees.

Sir Walter Raleigh bears a subsequent and incontrovertible evidence to the same fact, for he says expressly in his description of Guiana, that on his second voyage in 1595, when in the mouth of the Orinoco, "he saw the fire of the Tivitites and Qua-rawetes" (so he calls the Guaranes), "high up in the trees."* There is a drawing of the fire in the Latin edition of this work,† and Raleigh was the first who brought to England the fruit of the Mauritia palm, which he very justly compared, on account of its scales, to fir-cones. Father José Gumilla, who twice visited the Guaranes as a missionary, says, indeed, that this tribe have their dwelling in the Palmares (palm groves) of the morasses; but while he speaks more definitely of pendent habitations supported by high pillars, makes no mention of platforms attached to still growing trees.‡ Hillhouse and Sir Robert Schomburgk§ are of opinion that Bembo, through the relations of others, and Raleigh, by his own observation, were deceived into this belief in consequence of the high tops of the palm trees being lighted up in such a manner by the fires below them, that those sailing by thought the habitations of the Guaranes were attached to the trees themselves. "We do not deny," says Schomburgk, "that in order to escape the attacks of the mosquitos, the Indian sometimes suspends his hammock from the tops of trees, but on such occasions no fires are made under the hammock."‖

According to Martius, the beautiful Palm, Moriche, *Mau-*

* Raleigh, *Discovery of Guiana*, 1596, p. 90.

† *Brevis et admiranda Descriptio regni Guianæ* (Norib. 1599), tab. 4.

‡ Gumilla, *Historia natural, civil y geografica de las Naciones situadas en las riveras del Rio Orinoco*, nueva impr., 1791, pp. 143, 145, 163.

§ See *Journal of the Royal Geogr. Society*, vol. xii. 1842, p. 175, and *Description of the Murichi, or Ita Palm, read in the meeting of the British Association held at Cambridge, June* 1845 (*published in Simond's Colonial Magazine*).

‖ See also Sir Robert Schomburgk's new edition of *Raleigh's Discovery of Guiana* (1848), p. 50.

*ritia flexuosa*, *Quieteva*, or *Ita* Palm,* belongs, together with Calamus, to the family of the Lepidocaryæ or Coryphee. Linnæus has described it very imperfectly, as he erroneously considered it to be devoid of leaves. The trunk is 26 feet high, but it probably does not attain this height in less than 120 or even 150 years. The Mauritia extends high up the declivity of the Duida, north of the Esmeralda mission, where I found it in great beauty. It forms, in moist places, fine groups of a fresh and shining verdure, reminding us of that of our alders. The trees preserve the moisture of the ground by their shade, and hence the Indians believe that the Mauritia draws water around its roots by some mysterious attraction. In conformity with an analogous theory they advise, that serpents should not be killed, because the destruction of these animals is followed by the drying up of the lagoons. Thus do the rude children of nature confound cause and effect! Gumilla calls the *Mauritia flexuosa* of the Guaranes the tree of life ("arbol de la vida"). It is found on the mountains of Ronaima, east of the sources of the Orinoco, as high as 4263 feet. On the unfrequented banks of the Rio Atabapo, in the interior of Guiana, we discovered a new species of Mauritia having a prickly stem; our *Mauritia aculeata*.†

(32) p. 13.—"*An American Stylite.*"

The founder of the sect of Stylites, the fanatical Pillar-saint, Simeon Sisanites of Syria, the son of a Syrian herdsman, is said to have passed thirty-seven years in holy contemplation, elevated on five columns, each higher than the preceding. He died in the year 461. The last of the pillars which he occupied was 40 ells in height. For seven hundred years there continued to be followers of this mode of life, who were called *Sancti Columnares*, or Pillar-saints. Even in Germany, in the see of Treves, attempts were made to found similar aërial cloisters; but the dangerous practice met with the constant opposition of the bishops.‡

* Bernau, *Missionary Labours in British Guiana*, 1847, pp. 34, 44.

† Humboldt, Bonpland, et Kunth, *Nova genera et species Plantarum*, t. i. p. 310.

‡ Mosheim, *Institut. Hist. Eccles.*, 1755, p. 215.

(33) p. 14.—"*Towns on the banks of the Steppe-rivers.*"

Families who live by raising cattle and do not take part in agricultural pursuits have congregated together in the middle of the Steppe, in small towns, which, in the cultivated parts of Europe, would scarcely be regarded as villages. Among these are Calabozo, which, according to my astronomical observations, is situated in 8° 56′ 14″ north lat., and 67° 43′ west long.; Villa del Pao (8° 38′ 1″ north lat., and 66° 57′ west long.); Saint Sebastian, and others.

(34) p. 14.—"*Funnel-shaped clouds.*"

The singular phenomenon of these sand-spouts, of which we see something analogous on the cross roads of Europe, is especially characteristic of the Peruvian sandy desert between Amotape and Coquimbo. Such dense clouds of sand may endanger the safety of the traveller who does not cautiously avoid them. It is remarkable that these partial and opposing currents of air should arise only when there is a general calm. The aërial ocean resembles the sea in this respect; for here, too, we find that the small currents (*filets de courant*) in which the water may frequently be heard to flow with a splashing sound, occur only in a dead calm (*calme plat*).

(35) p. 14.—"*Increases the stifling oppression.*"

I have observed in the Llanos de Apure, at the cattle farm of Guadalupe, that the thermometer rose from 92°.7 to 97°.2 Fahr. whenever the hot wind began to blow from the desert, which was covered either with sand or short withered grass. In the middle of the sand-cloud the thermometer stood for several minutes together at 111° Fahr. The dry sand in the village of San Fernando de Apure had a temperature of 126° Fahr.

(36) p. 15.—"*The phantom of a moving undulating surface.*"

The well known phenomenon of the *mirage* is called in Sanscrit "the thirst of the gazelle."* All objects appear to float in the air, while their forms are reflected in the lower stratum of the atmosphere. At such times the whole desert

* See my *Rélat. hist.*, t. i. pp. 296, 625; t. ii. p. 161.

resembles a vast lake, whose surface undulates like waves, Palm trees, cattle, and camels sometimes appear inverted in the horizon. In the French expedition to Egypt, this optical illusion often nearly drove the faint and parched soldiers to distraction. This phenomenon has been observed in all quarters of the world. The ancients were alsoacquainted with the remarkable refraction of the rays of light in the Lybian Desert. We find mention made in Diodorus Siculus of strange illusive appearances, an African *Fata Morgana*, together with still more extravagant explanations of the conglomeration of the particles of air.*

(37) p. 15.—" *The Melocactus.*"

The *Cactus melocactus* is frequently from 10 to 12 inches in diameter, and has generally 14 ribs. The natural group of the Cactaceæ, the whole family of the Nopaleæ of Jussieu, belongs exclusively to the New Continent. The Cactus assumes a variety of shapes, being ribbed and melon-like (*Melocacti*); articulated (*Opuntiæ*); upright-like columns (*Cerei*); of a serpentine or creeping form (*Rhipsalides*); or provided with leaves (*Pereskiæ*). Many extend high up the slopes of the mountains. Near the foot of the Chimborazo, in the sandy table-land around Riobamba, I found a new species of Pitahaya (*Cactus sepium*), even at an elevation of 10,660 feet.†

(38) p. 16.—" *The scene suddenly changes in the Steppe.*"

I have endeavoured to describe the approach of the rainy season, and the signs by which it is announced. The deep blue of the heavens in the tropics is occasioned by the imperfect solution of vapour. The cyanometer indicates a lighter shade of blue as soon as the vapours begin to fall. The dark spot in the constellation of the Southern Cross becomes indistinct in proportion as the transparency of the atmosphere decreases. and this change announces the approach of rain. The bright radiance of the Magellanic clouds (*Nubecula major* and *Nubecula minor*) then gradually fades away. The fixed stars which had before been shining with a calm, steady, planet-

* Lib. iii. p. 184, Rhod., p. 219, Wessel.

† Humboldt, Bonpland, et Kunth, *Synopsis Plantarum æquinoct. Orbis Novi*, t. iii. p. 370.

like light, are now seen to scintillate in the zenith.* All these phenomena are the result of the increased quantity of aqueous vapour floating in the atmosphere.

(39) p. 16.—" *The humid clay soil is seen to rise slowly in a broad flake.*"

Drought produces the same phenomena in animals and plants as the abstraction of heat. During the dry season many tropical plants lose their leaves. The crocodile and other amphibious animals conceal themselves in the mud and lie apparently dead, like animals in cold regions who are thrown into a state of hybernation.†

(40) p. 17.—" *A vast inland sea.*"

Nowhere are these inundations on a larger scale than in the network of streams formed by the Apure, the Arachuna, the Pajara, the Arauca, and the Cabuliare. Large vessels sail across the country over the Steppe for 40 or 50 miles.

(41) p. 17.—" *To the mountainous plain of Antisana.*"

The great mountain plateau which surrounds the volcano of Antisana is 13,473 feet above the level of the sea. The pressure of the atmosphere is so inconsiderable at this height, that blood will flow from the nostrils and mouth of the wild bull when hunted with dogs.

(42) p. 17.—" *The marshy waters of Bera and Rastro.*"

I have elsewhere more circumstantially described the capture of the gymnotus.‡ Mons. Gay Lussac and myself were perfectly successful in the experiments we conducted without a chain on a living gymnotus, which was still very vigorous when it reached Paris. The discharge of electricity is entirely dependent on the will of the animal. We did not observe any electric sparks, but other physicists have done so on numerous occasions.

(43) p. 18.—"*Awakened by the contact of moist and dissimular particles.*"

In all organic bodies dissimilar substances come into

* Compare Arago in my *Rélation hist.*, t. i. p. 623.

† See my *Rélat. histor.*, t. ii. pp. 196, 626.

‡ *Observations de Zoologie et d'Anatomie comparée*, t. i. pp. 83–87, and *Rélat. hist.*, t. ii. pp. 173–190.

contact with each other, and solids are associated with fluids. Wherever there is organization and life, there must be electric tension, or, in other words, a voltaic pile must be brought into play, as the experiments of Nobili and Matteucci, and more especially the late most admirable labours of Emil Dubois, teach us. The last-named physicist has succeeded in "manifesting the presence of the electric muscular current in living and wholly uninjured animal bodies:" he shows that "the human body, through the medium of a copper wire, can at will cause the magnetic needle at a distance to deflect first in one direction and then in another."* I have myself witnessed these movements produced at will, and have thus unexpectedly seen much light thrown on phenomena, to which I had laboriously and ardently devoted so many years of my earlier life.

(44) p. 19.—"*The myth of Osiris and Typhon.*"

Respecting the struggle of two human races, the Arabian shepherd tribes of Lower Egypt and the cultivated agricultural races of Upper Egypt; on the subject of the fair-haired Prince *Baby* or *Typhon*, who founded Pelusium; and on the dark-complexioned Dionysos or Osiris; I would refer to Zoëga's older and almost universally discarded views as set forth at p. 577 of his masterly work "*De origine et usu obeliscorum.*"

(45) p. 19—"*The boundaries of European semi-civilization.*"

In the Capitania General de Caracas, as well as in all the eastern part of America, the civilization formerly introduced by Europeans is limited to the narrow strip of land which skirts the shore. In Mexico, New Granada, and Quito on the other hand, European civilization has penetrated far into the interior of the country and advanced up to the ridges of the Cordilleras. There existed already in the fifteenth century an earlier stage of civilization among the inhabitants of the last-named region. Wherever the Spaniards perceived this culture they pursued its track, regardless whether the seat of it was at a distance from the sea, or in its vicinity. The ancient cities were enlarged and their former significant

* *Untersuchungen über thierische Electricität*, von Emil du Bois-Raymond, 1848, bd. i. s. xv.

Indian names mutilated, or exchanged for those of Christian saints.

(46) p. 19—"*Huge masses of leaden-coloured granite.*"

In the Orinoco, and more especially at the cataracts of Maypures and Atures (not in the Black River or Rio Negro), all blocks of granite, even pieces of white quartz, wherever they come in contact with the water, acquire a grayish black coating, which does not penetrate beyond 0·01 of a line into the interior of the rock. The traveller might almost suppose that he was looking at basalt, or fossils coloured with graphite. Indeed, the crust does actually appear to contain manganese and carbon. I say "appears" to do so, because the phenomenon has not yet been thoroughly investigated. Something perfectly analogous to this was observed by Rozier in the syenitic rocks of the Nile (near Syene and Philæ); by the unfortunate Captain Tuckey on the rocky banks of the Zaire; and by Sir Robert Schomburgk at Berbice.* On the Orinoco these leaden-coloured rocks are supposed when wet to give forth noxious exhalations, and their vicinity is believed to be conducive to the generation of fevers.† It is also remarkable that the South American rivers generally, which have black waters (*aguas negras*), or waters of a coffee brown or wine yellow tint, do not darken the granite rocks; that is to say, they do not act upon the stone in such a manner as to form from its constituent parts a black or leaden-coloured crust.

(47) p. 20—"*The rain-foreboding howl of the bearded ape.*"

Some hours before the commencement of rain, the melancholy cries of various apes, as *Simia seniculus*, *Simia beelzebub*, &c., fall on the ear like a storm raging in the distance. The intensity of the noise produced by such small animals can only be explained by the circumstance that one tree often contains a herd of seventy or eighty apes. I have elsewhere spoken of the laryngeal sac, and the ossification of the larynx of these animals.‡

* *Reisen in Guiana und am Orinoko*, s. 212.

† See my *Rélat. hist.*, t. ii. pp. 299-304.

‡ See my anatomical treatise in *Recueil d'Observations de Zoologie*, vol. i. p. 18.

(48) p. 20—"*Its uncouth body often covered with birds.*"

The crocodiles lie so motionless, that I have often seen flamingoes (*Phœnicopterus*) resting on their heads, while the other parts of the body were covered, like the trunk of a tree, with aquatic birds.

(49) p. 20—"*Down its dilating throat.*"

The saliva with which the boa covers its prey tends to promote rapid decomposition. The muscular flesh is rendered gelatinously soft under its action, so that the animal is able to force entire limbs of its slain victim through its swelling throat. The Creoles call the giant boa *Tragavenado* (*stag-swallower*), and fabulously relate that the antlers of a stag which could not be swallowed by the snake have been seen fixed in its throat. I have frequently observed the boa constrictor swimming in the Orinoco, and in the smaller forest streams, the Tuamini, the Temi, and the Atabapo. It holds its head above water like a dog. Its skin is beautifully speckled. It has been asserted, that the animal attains a length of 48 feet, but the longest skins which have as yet been carefully measured in Europe do not exceed from 21 to 23 feet. The South American boa (a Python) differs from the East Indian.*

(50) p. 20—"*Living on gums and earth.*"

It is currently reported throughout the coasts of Cumana, New Barcelona, and Caracas (which the Franciscan monks of Guiana are in the habit of visiting on their return from the missions,) that there are men living on the banks of the Orinoco who eat earth. On the 6th of June, 1800, on our return from the Rio Negro, when we descended the Orinoco in thirty-six days, we spent the day at the mission inhabited by these people (the Otomacs). Their little village, which is called La Concepcion de Uruana, is very picturesquely built against a granite rock. It is situated in 7° 8′ 3″ north lat.; and according to my chronometrical determination, in 67° 18′ west longitude. The earth which the Otomacs eat, is an unctuous, almost tasteless clay, true potter's earth, of a yellowish grey

* On the Ethiopian Boa, see Diodor. Sicul., lib. iii. p. 204, ed. Wesseling.

colour, in consequence of a slight admixture of oxide of iron. They select it with great care, and seek it in certain banks on the shores of the Orinoco and Meta. They distinguish the flavour of one kind of earth from that of another; all kinds of clay not being alike acceptable to their palate. They knead this earth into balls measuring from four to six inches in diameter, and bake them before a slow fire, until the outer surface assumes a reddish colour. Before they are eaten, the balls are again moistened. These Indians are mostly wild, uncivilized men, who abhor all tillage. There is a proverb current among the most distant of the tribes living on the Orinoco, when they wish to speak of anything very unclean, "so dirty that the Otomacs eat it."

As long as the waters of the Orinoco and the Meta are low, these people live on fish and turtles. They kill the former with arrows, shooting the fish as they rise to the surface of the water with a skill and dexterity that has frequently excited my admiration. At the periodical swelling of the rivers, the fishing is stopped, for it is as difficult to fish in deep river water as in the deep sea. It is during these intervals, which last from two to three months, that the Otomacs are observed to devour an enormous quantity of earth. We found in their huts considerable stores of these clay balls piled up in pyramidal heaps. An Indian will consume from three-quarters of a pound to a pound and a quarter of this food daily, as we were assured by the intelligent monk, Fray Ramon Bueno, a native of Madrid, who had lived among these Indians for a period of twelve years. According to the testimony of the Otomacs themselves, this earth constitutes their main support in the rainy season. In addition, they however eat, when they can procure them, lizards, several species of small fish, and the roots of a fern. But they are so partial to clay, that even in the dry season, when there is an abundance of fish, they still partake of some of their earth-balls, by way of a *bonne bouche* after their regular meals.

These people are of a dark, copper-brown colour, have unpleasant Tartar-like features, and are stout, but not protuberant. The Franciscan who had lived amongst them as a missionary, assured us that he had observed no difference in the condition and well-being of the Otomacs during the periods in which they lived on earth. The simple facts are therefore as follows:—The Indians undoubtedly consume large quantities of clay without injuring their health; they regard this earth as a nutritious article of food, that is to say, they feel that it will satisfy their hunger for a long time. This property they ascribe exclusively to the clay, and not to the other articles of food which they contrive to procure from time to time in addition to it. If an Otomac be asked what are his winter provisions—the term winter in the torrid parts of South America implying the rainy season—he will point to the heaps of clay in his hut. These simple facts do not, however, by any means decide the questions: whether clay can actually be a nutritious substance; whether earths can be assimilated in the human body; whether they only serve as ballast; or merely distend the walls of the stomach, and thus appease the cravings of hunger? These are questions which I cannot venture to decide.* It is singular, that Father Gumilla, who is generally so credulous and uncritical, should have denied the fact of earth being eaten by and for itself.† He maintains that the clay-balls are largely mixed with maize-flour, and crocodile's fat. But the missionary Fray Ramon Bueno, and our friend and fellow-traveller, the lay-brother Fray Juan Gonzales, who perished at sea off the coast of Africa (at the time we lost a portion of our collections), both assured us, that the Otomacs never mix their clay cakes with crocodile's fat, and we heard nothing in Uruana of the admixture of flour.

The earth which we brought with us, and which was chemically investigated by M. Vauquelin, is quite pure and unmixed. May not Gumilla, by confounding heterogeneous facts, have intended to allude to a preparation of bread from the long pod of a species of Inga? as this fruit is certainly buried in the earth, in order to hasten its decomposition. It appears to me especially remarkable, that the Otomacs should not lose their health by eating so much earth. Has this tribe been habituated for generations to this stimulus?

In all tropical countries men exhibit a wonderful and almost irresistible desire to devour earth, not the so-called alkaline or calcareous earth, for the purpose of neutralizing acidity, but unctuous, strong-smelling clay. It is often found

* *Rélat. hist.*, t. ii. pp. 618–620.

† *Historia del Rio Orinoco*, nueva impr., 1791, t. i. p. 179.

necessary to shut children up in order to prevent their running into the open air to devour earth after recent rain. The Indian women who are engaged on the river Magdalena, in the small village of Banco, in turning earthenware pots, continually fill their mouths with large lumps of clay, as I have frequently observed, much to my surprise.* Wolves eat earth, especially clay, during winter. It would be very important, in a physiological point of view, to examine the excrements of animals and men that eat earth. Individuals of all other tribes, excepting the Otomacs, lose their health if they yield to this singular propensity for eating clay. In the mission of San Borja we found the child of an Indian woman, which, according to the statement of its mother, would hardly eat anything but earth. It was, however, much emaciated, and looked like a mere skeleton.

Why is it that in the temperate and cold zones this morbid eagerness for eating earth is so much less frequently manifested, and is indeed limited almost entirely to children and pregnant women, whilst it would appear to be indigenous to the tropical lands of every quarter of the earth? In Guinea the negroes eat a yellowish earth, which they call *caouac;* and when they are carried as slaves to the West Indies they even endeavour there to procure for themselves some similar species of food, maintaining that the eating of earth is perfectly harmless in their African home. The *caouac* of the American islands, however, deranges the health of the slaves who partake of it; for which reason the eating of earth was long since forbidden in the West Indies, notwithstanding which a species of red or yellowish tuff (*un tuf rouge jaunâtre*) was secretly sold in the public market of Martinique in the year 1751.

"The negroes of Guinea say that in their own country they *habitually* eat a certain earth, the flavour of which is most agreeable to them, and which does not occasion them any inconvenience. Those who have addicted themselves to the excessive use of *caouac* are so partial to it, that no punishment can prevent them from devouring this earth."† In the island of Java, between Sourabaya and Samarang, Labillardière saw

* This was also observed by Gilj, *Saggio di Storia Americana,* t. ii. p. 311.

† Thibault de Chanvalon, *Voyage à la Martinique,* p. 85.

small square reddish cakes publicly sold in the villages. The natives called them *tana ampo* (*tanah* signifies earth in Malay and Javanese); and on examining them more closely, he found that they were cakes made of a reddish clay, and intended for eating.* The edible clay of Samarang has recently (1847) been sent, by Mohnike, to Berlin in the shape of rolled tubes like cinnamon, and has been examined by Ehrenberg. It is a fresh-water formation deposited in tertiary limestone, and composed of microscopic polygastrica (Gallionella, Navicula) and of Phytolitharia.† The natives of New Caledonia, to appease their hunger, eat lumps as large as the fist of friable steatite, in which Vauquelin detected an appreciable quantity of copper.‡ In Popayan and many parts of Peru calcareous earth is sold in the streets as an article of food for the Indians. This is eaten together with the Coca (the leaves of the *Erythroxylon peruvianum*). We thus find that the practice of eating earth is common throughout the whole of the torrid zone among the indolent races who inhabit the most beautiful and fruitful regions of the earth. But accounts have also come from the north, through Berzelius and Retzius, from which we learn, that in the most remote parts of Sweden hundreds of cartloads of earth containing infusoria are annually consumed by the country people as bread-meal, more from fancy (like the smoking of tobacco) than from necessity. In some parts of Finland a similar kind of earth is mixed with the bread. It consists of empty shells of animalcules, so small and soft, that they break between the teeth without any perceptible noise, filling the stomach without yielding any actual nourishment. Chronicles and archives often make mention during times of war of the employment as food of infusorial earth, which is spoken of under the indefinite and general term of "mountain meal." Such, for instance, was the case in the Thirty Years' War, at Camin in Pomerania, Muskau in the Lausitz, and Kleiken in the Dessau territory; and subsequently in 1719 and 1733, at the fortress of Wittenberg.§

* *Voyage à la Recherche de La Pérouse*, t. ii. p. 322.

† *Bericht über die Verhandl. der Akad. d. Wiss. zu Berlin aus dem J.* 1848, s. 222—225.

‡ *Voy. à la Rech. de La Pérouse*, t. ii. p. 205.

§ See Ehrenberg, *Ueber das unsichtbar wirkende organiche Leben*, 1842, s. 41.

(51) p. 20.—" *Images graven in rocks.*"

In the interior of South America, between the parallels of 2° and 4° north lat., lies a wooded plain inclosed by four rivers, the Orinoco, the Atabapo, the Rio Negro, and the Cassiquiare. Here we find granitic and syenitic rocks, which, like those of Caicara and Uruana, are covered with colossal symbolical figures of crocodiles, tigers, utensils of domestic use, signs of the sun and moon, &c. This remote portion of the earth is at present wholly uninhabited throughout an extent of more than 8000 square miles. The neighbouring tribes, who occupy the lowest place in the scale of humanity, are naked wandering savages, who could not possibly have carved hieroglyphics in stone. A whole range of these rocks covered with symbolical signs may be traced from Rupunuri, Essequibo, and the mountains of Pacaraima, to the banks of the Orinoco and of the Yupura, extending over more than eight degrees of longitude.

These carvings may belong to very different periods of time, for Sir Robert Schomburgk even found on the Rio Negro representations of a Spanish galliot,* which must necessarily have been of a date subsequent to the beginning of the sixteenth century, and that in a wilderness where the inhabitants were probably as rude then as they now are. But it must not be forgotten, as I have already elsewhere observed, that nations of very different descent, but in similarly uncivilized conditions, possessed of the same disposition to simplify and generalize outlines, and urged by identical inherent mental tendencies, may be led to produce similar signs and symbols.†

At the meeting of the Society of Antiquaries in London a memoir was read on the 17th of November, 1836, by Sir Robert Schomburgk, " On the religious traditions of the Macusi Indians, who inhabit the Upper Mahu, and a portion of the Pacaraima mountains," and who have therefore not changed their habitation for a century (since the journey of the intrepid Hortsmann). "The Macusis," says Sir Robert Schomburgk, " believe that the only being who survived a

* *Reisen in Guiana und am Orinoko* übersetzt von Otto Schomburgk, 1841, s. 500.

† Compare *Rélation historique,* t. ii. p. 589, with Martius, *Ueber die Physiognomie des Pflanzenreichs in Brasilien,* 1824, s. 14.

general deluge, repeopled the earth by converting stones into human beings." This myth, which is the fruit of the lively imagination of these tribes, and which reminds us of that of Deucalion and Pyrrha, shows itself in a somewhat modified form among the Tamanacs of the Orinoco. When these people are asked how the human race survived this great flood, *the age of waters* of the Mexicans, they unhesitatingly reply, "that one man and one woman were saved by taking refuge on the summit of the lofty mountain of Tamanacu, on the banks of the Asiveru, and that they then threw over their heads the fruits of the Mauritia palm, from the kernels of which sprang men and women, who again peopled the earth." Some miles from Encaramada there rises in the midst of the savannah the rock of Tepu-Mereme; *i.e.*, the "painted rock," which exhibits numerous figures of animals and symbolical signs, having much resemblance to those which we observed at some distance above Encaramada, near Caycara, (7° 5′ to 7° 40′ north lat., and 66° 28′ to 67° 23′ west long.) Similarly carved rocks are found between the Cassiquiare and the Atabapo (2° 5′ to 3° 20′ lat.); and what is most striking, also 560 miles further eastward in the solitudes of the Parime. The last-named fact is proved beyond a doubt, by the journal of Nicolas Hortsmann of Hildesheim, of which I have seen a copy in the handwriting of the celebrated d'Anville. That simple and modest traveller wrote down every day on the spot whatever had struck him as worthy of notice; and his narrative deserves perhaps the more confidence from the fact that the great disappointment he experienced in having failed in the object of his researches, which was the discovery of the Lake of Dorado, with its lumps of gold and a diamond mine (which proved to be merely rock crystal of a very pure kind), led him to look with a certain degree of contempt on all that fell in his way. On the bank of the Rupunuri, at the point where the river, winding between the Macarana mountains, forms several small cascades; and before reaching the country immediately surrounding the Lake of Amucu, he found, on the 16th of April, 1749, "rocks covered with figures," or, as he says in Portuguese, "*de varias letras*" (with various letters or characters). We were shown, at the rock of Culimacari, on the banks of the Cassiquiare, signs said to be characters drawn by line and rule;

but they were merely ill-formed figures of the heavenly bodies. crocodiles, boa-constrictors, and utensils used in the preparation of manioc-meal. I found among these painted rocks (*piedras pintadas*) neither a symmetrical arrangement nor any trace of characters drawn with a regard to regularity in space and size. The word "*letras*" in the journal of the German Surgeon (Hortsmann) must not, therefore, I am disposed to think, be taken in the strictest sense.

Schomburgk did not succeed in finding the rocks observed by Hortsmann, but he has described others which he saw on the bank of the Essequibo, near the cascade of Waraputa. "This cascade," he says, "is celebrated not only for its height, but also for the great number of figures hewn in the rock, which bear a great resemblance to those that I have seen on the island of St. John, (one of the Virgin Islands,) and which I consider to be without doubt the work of the Caribs, by whom this part of the Antilles was peopled in former times. I made the most strenuous efforts to hew away a portion of the rock carved with inscriptions, which I was desirous of taking with me; but the stone was too hard, and my strength had been wasted by fever. Neither threats nor promises could prevail on the Indians to aim a single stroke of the hammer against these rocks—the venerable monuments of the culture and superior skill of their forefathers. They regard them as the work of the Great Spirit; and all the different tribes we met were acquainted with them, although living at a great distance. Terror was painted on the faces of my Indian companions who seemed to expect every moment that the fire of heaven would fall on my head. I now saw clearly that all my efforts were fruitless, and I was therefore obliged to content myself with bringing away a complete drawing of these monuments."

The last resolution was undoubtedly the best, and the editor of the English journal, to my great satisfaction, subjoins in a note the remark, "that it is to be wished that others may succeed no better than Schomburgk, and that no traveller belonging to a civilized nation will in future attempt the destruction of these monuments of the unprotected Indians."

The symbolical signs which Sir Robert Schomburgk found in the fluvial valley of the Essequibo, near the rapids of Waraputa,* resemble, indeed, according to his observation,

* Richard Schomburgk, *Reisen in Britisch Guiana*, th. i. s. 320.

the genuine Carib carvings of one of the smaller Virgin Islands (St. John); but notwithstanding the wide extent of the Carib invasions, and the ancient power of that fine race, I cannot believe that this vast belt of carved rocks which intersects a great portion of South America from west to east, is actually to be ascribed to the Caribs. These remains seem rather to be traces of an ancient civilization, which may have belonged to an epoch when the tribes, whom we now distinguish by various names and races, were still unknown. The veneration which is everywhere shown by the Indians for these rude carvings of their predecessors, proves that the present races have no idea of the execution of similar works. Nay, more than this, between Encaramada and Caycara, on the banks of the Orinoco, many of these hieroglyphic figures are found sculptured on the sides of rocks at a height which can now only be reached by means of extremely high scaffolding. When asked who can have carved these figures, the natives answer with a smile, as if it were a fact of which none but a white man could be ignorant, that "in the days of the great waters their fathers sailed in canoes at this height." Here we find a geological dream serving as a solution of the problem presented by a long extinct civilization.

I would here be permitted to subjoin a remark, which I borrow from a letter addressed to me by Sir Robert Schomburgk, the distinguished traveller already mentioned. "The hieroglyphic figures are much more widely extended than you probably have conjectured. During my expedition, the object of which was the exploration of the river Corentyn, I not only observed several gigantic figures on the rock of Timeri (4° 30′ north lat. and 57° 30′ west long.), but I also discovered similar ones in the vicinity of the great cataracts of the river Corentyn (in 4° 21′ 30″ north lat. and 57° 55′ 30″ west long.) These figures have been executed more carefully than any others which I met with in Guiana. They are about 12 feet in height and appear to represent human figures. The head-gear is extremely remarkable; it surrounds the entire head, spreads far out, and is not unlike the glory represented round the heads of Saints. I left drawings of these images in the colony, which I hope some day to be able to lay collectively before the public. I have seen less complete figures on the Cuyuwini, a river which, flowing from the north-west, empties itself into the Essequibo in 2° 1·6′ north lat.; and I subsequently found similar figures on the Essequibo itself in 1° 40′ north lat. These figures, therefore, as appears from actual observations, extend from 7° 10′ to 1° 40′ north lat., and from 57° 30′ to 66° 30′ west long. The zone (or belt) of the sculptured rocks (as far as it has yet been investigated) thus extends over an area of 192,000 square miles, and includes within its circuit the basins of the Corentyn, Essequibo, and Orinoco—a circumstance that enables us to judge of the former population of this portion of the continent."

Remarkable relics of a former culture, consisting of granitic vessels ornamented with beautiful representations of labyrinths, and the earthenware forms resembling the Roman masks, have been discovered among the wild Indians on the Mosquito coast.* I had them engraved in the picturesque Atlas appended to the historical portion of my travels. Antiquarians are astonished at the resemblance of these *al-greco* vessels to those which embellish the Palace of Mitla (near Oaxaca, in New Spain). The large-nosed race, who are so frequently sculptured in relief on the Palenque of Guatimala and in Aztec pictures, I have never observed in Peruvian carvings. Klaproth recollects having noticed that the Chalkas, a horde of Northern Mongolia, had similar large noses. It is universally known, that many races of the North American, Canadian, and copper-coloured Indians, have fine aquiline noses, which constitute an essential physiognomical mark of distinction between them and the present inhabitants of New Granada, Quito, and Peru. Are the large-eyed, fair-skinned natives of the north-west coast of America, of whom Marchand speaks as living in 54° and 58° north lat., descended from the Usuns, an Alano-Gothic race of Central Asia?

(52) p. 20.—"*Deal certain death with a poisoned thumb-nail.*"

The Otomacs frequently poison their thumb-nails with *curare*. The mere impress of the nail proves fatal, should the curare become mixed with the blood. We have in our possession the creeping plant, from the juice of which the curare is prepared, in the Esmeralda Mission, on the Upper Orinoco,

* *Archæologia Britannica,* vol. v. 1779, pp. 318-324; and vol. vi. 1782, p. 107.

but, unfortunately, we did not find the plant when in blossom. From its physiognomy, it seems to be allied to *Strychnos*.*

Since I wrote the above notice of the *Curare*, or *Urari*, as the plant and poison were called by Raleigh, the brothers Robert and Richard Schomburgk have rendered important service to science by making us accurately acquainted with the nature and mode of preparing this substance, which I was the first to bring to Europe in any considerable quantity. Richard Schomburgk found this creeping plant in flower in Guiana, on the banks of the Pomeroon and Sururu, in the territory of the Caribs, who are, however, ignorant of the mode of preparing the poison. His instructive work† gives the chemical analysis of the juice of the *Strychnos toxifera*, which, notwithstanding its name and organic structure, contains, according to Boussingault, no trace of strychnine. Virchow's and Münter's interesting physiological experiments show that the curare or urari poison does not appear to destroy by resorption from without, but chiefly when it is absorbed by the animal substance after the separation of the continuity of the latter; that curare does not belong to tetanic poisons; and that it especially produces paralysis, *i.e.*, a cessation of voluntary muscular movement, while the function of the involuntary muscles (as the heart and intestines) continues unimpaired.‡

* See my *Rêlat. historique*, t. ii. pp. 547—556.

† *Reisen in Britisch Guiana*, th. i. s. 441—461.

‡ Compare also the older chemical analysis of Boussingault, in the *Annales de Chimie et Physique*, t. xxxix. 1828, pp. 24—37.

# ON THE CATARACTS OF THE ORINOCO,

*Near Atures and Maypures.*

---

In the preceding section, which I made the subject of an Academical Lecture, I have delineated those boundless plains, whose natural character is so variously modified by climatic relations, that what in one region appear as barren treeless wastes or deserts, in another are Steppes or far-stretching Prairies. With the Llanos of the southern portion of the New Continent, may be contrasted the fearful sandy deserts in the interior of Africa; and these again with the Steppes of Central Asia, the habitation of those world-storming herdsmen, who, once pouring forth from the east, spread barbarism and devastation over the face of the earth.

While on that occasion (1806), I ventured to combine many massive features in one grand picture of nature, and endeavoured to entertain a public assembly with subjects, somewhat in accordance with the gloomy condition of our minds at that period, I will now, confining myself to a more limited circle of phenomena, pourtray in brighter tints the cheerful picture of a luxuriant vegetation, and fluvial valleys with their foaming mountain torrents. I will describe two scenes of Nature from the wild regions of Guiana,—Atures and Maypures, the far-famed Cataracts of the Orinoco,—which, previously to my own travels, had been visited by few Europeans.

The impression which is left on the mind by the aspect of natural scenery is less determined by the peculiar character of the region, than by the varied nature of the light through which we view, or mountain or plain, sometimes beaming beneath an azure sky, sometimes enveloped in the gloom of lowering clouds. Thus, too, descriptions of nature affect us more or less powerfully, in proportion as they harmonize with the condition of our own feelings. For the physical world is reflected with truth and animation on the inner susceptible world of the mind. Whatever marks the character of a landscape: the profile of mountains, which in the far and hazy distance bound the horizon; the deep gloom of pine forests; the mountain torrent, which rushes headlong to its fall through overhanging cliffs: all stand alike in an ancient and mysterious communion with the spiritual life of man.

From this communion arises the nobler portion of the enjoyment which nature affords. Nowhere does she more deeply impress us with a sense of her greatness, nowhere does she speak to us more forcibly than in the tropical world, beneath the "Indian sky," as the climate of the torrid zone was called in the early period of the Middle Ages. While I now, therefore, venture to give a delineation of these regions, I am encouraged to hope that the peculiar charm which belongs to them will not be unfelt. The remembrance of a distant and richly endowed land, the aspect of a free and powerful vegetation, refreshes and strengthens the mind; even as our soaring spirit, oppressed with the cares of the present, turns with delight to contemplate the early dawn of mankind and its simple grandeur.*

Western currents and tropical winds favour the passage over that pacific arm of the sea (1) which occupies the vast valley stretching between the New Continent and Western Africa. Before the shore is seen to emerge from the highly curved expanse of waters, a foaming rush of conflicting and

* Humboldt, in this and other pages of his lecture, addressed, it should be remembered, to the citizens of Berlin, in 1806, evidently alludes to the troubles of the times.—Ed.

intermingling waves is observed. The mariner who is unacquainted with this region would suspect the vicinity of shoals, or a wonderful burst of fresh springs, such as occur in the midst of the Ocean among the Antilles (2).

On approaching nearer to the granitic shores of Guiana, he sees before him the wide mouth of a mighty river, which gushes forth like a shoreless sea, flooding the ocean around with fresh water. The green waves of the river, which assume a milky white hue as they foam over the shoals, contrast with the indigo-blue of the sea, which marks the waters of the river in sharp outlines.

The name Orinoco, which the first discoverers gave to this river, and which probably owes its origin to some confusion of language, is unknown in the interior of the country. For in their condition of animal rudeness, savage tribes only designate by peculiar geographical names, those objects which might be confounded with others. Thus the Orinoco, the Amazon, and the Magdalena, are each simply termed *The River*, the *Great River*, and *The Great Water;* whilst, those who dwell on the banks of even the smallest streams distinguish them by special names.

The current produced by the Orinoco between the South American Continent and the asphaltic island of Trinidad is so powerful, that ships, with all their canvass spread, and a westerly breeze in their favour, can scarcely make way against it. This desolate and fearful spot is called the Bay of Sadness (*Golfo Triste*), and its entrance the *Dragon's Mouth* (*Boca del Drago*). Here isolated cliffs rise tower-like in the midst of the rushing stream. They seem to mark the old rocky barrier (3) which, before it was broken through by the current, connected the island of Trinidad with the coast of Paria.

The appearance of this region first convinced the bold navigator Columbus of the existence of an American continent. "Such an enormous body of fresh water," concluded this acute observer of nature, "could only be collected from a river having a long course; the land, therefore, which supplied it must be a continent, and not an island." As, according to Arrian, the companions of Alexander, when they penetrated across the snow-crowned summits of Paropanisus (4), believed that they recognized in the crocodile-teeming Indus a part of the Nile,* so Columbus, in his ignorance of the similarity of physiognomy which characterises all the products of the climate of palms, imagined that the New Continent was the eastern coast of the far projecting Asia. The grateful coolness of the evening air, the ethereal purity of the starry firmament, the balmy fragrance of flowers, wafted to him by the land breeze—all led him to suppose, (as we are told by Herrera, in the Decades (5),) that he was approaching the garden of Eden, the sacred abode of our first parents. The Orinoco seemed to him one of the four rivers, which, according to the venerable tradition of the ancient world, flowed from Paradise, to water and divide the surface of the earth, newly adorned with plants. This poetical passage in the Journal of Columbus, or rather in a letter to Ferdinand and Isabella, written from Haiti in October, 1498, presents a peculiar psychological interest. It teaches us anew, that the creative fancy of the poet manifests itself in the discoverer of a world, no less than in every other form of human greatness.

When we consider the great mass of water poured into the Atlantic Ocean by the Orinoco, we are naturally led to ask which of the South American rivers is the greatest—the Orinoco, the Amazon, or the La Plata? The question is as indeterminate as the idea of greatness itself. The Rio de la Plata has undoubtedly the widest mouth, its width measuring 92 miles across; but this river, like those of Great Britain, is comparatively of but inconsiderable length. Its shallowness, too, is so great as to impede navigation at

* *Hist.*, lib. vi., initio.

Buenos Ayres. The Amazon, which is the longest of all rivers, measures 2880 miles from its rise in the Lake of Lauricocha to its estuary. Yet its width in the province of Jaen de Bracamoros, near the cataract of Rentama, where I measured it at the foot of the picturesque mountain of Patachuma, is scarcely equal to that of the Rhine at Mayence.

The Orinoco is narrower at its mouth than either the La Plata or the Amazon, while its length, according to my astronomical observations, does not exceed 1120 geographical miles. But in the interior of Guiana, 560 miles from its estuary, I found that at high water the width of the river measured upwards of 17,265 feet. Its periodical swelling here raises the level of the waters every year from 30 to 36 feet above the lowest water-mark. We are still without sufficient data for an accurate comparison between the enormous rivers which traverse the South American Continent. For such a comparison it would be necessary to ascertain the profile of the river-bed, as well as the velocity of the water, which varies very considerably at different points.

If the Orinoco, in the Delta formed by its variously divided and still unexplored branches, as well as in the regularity of its rise and fall, and in the number and size of its crocodiles, exhibits numerous points of resemblance to the Nile; there is this further analogy between the two rivers, that they for a long distance wind their impetuous way, like forest torrents, between granitic and syenitic rocks, till, slowly rolling their waters over an almost horizontal bed, skirted by treeless banks, they reach the sea.

An arm of the Nile (the Green Nile, Bahr-el-Azrek), from the celebrated mountain lake, near Gondar, in the Gojam Alps, in Abyssinia, to Syene and Elephantis, winds its way through the mountain range of Schangalla and Sennar; and in like manner the Orinoco rises on the southern slope of a mountain chain, which stretches between the parallels of 4° and 5° north lat., from French Guiana, in a westerly direction towards the Andes of New Granada. The sources of the Orinoco have never been visited by any European (6), nor even by any natives who have held intercourse with Europeans.

When, in the summer of 1800, we ascended the Upper Orinoco, we passed the mission of Esmeralda, and reached the mouths of the Sodomoni and the Guapo. Here soars high above the clouds, the mighty peak of the Yeonnamari or Duida; a mountain which presents one of the grandest spectacles in the natural scenery of the tropical world. Its altitude, according to my trigonometrical measurement, is 8278 (8823 English) feet above the level of the sea. Its southern slope is a treeless grassy plain, redolent with the odour of pine-apples, whose fragrance scents the humid evening air. Among lowly meadow plants rise the juicy stems of the *anana*, whose golden yellow fruit gleams from the midst of a bluish green diadem of leaves. Where the mountain springs break forth from beneath the grassy covering, rise isolated groups of lofty fan-palms, whose leaves, in this torrid region, are never stirred by a cooling breeze.

To the east of the Duida mountain, begins a thicket of wild cacao trees, among which are found the celebrated almond tree, *Bertholletia excelsa*, the most luxurious product of a tropical vegetation (7). Here the Indians collect colossal stalks of grass, whose joints measure upwards of 18 feet from knot to knot, which they use as blow-pipes for the discharge of their arrows (8). Some Franciscan monks have penetrated as far as the mouth of the Chiguire, where the river is already so narrow that the natives have suspended over it, near the waterfall of the Guaharibes, a bridge woven of the stems of twining plants. The Guaicas, of palish complexion and short stature, armed with poisoned arrows, oppose all further progress eastward.

Therefore, all that has been advanced to prove that the

Orinoco derives its source from a lake must be regarded as a fable (9). In vain the traveller seeks to discover the Lake of El Dorado, which, in Arrowsmith's maps, is set down as an inland sea measuring upwards of 20 geographical (80 English) miles. Can the little reed-covered lake of Amucu, near which rises the Pirara (a branch of the Mahu), have given rise to this myth? This swamp lies, however, 4° to the east of the region in which we may suppose the sources of the Orinoco to be situated. Here tradition placed the island of Pumacena, a rock of micaceous schist, whose shining brightness has played a memorable, and, for the deluded adventurers, often a fatal, part in the fable of *El Dorado*, current since the sixteenth century.

According to the belief of many of the natives, the Magellanic clouds of the southern sky, and even the glorious nebulæ in the constellation Argo, are mere reflections of the metallic brilliancy of these silver mountains of the Parime. It was besides an ancient custom of dogmatising geographers to make all the most considerable rivers of the world originate in lakes.

The Orinoco is one of those remarkable rivers which, after numerous windings, first towards the west and then to the north, finally return towards the east in such a manner as to bring both its estuary and its source into nearly the same meridian. From the Chiguire and the Gehette as far as the Guaviare, the course of the Orinoco inclines westward, as if it would pour its waters into the Pacific. Here branches off to the south, the Cassiquiare, a remarkable river, but little known to Europeans, which unites with the Rio Negro, or as the natives call it, the Guainia: furnishing the only example of a bifurcation which forms in the very interior of a continent a natural connection between two great river valleys.

The nature of the soil, and the junction of the Guaviare and Atabapo with the Orinoco, cause the latter to deflect suddenly northwards. From a want of correct geographical data, the Guaviare, flowing in from the west, was long regarded as the true source of the Orinoco. The doubts advanced since 1797 by an eminent geographer, M. Buache, regarding the possibility of a connection with the Amazon, have, I trust, been completely set at rest by my expedition. In an uninterrupted voyage of 920 miles, I penetrated through a remarkable net-work of rivers, from the Rio Negro, along the Cassiquiare, into the Orinoco; across the interior of the continent, from the Brazilian boundary to the coast of Caracas.

In the upper portion of this fluvial district, between 3° and 4° north lat., nature has exhibited, at many different points, the puzzling phenomenon of the so-called *black waters*. The Atabapo, whose banks are adorned with *Carolinias* and arborescent *Melastomas*, the Temi, Tuamini, and Guainia, are all rivers of a brown or coffee colour, which, under the deep shade of the palms, assumes a blackish, inky tint. When placed in a transparent vessel, the water appears of a golden yellow colour. These black streams reflect the images of the southern stars with the most remarkable clearness. Where the waters flow gently they afford the astronomer, who is making observations with reflecting instruments, a most excellent artificial horizon.

An absence of crocodiles as well as of fish—greater coolness—less torment from stinging mosquitoes—and salubrity of atmosphere, characterize the region of the black rivers. They probably owe their singular colour to a solution of carburetted hydrogen, to the rich luxuriance of tropical vegetation, and to the abundance of plants on the soil over which they flow. Indeed, I have observed that on the western declivity of the Chimborazo, towards the shores of the Pacific, the overflowing waters of the Rio de Guayaquil gradually assume a golden yellow, approaching to a coffee colour, after they have covered the meadows for several weeks.

Near the mouths of the Guaviare and Atapabo grows one

of the noblest forms of the palm-tree, the Piriguao (10), whose smooth stem, which is nearly 70 feet in height, is adorned with delicate flag-like leaves having curled margins. I know no palm which bears equally large and beautifully coloured fruits. They resemble peaches in their blended tints of yellow and crimson. Seventy or eighty of these form one enormous cluster, of which each stem annually ripens three. This noble tree might be termed the peach-palm. Its fleshy fruit, owing to the extreme luxuriance of vegetation, is generally devoid of seed; and it yields the natives a nutritious and farinaceous article of food which, like the banana and the potato, is capable of being prepared in many different ways.

To this point, that is, as far as the mouth of the Guaviare, the Orinoco flows along the southern declivity of the chain of the Parime. From its left bank, across the equator, and as far as the parallel of 15° south lat., extends the boundless wooded plain of the river Amazon. At San Fernando de Atabapo the Orinoco, turning off abruptly in a northerly direction, intersects a portion of the mountain chain itself. Here are the great waterfalls of Atures and Maypures, and here the bed of the river is everywhere contracted by colossal masses of rocks, which give it the appearance of being divided by natural dams into separate reservoirs.

At the entrance of the Meta stands, in the midst of an enormous whirlpool, an isolated rock, which the natives very aptly term the "Rock of Patience," because when the waters are low, it sometimes retards for two whole days the ascent of the navigator. Here the Orinoco, biting deep into its shores, forms picturesque rocky bays. Opposite the Indian mission of Carichana, the traveller is surprised by a most remarkable prospect. Involuntarily his eye is arrested by a steep granite rock, "El Mogote de Cocuyza," a cubiform mass, which rises precipitously to a height of more than 200 feet; and whose summit is crowned with a luxuriant forest.

Like a Cyclopic monument of simple grandeur, this bold promontory towers high above the tops of the surrounding palms, cutting the deep azure of the sky with its strongly marked outlines, and lifting, as it were, forest upon forest.

On descending beyond Carichana, the traveller arrives at a point where the river has opened itself a passage through the narrow pass of Baraguan. Here we everywhere recognise traces of chaotic devastation. To the north, towards Uruana and Encaramada, rise granite rocks of grotesque appearance, which, in singularly formed crags of dazzling whiteness, gleam brightly from amidst the surrounding groves.

At this point, near the mouth of the Apure, the stream leaves the granitic chain, and flowing eastward, separates as far as the Atlantic, the impenetrable forests of Guiana from the Savannahs, on whose far distant horizon the vault of heaven seems to rest. Thus the Orinoco surrounds on the south, west, and north, the high mountain chain of the Parime, which occupies the vast space between the sources of the Jao and of the Caura. No cliffs or rapids obstruct the course of the river from Carichana to its mouth, excepting, indeed, the "Hell's Mouth" (Boca del Inferno) near Muitaco, a whirlpool occasioned by rocks, as at Atures and Maypures, which does not, however, block up the whole breadth of the stream. In this district, which is contiguous to the sea, the only dangers encountered by the boatmen arise from the natural timber-floats, against which canoes are often wrecked at night. These floats consist of forest trees which have been uprooted and torn away from the banks by the rising of the waters. They are covered, like meadows, with blooming water-plants, and remind us of the floating gardens of the Mexican lakes.

After this brief glance at the course of the Orinoco and its general features, I pass to the waterfalls of Maypures and Atures.

From the high mountain-group of Cunavami, between the sources of the rivers Sipapo and Ventuari, a granite ridge projects to the far west towards the mountain of Uniama. From this ridge descend four streams, which mark, as it were, the limits of the cataracts of Maypures; two bound Sipapo and Sanariapo, on the eastern shore of the Orinoco; and two the Cameji and Toparo, on the western side. At the site of the missionary village of Maypures the mountains form a wide bay opening towards the south-west.

Here the stream rushes foaming down the eastern declivity of the mountain, while far to the west traces remain of the ancient and now forsaken bank of the river. An extensive Savannah stretches between the two chains of hills, at an elevation of scarcely 30 feet above the upper water-level of the river, and here the Jesuits have erected a small church formed of the trunks of palms.

The geognostical aspect of this region, the insular form of the rocks of Keri and Oco, the cavities worn in the former by the current, and which are situated at exactly the same level as those in the opposite island of Uivitari; all these indications tend to prove that the Orinoco once filled the whole of this now dried-up bay. It is probable that the waters formed a wide lake, as long as the northern dam withstood their passage. When this barrier gave way, the Savannah now inhabited by the Guareke Indians emerged as an island. The river may perhaps long after this have continued to surround the rocks of Keri and Oco, which now picturesquely project, like castellated fortresses, from its ancient bed. After the gradual diminution of the waters, the river withdrew wholly to the eastern side of the mountain chain.

This conjecture is confirmed by various circumstances. Thus, for instance, the Orinoco, like the Nile at Philæ and Syene, has the singular property of colouring black the reddish-white masses of granite, over which it has flowed for thousands of years. As far as the waters reach one observes on the rocky shore a leaden-coloured manganeseous and perhaps carbonaceous coating which has penetrated scarcely one tenth of a line into the stone. This black coloration, and the cavities already alluded to, show the former water level of the Orinoco.

These black cavities may be traced at elevations of from 160 to 192 feet above the present level of the river on the rocks of Keri, in the islands of the cataracts; in the gneiss-like hills of Cumadanimari, which extend above the island of Tomo; and lastly at the mouth of the Jao. Their existence proves, what indeed we learn from all the river-beds of Europe, that those streams which still excite our admiration by their magnitude, are but inconsiderable remains of the immense masses of water belonging to a former age.

These simple facts have not escaped even the rude natives of Guiana. Everywhere the Indians drew our attention to these traces of the ancient water-level. Nay, in a Savannah near Uruana there rises an isolated rock of granite, which, according to the testimony of persons worthy of credit, exhibits at an elevation of between 80 and 90 feet, a series of figures of the sun and moon, and of various animals, especially crocodiles and boa-constrictors, graven, almost in rows. At the present day this perpendicular rock, which well deserves the careful examination of future travellers, cannot be ascended without the aid of scaffolding. In a similarly remarkable elevated position, the traveller can trace hieroglyphic characters carved on the mountains of Uruana and Encaramada.

If the natives are asked how these characters could have been graven there, they answer that it was done in former times, when the waters were so high that their fathers' canoes floated at that elevation. Such lofty condition of the water level must therefore have been coeval with these rude memorials of human skill. It indicates an ancient distribution of land and water over the surface of the globe widely different from that which now exists; but which must not be confounded with that condition when the primeval vegetation of our planet, the colossal remains of extinct terrestrial animals, and the oceanic creatures of a chaotic world, found one common grave in the indurating crust of our earth.

At the most northern extremity of the cataracts our attention is attracted by what are called the natural representations of the Sun and Moon. The rock of Keri, to which I have more than once referred, derives its name from a glistening white spot seen at a considerable distance, and in which the Indians profess to recognize a striking resemblance to the disc of the full moon. I was not myself able to climb this precipitous rock, but it seems probable that the white spot is a large knot of quartz, formed by a cluster of veins in the greyish-black granite.

Opposite to the Keri rock, on the twin mountain of the island of Uivitari, which has a basaltic appearance, the Indians point, with mysterious admiration, to a similar disc, which they venerate as the image of the Sun, *Camosi.* The geographical position of these two rocks may have contributed to their respective appellations, for I found that Keri was turned towards the west, and Camosi towards the east. Some etymological inquirers have thought they could recognize an analogy between the American word Camosi and the word Camosh, a name applied in one of the Phœnician dialects to the sun, and identical with the Apollo Chomeus or Beelphegor and Amun.

The lofty falls of Niagara, which are 150 feet in height, derive their origin, as is well known, from the combined precipitation of one enormous mass of water. Such, however, is not the case with respect to the cataracts of Maypures, nor are they narrow straits or passes through which the stream rushes with increasing velocity, like the Pongo of Manseriche on the Amazon, but rather to be regarded as a countless number of small cascades succeeding each other like steps. The *Raudal,* (as the Spaniards term this kind of cataract,) is formed by an archipelago of islands and rocks, which so contract the bed of the river that its natural width of more than 8500 feet is often reduced to a channel scarcely navigable to the extent of 20 feet. At the present day the eastern side is far less accessible and far more dangerous than the western.

At the mouth of the Cameji the boatmen unload their cargo that they may leave the empty canoe, or, as it is here called, the *Piragua*, to be piloted by Indians well acquainted with the Raudal, as far as the mouth of the Toparo, where all danger is supposed to be past. Where the rocks or shelvy ledges, (each of which has its particular name,) are not above two or three feet in height, the natives venture to shoot the rapid with their canoes. When, however, they have to ascend the stream, they swim in advance of the piragua, and after much labour, and, perhaps, many unsuccessful efforts, succeed in throwing a rope round a point of rock projecting above the breakers, and by this means draw the canoe against the stream, which, in this arduous operation, is often water-logged, or upset.

Sometimes the canoe is dashed to pieces on the rock, and this is the only danger the natives fear. With bleeding bodies they then strain every nerve to escape the fury of the whirlpool and swim to land. Where the rocky ledges are very high and form a barrier by extending across the entire bed of the river, the light canoe is hauled to land and dragged for some distance along the shore on branches of trees which serve the purpose of rollers.

The most celebrated and most perilous ledges are those of Purimarimi and Manimi, which are between nine and ten feet in height. It was with surprise I found, by barometrical measurements, that the entire fall of the Raudal, from the mouth of the Cameji to that of the Toparo, scarcely amounted to more than 30 or 32 feet. (A geodesic levelling is not practicable, owing to the inaccessibility of the locality and the pestiferous atmosphere, which swarms with mosquitoes.) I say with surprise, for I hence discovered that the tremendous roar and wild dashing of the stream arose from the contraction of its bed by numerous rocks and islands, and the counter-currents produced by the form and position of the masses of rock. The truth of my assertion regarding the inconsiderable height of the whole fall will be best verified by observing the cataracts, in descending to the bed of the river, from the village of Maypures, across the rocks of Manimi.

At this point the beholder enjoys a most striking and wonderful prospect. A foaming surface, several miles in length, intersected with iron-black masses of rock projecting like battlemented ruins from the waters, is seen at one view. Every islet and every rock is adorned with luxuriant forest trees. A perpetual mist hovers over the watery mirror, and the summits of the lofty palms pierce through the clouds of vapoury spray. When the rays of the glowing evening sun are refracted in the humid atmosphere, an exquisite optical illusion is produced. Coloured bows appear, vanish, and re-appear, while the ethereal picture dances, like an ignis fatuus, with every motion of the sportive breeze.

During the long rainy seasons, the falling waters carry down quantities of vegetable mould, which accumulating form islands of the naked rocks; adorning the barren stone with blooming beds of Melastomes and Droseras, silver-leaved *Mimosæ*, and a variety of ferns. They recal to the mind of the European those groups of vegetation which the inhabitants of the Alps term *courtils*, blocks of granite bedecked with flowers which project solitarily amid the Glaciers of Savoy.

In the blue distance the eye rests on the mountain chain of Cunavami, a far-stretching chain of hills which terminates abruptly in a sharply truncated cone. We saw this conical hill, called by the Indians Calitamini, glowing at sunset as if in crimson flames. This appearance daily returns. No one has ever been in the immediate neighbourhood of this mountain. Possibly its dazzling brightness is produced by the reflecting surface of decomposing talc, or mica schist.

During the five days that we passed in the neighbourhood of the cataracts, we were much struck by the fact that the roar of the rushing torrent was three times as great by night as by day. The same phenomenon is observed in all European waterfalls. To what can we ascribe this effect in a solitude where the repose of nature is undisturbed? Probably to ascending currents of warm air, which producing an unequal density of the elastic medium, obstruct the propagation of sound by displacing its waves; causes which cease after the nocturnal cooling of the earth's surface.

The Indians showed us traces of ruts caused by wheels. They speak with wonder of the horned cattle, (oxen,) which at the period of the Jesuit missions used to draw the trucks, that conveyed the canoes, along the left shore of the Orinoco, from the mouth of the Cameji to that of the Toparo. The canoes at that time were transported without the discharge of their cargoes, and were not as now injured by being constantly dragged over sharp-pointed rocks, or stranded.

The topographical plan which I have sketched of the locality, shews that a canal might be opened between the Cameji and the Toparo. The valley in which these two abundantly watered rivers flow is a gentle level; and the canal, of which I suggested a plan to the Governor-General of Venezuela, would become a navigable arm of the Orinoco, and supersede the old and dangerous bed of the river.

The Raudal of Atures is exactly similar to that of Maypures, like which it consists of a cluster of islands between which the river forces itself a passage extending from 18,000 to 24,000 feet. Here too a forest of palm trees rises from the midst of the foaming surface of the waters. The most celebrated ledges of the cataract are situated between the islands of Avaguri and Javariveni, between Suripamana and Uirapuri.

When M. Bonpland and myself were returning from the banks of the Rio Negro, we ventured to pass the latter, that is the lower half, of the Raudal of Atures in our loaded canoe. We several times disembarked to climb over rocks, which, like dykes, connected one island with another. At one time the water shoots over these dykes; at another it falls into their cavities with a deafening hollow sound. In some places considerable portions of the bed of the river are perfectly dry, in consequence of the stream having opened for itself a subterranean passage. In this solitude the golden-coloured Rock Manakin (*Pipra rupicola*) builds its nest. This bird, which is as pugnacious as the East India cock, is one of the most beautiful birds of the tropics, and is remarkable for its double moveable crest of feathers with which its head is decorated.

In the Raudal of Canucari the dyke is formed of piled-up granitic boulders. We crept into the interior of a cavern, whose humid walls were covered with confervæ and phosphorescent Byssus. The river rushed over our heads with a terrible and stunning noise. By accident we had an opportunity of contemplating this grand scene longer than we desired. The Indian boatmen had left us in the middle of the cataract, to take the canoe round a small island, at the other extremity of which, after a considerable circuit, we were to re-embark. For an hour and a half we remained exposed to a fearful thunder-storm. Night was approaching, and we in vain sought shelter in the fissures of the rocks. The little apes which we had carried with us for months in wicker cages, attracted by their plaintive cries large crocodiles, whose size and leaden-grey colour indicated their great age. I should not have alluded to the appearance of these animals in the Orinoco, where they are of such common occurrence, were it not that the natives had assured us that no crocodiles had ever been seen among the cataracts; indeed, on the strength of that assertion, we had repeatedly ventured to bathe in this portion of the river.

Meanwhile our anxiety increased every moment, lest, drenched as we were and deafened by the thundering roar of the falling waters, we should be compelled to spend the long tropical night in the midst of the Raudal. At length, however, the Indians made their appearance with our canoe. Their delay had been occasioned by the inaccessibility of the steps they had to descend, owing to the low state of the water; which had obliged them to seek in the labyrinth of channels a more practicable passage.

Near the southern entrance of the Raudal of Atures, on the right bank of the river, lies the cavern of Ataruipe, so celebrated among the Indians. The surrounding scenery has a grand and solemn character, which seems to mark it as a national burial-place. With difficulty, and not without danger of being precipitated into the depths below, we clambered a steep and perfectly bare granite rock, on whose smooth surface it would be hardly possible to keep one's footing were it not for large crystals of feldspar, which, defying the action of weather, project an inch or more from the mass.

On gaining the summit, a wide prospect of the surrounding country astonishes the beholder. From the foaming bed of the river rise hills richly crowned with woods, while beyond its western bank the eye rests on the boundless Savannah of the Meta. On the horizon loom like threatening clouds the mountains of Uniama. Such is the distant view; but immediately around all is desolate and contracted. In the deep ravines of the valley moves no living thing save where the vulture and the whirring goat-sucker wing their lonely way, their heavy shadows gleaming fitfully past the barren rock.

The cauldron-shaped valley is encompassed by mountains, whose rounded summits bear huge granite boulders, measuring from 40 to more than 50 feet in diameter. They appear poised on only a single point of their surface, as if the slightest shock of the earth would hurl them down.

The further side of this rocky valley is thickly wooded. It is in this shady spot that the cave of the Ataruipe is situated; properly speaking, however, it is not a cave, but a vault formed by a far projecting and overhanging cliff,—a kind of bay hollowed out by the waters when formerly at this high level. This spot is the grave of an extinct tribe (11). We counted about six hundred well preserved skeletons, placed in as many baskets, formed of the stalks of palm-leaves. These baskets, called by the Indians *mapires*, are a kind of square sack varying in size according to the age of the deceased. Even new-born childre have each their own mapire. These skeletons are so perfect, that not a rib or a finger is wanting.

The bone are prepared in three different ways: some are bleached, some dyed red with onoto, the pigment of the *Bixa Orellana;* others like mummies, are anointed with fragrant resin and wrapped in banana leaves.

The Indians assured me that the corpse was buried during several months in a moist earth, which gradually destroyed the flesh; and that after being disinterred, any particles of flesh still adhering to the bones were scraped off with sharp stones. This practice is still continued among many tribes of Guiana. Besides these baskets or mapires, we saw many urns of half-burnt clay, which appear to contain the bones of whole families. The largest of these urns are upwards of three feet in height and nearly six feet in length, of an elegant oval form, and greenish colour; with handles shaped like crocodiles and serpents, and the rims bordered with flowing scrolls and labyrinthine figures. These ornaments are precisely similar to those which cover the walls of the Mexican palace at Mitla. They are found in every clime and every stage of human culture,—among the Greeks and Romans, no less than on the shields of Otaheitans, and other South Sea islanders,—in all regions where a rhythmical repetition of regular forms delights the eye. The causes of these resemblances, as I have explained elsewhere, are rather to be referred to psychical conditions, and to the inner nature of our mental qualifications, than as affording evidence in favour of a common origin and the ancient intercourse of nations.*

Our interpreters could give us no certain information regarding the age of these vessels; but that of the skeletons did not in general appear to exceed a hundred years. There is a legend amongst the Guareke Indians, that the brave Atures, when closely pursued by the cannibal Caribs, took refuge on the rocks of the cataracts,—a mournful place of abode, in which this oppressed race perished, together with its language! (12) In the most inaccessible portion of the Raudal other graves of the same character are met with; indeed it is probable that the last descendants of the Atures did not become extinct until a much more recent period. There still lives and it is a singular fact, an old parrot in Maypures which cannot be understood, because, as the natives assert, it speaks the language of the Atures!

We left the cave at nightfall, after having collected, to the extreme annoyance of our Indian guides, several skulls and the perfect skeleton of an aged man. One of these skulls has been delineated by Blumenbach in his admirable craniological work;† but the skeleton, together with a large portion of our natural history collections, especially the entomological, was lost by shipwreck off the coast of Africa on the same occasion when our friend and former travelling companion, the young Franciscan monk, Juan Gonzalez, lost his life.

As if with a presentiment of this painful loss, we turned from the grave of a departed race with feelings of deep emo-

* This subject is elaborately discussed in Heeren's various works.—Ed.

† *Blumenbach, Collectiones suæ Craniorum diversarum gentium,* &c., 4to, Götting., 1798–1828.—Ed.

tion. It was one of those clear and deliciously cool nights so frequent beneath the tropics. The moon stood high in the zenith, encircled by a halo of coloured rings, her rays gilding the margins of the mist, which in well defined outline hovered like clouds above the foaming flood. Innumerable insects poured their red phosphorescent light over the herb-covered surface, which glowed with living fire, as though the starry canopy of heaven had sunk upon the grassy plain. Climbing Bignonia, fragrant Vanillas, and golden-flowered Banisterias, adorned the entrance of the cave, while the rustling palm-leaves waved over the resting-place of the dead.

Thus pass away the generations of men!—thus perish the records of the glory of nations! Yet when every emanation of the human mind has faded—when in the storms of time the monuments of man's creative art are scattered to the dust—an ever new life springs from the bosom of the earth. Unceasingly prolific nature unfolds her germs,—regardless though sinful man, ever at war with himself, tramples beneath his foot the ripening fruit!

---

## ILLUSTRATIONS AND ADDITIONS.

(1) p. 154—"*Across that pacific arm of the sea.*"

The Atlantic Ocean, between the parallels of 23° south lat. and 70° north lat., has the form of a furrowed longitudinal valley, in which the advancing and receding angles are opposite to each other. I first developed this idea in my work entitled *Essai d un Tableau Géologique de l'Amérique méridionale*, which was published in the *Journal de Physique*, t. liii. p. 61.* From the Canary Isles, especially from 21° north lat., and 23° west long., to the north-east coast of South America, the surface of the ocean is so calm, and the waves so gentle, that an open boat might navigate it in safety.

(2) p. 155—"*Fresh springs among the Islands of the Antilles.*"

On the southern coast of the Island of Cuba, south-west of the harbour of Batabano, in the Gulf of Xagua, at a distance of eight to twelve miles from the shore, springs of fresh water gush from the bed of the ocean, probably from the action of hydrostatic pressure. The jet is propelled with such force that boats use extreme caution in approaching this spot, which is well known for its counter current producing a heavy swell. Trading vessels sailing along the coast, which do not purpose putting into port, sometimes visit these springs, in order to provide themselves, in the midst of the ocean, with a supply of fresh water. The freshness of the water increases with the depth from which it is drawn. River cows (*Trichecus manati*), which do not generally inhabit salt water, are frequently killed here. This singular phenomenon (the fresh springs), of which no mention had hitherto been made, was most accurately investigated by my friend, Don Francisco Lemaur, who made a trigonometrical survey of the Bahia de Xagua. I did not myself visit Xagua, but remained in the insular group situated further to the south (the so-called *Jardines del Rey*), to make astronomical determinations of their latitude and longitude.

(3) p. 155—"*Ancient rocky barrier.*"

Columbus, whose unwearied spirit of observation was di-

* Gilbert's *Annalen der Physik*, bd. xvi. 1804, s. 394—449.

rected on every side, proposes in his letters to the Spanish monarchs, a geognostic hypothesis regarding the configuration of the larger Antilles. Being fully impressed with the idea of the strength of the Equinoctial current, which has often a westerly direction, he ascribes to it the disintegration of the group of the smaller Antilles, and the singularly lengthened configuration of the southern coasts of Porto Rico, Haiti, Cuba, and Jamaica, all of which follow almost exactly the direction of parallels of latitude. On his third voyage (from the end of May, 1498, to the end of November, 1500), when, from the Boca del Drago to the Island of Margarita, and afterwards from that island to Haiti, he felt the whole force of the equinoctial current, "that movement of the waters which accords with the movement of the heavens—*movimiento de los cielos*," he says expressly that the violence of the current has torn the Island of Trinidad from the mainland. He refers the sovereigns to a chart which he sends them—a "*pintura de la tierra*," drawn by himself, to which frequent reference is made in the celebrated lawsuit against Don Diego Colon respecting the rights of the first Admiral. "Es la carta de marear y figura que hizo el Almirante señalando los rumbos y vientos por los quales vino á Paria, que dicen parte del Asia."*

(4) p. 156—"*Across the snow-crowned Paropanisus.*"

In Diodorus' description of the Paropanisus,† we seem to recognise a delineation of the Peruvian chain of the Andes. The army passed through inhabited districts in which snow daily fell!

(5) p. 156—"*Herrera in his Decades.*"

*Historia general de las Indias Occidentales*, Dec. i. lib. iii. cap. 12 (ed. 1601, p. 106); Juan Batista Muñoz, *Historia del Nuevo Mundo*, lib. vi. c. 31, p. 301; Humboldt, *Examen Crit.*, t. iii. p. 111.

(6) p. 158—"*The Sources of the Orinoco have never been visited by any European.*"

Thus I wrote respecting these sources in the year 1807, in

* Navarrete, *Viages y Descubrimientos que hiciéron por mar los Españoles*, t. i. pp. 253, 260; t. iii. pp. 539, 587.

† Diodor. Sicul., lib. xvii. p. 553 (Rhodom.)

the first edition of the *Ansichten der Natur*, and I repeat with equal truth the same statement after an interval of forty-one years. The travels of the brothers Robert and Richard Schomburgk, so important in reference to all departments of natural science and geography, have established other and more interesting facts; but the problem of the situation of the sources of the Orinoco has been only partially solved by Sir Robert Schomburgk. M. Bonpland and myself advanced from the west as far as Esmeralda, or the confluence of the Orinoco with the Guapo; and I was enabled, by the aid of well-attested information, to describe the upper course of the Orinoco to above the mouth of the Gehette, and to the small waterfall (Raudal) de los Guaharibos. From the east Sir Robert Schomburgk, proceeding from the mountains of the Majonkong Indians, the inhabited portion of which he estimated by the boiling point of water to be 3517 feet in height, succeeded in reaching the Orinoco by the Padamo River, which the Majonkongs and Guinaus (Guaynas?) call Paramu.* I had placed this confluence of the Padamo with the Orinoco in my Atlas, in 3° 12′ N. lat, and 65° 46′ W. long.. but Schomburgk found it by direct observation in 2° 53′ lat. and 65° 48′ W. long. The main object of this traveller's journey was not 'natural history,' but the solution of the prize question proposed by the Royal Geographical Society of London, in November, 1834,—on the connection of the coast of British Guiana with the easternmost point which I had reached on the Upper Orinoco. After undergoing many sufferings, this object was thoroughly achieved. Robert Schomburgk reached Esmeralda, with his instruments, on the 22nd of February, 1839. His determinations of the latitude and longitude of the place agreed more closely with mine than I had anticipated. Let us here allow the observer to speak for himself:—"Words are inadequate to describe the feelings which overwhelmed me when I sprang on shore. My object was attained; my observations, begun on the coast of Guiana, were brought into connection with those of Humboldt at Esmeralda, and I freely admit that at a time when my physical powers had almost entirely deserted me, and when I was surrounded by dangers and difficulties of no ordinary kind, the recognition which I hoped

* *Reisen in Guiana,* 1841, s. 448.

for from him, was the sole inducement which inspired me with a fixed determination to press forward towards the goal which I had now reached. The emaciated figures of my Indian companions and my faithful guides proclaimed more fully than any words could do, what difficulties we had had to surmount, and had surmounted." After citing expressions so gratifying, I must be permitted to subjoin the opinions I expressed regarding this great undertaking promoted by the Royal Geographical Society of London, in my Preface to the German edition of Robert Schomburgk's Account of his Travels, published in 1841. "Immediately after my return from Mexico, I indicated the direction and the routes by which the unknown portion of the South American Continent between the sources of the Orinoco, the mountain chain of Pacaraima, and the sea-shore near Essequibo, might be explored. These wishes, so strongly expressed in the personal narrative of my journey, have at length, after the lapse of nearly half a century, been for the most part fulfilled. I rejoice that I have been spared to see so important an enlargement of our geographical knowledge; I rejoice too in seeing a courageous and well-conducted enterprise, requiring the most devoted perseverance, executed by a young man, to whom I feel bound no less by the ties of similarity of pursuits than those of country. These circumstances were alone able to overcome the aversion and disinclination which I entertain, perhaps unjustly, for introductory prefaces by a different hand than that of the author himself. But I could not resist the impulse of expressing thus publicly my sincere esteem for the accomplished traveller who, led on by the meritorious idea of penetrating from east to west, from the Valley of the Essequibo to Esmeralda, has succeeded, after five years of efforts and of sufferings (the extent of which I well appreciate from my own experience), in attaining the object of his ambition. Courage for the sudden accomplishment of a hazardous undertaking is easier to find, and implies less inward strength, than the resolution to endure with resignation long-continued physical sufferings, excited by absorbing mental interest; and still to press forward, undismayed by the certainty of having to retrace his steps under equally great privations and with enfeebled powers. Serenity of mind, which is almost the first requisite for an enterprise in inhospitable regions, a passionate love for any department of scientific labour (be it natural history, astronomy, hypsometrics, or magnetism), a pure feeling for the enjoyment which nature is capable of imparting, are elements which, when they combine together in one individual, ensure valuable results from a great and important journey."

I will preface my consideration of the question of the sources of the Orinoco with my own conjectures in relation to the subject. The perilous route travelled in 1739 by the surgeon Nicolas Hortsmann, of Hildesheim; in 1775 by the Spaniard Don Antonio Santos, and his friend Nicolas Rodriguez; in 1793 by the Lieutenant-Colonel of the 1st Regiment of the Line of Para, Don Francisco José Rodriguez Barata; and (according to manuscript maps, for which I am indebted to the former Portuguese Ambassador in Paris, Chevalier de Brito) by several English and Dutch settlers, who in 1811 travelled from Surinam to Para by the portage of the Rupunuri and by the Rio Branco;—divides the *terra incognita* of the Parime into two unequal parts, and serves to mark the position of a very important point in the geography of those regions—viz., the sources of the Orinoco, which it is no longer possible to remove to an indefinite distance towards the east, without intersecting the bed of the Rio Branco, which flows from north to south through the fluvial district of the Upper Orinoco; while this portion of the great river itself pursues for the most part a direction from east to west. The Brazilians, since the beginning of the present century, have from political motives manifested a vivid interest in the extensive plains east of the Rio Branco.* Owing to the position of Santa Rosa on the Uraricapara, whose course appears to have been pretty accurately determined by Portuguese engineers, the sources of the Orinoco cannot be situated east of the meridian of 63° 8′ west long. This is the eastern limit beyond which they cannot be placed, and taking into consideration the state of the river at the Raudal de los Guaharibos (above Caño Chiguire, in the country of the strikingly fair-skinned Guaycas Indians, and 52′ east of the great Cerro

* See the Memoir which I drew up at the request of the Portuguese Government, in 1817, "*Sur la fixation des limites des Guyanes Francaise et Portuguaise.*" Schoell, *Archives historiques et politiques, ou Recueil de Pièces officielles, Mémoires, &c.* t. i. 1818, pp. 48—58.

Duida), it appears to me probable that the Orinoco in its upper part does not extend, at the utmost, beyond the meridian of 64° 8′ west long. This point is, according to my combinations, 4° 12′ west of the little lake of Amucu, which was reached by Sir Robert Schomburgk.

I will now detail the conjectures of that traveller, after having first given my own earlier ones. According to him the course of the Upper Orinoco, to the east of Esmeralda, is directed from south-east to north-west; my estimations of latitude for the mouths of the Padamo and the Gehette appear to be respectively 19′ and 36′ too small. Schomburgk conjectures that the sources of the Orinoco are situated in lat. 2° 30′, and the fine "Map of Guayana, to illustrate the route of R. H. Schomburgk," which accompanies the splendid English work entitled *Views in the Interior of Guiana*, places its geographical sources in 64° 56′ west long., *i.e.*, 1° 6′ west of Esmeralda, and only 48′ of longitude nearer to the Atlantic than I had determined the position of this point. Astronomical combinations led Schomburgk to place the mountain of Maravaca, which is about ten thousand feet high, in 3° 41′ lat. and 65° 48′ west long. The Orinoco was scarcely three hundred yards wide near the mouth of the Padamo or Paramú, and more to the west, where it expands to a width of from four to six hundred yards, it was so shallow, and so full of sandbanks, that the expedition was obliged to dig channels, as the river bed was only fifteen inches deep. Fresh-water dolphins were still to be seen in great numbers everywhere—a phenomenon which the zoologists of the eighteenth century would not have expected to find in the Orinoco and the Ganges.

(7) p. 158—"*The most luxurious product of a tropical climate.*"

The *Bertholletia excelsa* (Juvia), of the family of Myrtaceæ (and placed in Richard Schomburgk's proposed division of Lecythideæ), was first described in *Plantes Equinoxiales*, t. i. 1808, p. 122, tab. 36. This colossal and magnificent tree offers, in the perfect development of its cocoa-like, round, close-grained, woody fruit, inclosing the three-cornered and also woody seed-vessels, the most remarkable example of luxuriant organic development. The Bertholletia grows in the forests of the Upper Orinoco, between the Padamo and the Ocamu, in the vicinity of the mountain of Mapaya, as well as between the rivers Amaguaca and Gehette.*

(8) p. 158—"*Grass stalks, whose joints measure upwards of eighteen feet from knot to knot.*"

Robert Schomburgk, when visiting the small mountainous country of the Majonkongs, on his route to Esmeralda, was fortunate enough to determine the species of Arundinaria, which furnishes the material for these blowing-tubes. He says of this plant: "It grows in large tufts, like the bambusa; the first joint rises, in the old cane, without a knot, to a height of from 16 to 17 feet before it begins to bear leaves. The entire height of the Arundinaria, growing at the foot of the great mountain-cluster of Maravaca, is from 30 to 40 feet, with a thickness of scarcely half an inch in diameter. The top is always inclined; and this species of grass is peculiar to the sandstone mountains between the Ventuari, the Paramu (Padamo), and the Mavaca. The Indian name is Curata, and, therefore, from the excellence of these celebrated long blowing-tubes, the Majonkongs and Guinaus of these districts have acquired the name of the Curata nation."†

(9) p. 159—"*Fabulous origin of the Orinoco from a lake.*"

The lakes of these regions (some of which are wholly imaginary, while the real size of others has been much exaggerated by theoretical geographers) may be divided into two groups. The first of these groups comprise those situate between Esmeralda (the most easterly mission on the Upper Orinoco), and the Rio Branco; to the second, belong the lakes presumed to exist in the district between the Rio Branco and French, Dutch, and British Guiana. This general view, of which travellers should never lose sight, proves that the question of whether there is another Lake Parime eastward of the Rio Branco, besides the Lake Amucu, seen by Hortsmann, Santos, Colonel Barata, and Schomburgk, has nothing whatever to do with the problem of the sources of the Orinoco. As the name of my distinguished friend the former Director of the Hydrographic Office at Madrid, Don Felipe Bauza, is

* *Relation historique*, t. ii. pp. 474—496, 558—562.
† *Reisen in Guiana und am Orinoko*, 451.

of great weight in questions of geography, the impartiality which ought to influence every scientific investigation makes it incumbent on me to mention that this learned man was inclined to the view that there must be lakes west of the Rio Branco, at no great distance from the sources of the Orinoco. He wrote to me from London shortly before his death, "I wish you were here that I might converse with you respecting the geography of the Upper Orinoco, which has occupied you so much. I have been fortunate enough to rescue from entire destruction the papers of the General of Marine, Don José Solano, father of the Solano who perished in so melancholy a manner at Cadiz. These documents relate to the settlement of the boundary line between the Spaniards and Portuguese, with which Solano had been charged since 1754, in conjunction with the Escadre Chef Yturriaga and Don Vicente Doz. In all these plans and sketches I find a Laguna Parime sometimes as a source of the Orinoco, and sometimes as wholly detached from it. Are we then to assume that there is another lake further eastward to the north-east of Esmeralda?"

Löffling, the celebrated pupil of Linnæus, accompanied the last-named expedition to Cumana in the capacity of botanist. He died on the 22nd of February. 1756, at the mission of Santa Eulalia de Murucuri (somewhat to the south of the confluence of the Orinoco and Caroni), after traversing the missions on the Piritu and Caroni. The documents of which Bauza speaks are the same as those on which the great map of De la Cruz Olmedilla is based. They have served as the foundation of all the maps of South America, which appeared in England, France, and Germany, before the end of the last century; and have also served for the two maps executed in 1756 by Father Caulin, the historiographer of Solano's expedition, and by M. de Surville, Keeper of the Archives in the Secretary of State's Office at Madrid, who was but an unskilful compiler. The contradictions abounding in these maps show the little reliance that can be placed on the results of this expedition. Nay more, Father Caulin, above referred to, acutely details the circumstances which gave rise to this fable of the lake of Parime; and the map of Surville, which accompanies his work, not only restores this lake, under the name of the White Lake, and the Mar Dorado, but indicates another smaller one, from which flow partly by means of collateral branches, the Orinoco, Siapa, and Ocamo. I was able to convince myself on the spot of the following facts well known in the missions; that Don José Solano did not do more than cross the cataracts of Atures and Maypures; that he did not reach the confluence of the Guaviare and the Orinoco in 4° 3′ north lat., and 68° 9′ west long.; and that the astronomical instruments of the boundary expedition were neither carried to the isthmus of the Pimichin and the Rio Negro, nor to the Cassiquiare; and even on the Upper Orinoco, not beyond the mouth of the Atabapo. This vast extent of territory was not made the scene of any accurate observations before my journey, and has subsequently to Solano's expedition been traversed only by some few soldiers who had been sent on exploring expeditions; while Don Apolinario de Fuente, whoss journal I obtained from the archives of the province of Quixos, has gathered without discrimination everything from the fallacious narratives of the Indians that could flatter the credulity of the Governor Centurion. No member of the expedition had seen a lake, and Don Apolinario was unable to advance beyond the Cerro Yumarique and Gehette.

Although a line of separation, formed by the basin of the Rio Branco, is now established throughout the whole extent of the country, to which we are desirous of directing the inquiring zeal of travellers, it must yet be admitted, that our geographical knowledge of the district west of this valley between 62° and 66° long., has made no advance whatever for at least a century. The repeated attempts made by the Government of Spanish Guiana since the expeditions of Iturria and Solano, to reach and to pass over the Pacaraima Mountains, have been attended by very unimportant results. When the Spaniards, in proceeding to the missions of the Catalonian capuchins of Barceloneta, at the confluence of the Caroni and the Rio Paragua, ascended the last-named river southward to its junction with the Paraguamusi, they founded at this point the mission of Guirion, which, at first, bore the pompous appellation of Ciudad de Guirion. I place it in about 4° 30′ north latitude. From thence the Governor Centurion, in consequence of the exaggerated accounts given by two Indian chiefs, Paranacare and Arimuicapi, respecting the powerful tribe of the Ipurucotos, was excited to search for 'El Dorado,' and in carrying what were then called spiritual conquests still further, founded, beyond the

Pacaraima Mountains, the two villages of Santa Rosa and San Bautista de Caudacacla. The former was situate on the upper eastern bank of the Uraricapara, a tributary of the Uraricuera, which I find in the journal of Rodriguez under the name of the Rio Curaricara; the latter, at from 24 to 28 miles further east-south-east. The astronomo-geographer of the Portuguese Boundary Commission, Captain Don Antonio Pires de Sylva Pontes Leme, and the Captain of Engineers, Don Ricardo Franco d Almeida de Serra, who between 1787 and 1804, surveyed with the greatest care the whole course of the Rio Branco and its upper tributaries, call the most western part of the Uraricapara, "The Valley of Inundation." They place the Spanish mission of Santa Rosa in 3° 46′ north lat., and mark the route that leads from thence northward across the mountain chain to the Caño Anocapra, a branch of the Paraguamusi, which forms a connecting passage between the basin of the Rio Branco and that of the Caroni. Two maps of these Portuguese officers, embracing all the details of the trigometrical survey of the bends of the Rio Branco, the Uraricuera, the Tacutu, and the Mahu, were most kindly communicated to Colonel Lapie and myself by the Count o Linhares. These valuable unpublished documents, of which I have availed myself, are still in the hands of the learned geographer, who long since began to have them engraved at his own expense. The Portuguese sometimes call the whole of the Rio Branco by the name of Rio Parime, and sometimes limit this appellation to one branch only, the Uraricuera, somewhat below the Caño Mayari and above the old mission of San Antonio. As the words *Paragua* and *Parime* alike imply water, great water, lake, and sea, we cannot wonder at finding them so often repeated among tribes living at great distances from each other; as, for instance, by the Omaguas on the Upper Marañon, by the Western Guaranis, and by the Caribs. In all parts of the world, as I have already remarked, large rivers are called by those who live on their banks "the River," without any specific denomination. Paragua, the name of a branch of the Caroni, is also the term applied by the natives to the Upper Orinoco. The name Orinucu is Tamanakish; and Diego de Ordaz first heard it used in the year 1531, when he ascended to the mouth of the Meta. Besides the Valley of Inundation above mentioned we find other large pieces of water between the Rio Xumuru and the Parime. One of these bays is a branch of the Tacutu, and the other of the Uraricuera. Even at the base of the Pacaraima Mountains the rivers are subject to great periodical overflowings; and the Lake Amucu, of which we shall subsequently speak more fully, exhibits exactly the same character at the commencement of the plains. The Spanish missions, Santa Rosa and San Bautista de Caudacacla, or Cayacaya, founded in the years 1770 and 1773, by the Governor Don Manuel Centurion, were destroyed before the close of the last century; and since that time, no new attempt has been made to advance from the basin of the Caroni to the southern declivity of the Pacaraima Mountains.

The territory east of the valley of the Rio Branco has of late years been made the subject of several successful explorations. Mr. Hillhouse navigated the Massaruni as far as the Bay of Caranang, whence, as he says, a path would lead the traveller, in two days, to the source of the Massaruni; and, in three days, to the tributaries of the Rio Branco. With respect to the windings of the great river Massaruni, described by Mr. Hillhouse, he himself observes, in a letter addressed to me from Demerara, 1st January, 1831, that "the Massaruni, reckoning from its sources, flows first to the west, then for one degree of latitude to the north; afterwards nearly 200 miles eastward; and, finally, to the north and north-north-east till it merges in the Essequibo." As Mr. Hillhouse was unable to reach the southern declivity of the Pacaraima chain, he was not acquainted with the Amucu Lake; and he says himself, in his printed report, that "from the accounts given him by the Accaouais, who are continually traversing the country between the shore and the Amazon River, he is convinced there is no lake in this district." This assertion occasioned me some surprise, as it was directly opposed to the views I had previously formed regarding the Lake Amucu, from which flows the Caño Pirara, according to the accounts given by the travellers Hortsmann, Santos, and Rodriguez (and which had inspired me with the more confidence, because they entirely coincide with the recent Portuguese manuscript charts). Finally, after five years of expectation, Schomburgk's journey has removed all farther doubt.

"It is difficult to believe," says Mr. Hillhouse, in his interesting memoir on the Massaruni, "that the tradition of a large inland sea is wholly unfounded According to my views, the following circumstance may have given rise to the belief in the existence of the fabulous lake of the Parime. At some distance from the rocky fall of Teboco the waters of the Massaruni present to the eye as little motion as the calm surface of a lake. If at a more or less remote period the horizontal granitic strata of Teboco had been totally compact and without fissures, the waters must have been at least 50 feet above their present level, and there would have been formed an immense lake 10 or 12 miles in width, and 1500 or 2000 miles in length."* The extent of this supposed inundation is not the only reason which prevents me from acceding to this explanation; for I have seen plains (Llanos), where, during the rainy season, the overflowing of the tributaries of the Orinoco annually covered a surface of 6400 square miles. The labyrinth of ramifications between the Apure, Arauca, Capanaparo, and Sinaruco (see maps 17 and 18 of my Physical Atlas), is then wholly lost sight of; the configuration of the river beds can no longer be traced, and the whole appears like one vast lake. But the locality of the fabulous Dorado, and of the Lake Parime, belongs historically to quite a different part of Guiana, namely, that lying south of the Pacaraima mountains. This myth of the White Sea and of the Dorado of the Parime, has arisen, as I endeavoured thirty years ago to show in another work, from the appearance of the micaceous rocks of the Ucucuamo, the name *Rio Parime* (Rio Branco) the inundations of the tributaries; and especially from the existence of the lake *Amucu*, which is in the neighbourhood of the Rio Rupunuwini (Rupunuri), and is connected by means of the Pirara with the Rio Parime.

I have had much pleasure in finding that the travels of Sir Robert Schomburgk have fully corroborated these early views. The section of his map which gives the course of the Essequibo and of the Rupunuri is quite new, and of great importance in a geographical point of view. It places the Pacaraima chain between 3° 52′ and 4° north lat., while I had given its mean direction from 4° to 4° 10′. The chain reaches the confluence of the Essequibo and Rupunuri in 3° 57′ north lat., and 58° 1′ west longitude; I had placed it half a degree too

* *Nouvelles Annales des Voyages*, 1836, Sept. p. 316.

far to the north. Schomburgk calls the last-named river Rupununi, according to the pronunciation of the Macusis; and gives as the synonymes Rupunuri, Rupunuwini and Opununy, which have arisen from the difficulty the Carib tribes of these districts find in pronouncing the letter "r." The position of the lake Amucu and its relations to the Mahu (Maou) and Tacutu (Tacoto) correspond perfectly with my map of Colombia drawn in 1825. We agree equally well regarding the latitude of the lake of Amucu, for while he places it in 3° 33′, I considered it to be in 3° 35′; the Caño Pirara (Pirarara) which connects the Amucu with the Rio Branco, flows from it towards the north, and not to the west as I had marked it. The Sibarana of my map, the sources of which Hortsmann placed to the north of the Cerro Ucucuamo near a fine mine of rock crystal, is the Siparuni of Schomburgk's map. His Waa-Ekuru is the Tavaricaru of the Portuguese geographer Pontes Leme, and is the branch of the Rupunuri which lies the nearest to the lake of Amucu.

The following remarks from the report of Sir Robert Schomburgk throw some light on the subject in question. "The lake of Amucu," says this traveller, "is without doubt the nucleus of the Lake of Parime and of the supposed White Sea. In December and January, when we visited it, it was scarcely a mile in length, and was half covered with reeds." The same observation occurs on D'Anville's map of 1748. "The Pirara flows from the lake to the W.N.W. of the Indian village of Pirara and falls into the Maou or Mahu. The last-named river rises, according to the information given me, north of the ridge of the Pacaraima mountains, which in their eastern portion do not attain a greater elevation than about 1600 feet. The sources of the river are on a plateau, from whence it is precipitated in a beautiful waterfall, known as the Corona. We were on the point of visiting this fall, when on the third day of our excursion to the mountains, the indisposition of one of my companions compelled me to return to the station at the lake Amucu. The Mahu has black coffee-coloured water, and its current is more impetuous than that of the Rupunuri. In the mountains through which it pursues its course it is about 60 yards in breadth. Its environs are here extremely picturesque. This valley as well as the bank of the Buroburo, which flows into the Siparuni, are inhabited by the Macusis.

In April the whole Savannahs are overflowed, and then present the peculiar phenomenon of the waters belonging to two different river basins commingling together. It is probable that the vast extent of this temporary inundation may have given rise to the fable of the lake of Parime. During the rainy season a water communication is formed in the interior of the country between the Essequibo, the Rio Branco, and the Gran Para. Some groups of trees, rising like Oases on the sand-hills of the Savannahs, present, at the time of the inundation, the appearance of islands scattered over a lake; and these are without doubt the Ipomucena islands of Don Antonio Santos."

In D'Anville's manuscripts, which his heirs kindly allowed me to examine, I find that Hortsmann of Hildesheim, who described these districts with great care, saw a second Alpine lake, which he places two day's journey above the confluence of the Mahu with the Rio Parime (Tacutu?). It is a black water lake, situated on the summit of a mountain. He explicitly distinguishes it from the lake of Amucu, which he describes as "covered with rushes." The descriptions given by Hortsmann and Santos coincide with the Portuguese manuscript maps of the Marine Bureau at Rio Janeiro, in not indicating the existence of an uninterrupted connection between the Rupunuri and the lake of Amucu. In D'Anville's maps of South America, the rivers are better drawn in the first edition published in 1748, than in the more extensively circulated one of 1760. Schomburgk's travels fully confirm the independence of the basin of the Rupunuri and Essequibo; but he draws attention to the fact that, during the rainy season, the Rio Waa-Ekuru, a tributary of the Rupunuri, is in connection with the Caño Pirara. Such is the condition of these river-channels, which are still but little developed, and almost entirely without separating ridges.

The Rupunuri and the village of Anai, 3° 56′ north latitude, 58° 34′ west longitude, are at present recognised as the political boundaries between the British and Brazilian domains in these desert regions. Sir Robert Schomburgk was compelled by severe illness to make a protracted stay at Anai. He bases his chronometrical determinations of the position of the lake of Amucu on the mean of many lunar distances, east and west, which he measured during his sojourn at Anai. His determinations of longitude for these points of the Parime are in general one degree more east than those in my map of Colombia. While I am far from calling in question the result of these lunar observations taken at Anai, I may be allowed to observe that the calculation of these distances is of importance, when it is desired to carry the comparison from the lake of Amucu to Esmeralda, which I found in 66° 19′ west longitude.

Thus then we see the great *Mar de la Parima*, (which it was so difficult to remove from our maps, that even after my return from America it was still supposed to be 160 miles in length,) reduced by recent investigations to the lake of Amucu, measuring only two or three miles in circumference. The illusions entertained for nearly two hundred years, and which in the last Spanish expedition, in 1775, for the discovery of El Dorado, cost several hundred lives, have finally terminated by enriching geography with some few results. In the year 1512 thousands of soldiers perished in the expedition, undertaken by Ponce de Leon, to discover the "Fountain of Youth," on one of the Bahama Islands, called Binimi, which is hardly to be found on any of our maps. This expedition led to the conquest of Florida, and to the knowledge of the great oceanic current, or gulf-stream, which flows through the Straights of Bahama. The thirst after gold, and the desire of rejuvenescence—the *Dorado* and the *Fountain of Youth*—stimulated, to an almost equal extent, the passions of mankind.

(10) p. 161—"*The Piriguao, one of the noblest forms of the Palm.*"

Compare Humboldt, Bonpland, and Kunth, *Nova Genera Plantarum*, and *Plant. æquinoct.*, t. i. p. 315.

(11) p. 171—"*The grave of an extinct race.*"

During my stay in the forests of the Orinoco, researches were being made, by royal command, in reference to these bone-caves. The missionary of the cataracts had been falsely accused of having discovered in these caves treasures which the Jesuits had concealed there prior to their flight.

(12) p. 172—"*When his language perished with him.*"

The parrot of the Atures has been made the subject of a

charming poem by my friend Professor Ernst Curtius, the tutor of the promising young Prince Friedrick Wilhelm of Prussia. The author will forgive me for closing the present section of the "Views of Nature" with this poem, which was not designed for publication, and was communicated to me by letter.

---

## THE PARROT OF ATURES.

Where, through deserts wild and dreary,
  Orinoco dashes on,
Sits a Parrot old and weary,
  Like a sculptur'd thing of stone.

Through its rocky barriers flowing,
  Onward rolls the foaming stream;
Waving palms on high are glowing
  In the sun's meridian beam.

Ceaselessly the waves are heaving,
  Sparkling up in antic play;
While the sunny rays are weaving
  Rainbows in the feathery spray.

Where yon billows wild are breaking,
  Sleeps a tribe for evermore,
Who, their native land forsaking,
  Refuge sought on this lone shore.

As they lived, free, dauntless ever,
  So the brave Aturians died;
And the green banks of the river
  All their mortal relics hide.

Yet the Parrot, ne'er forgetting
  Those who loved him, mourns them still,
On the stone his sharp beak whetting,
  While the air his wailings fill.

Where are now the youths who bred him,
  To pronounce their mother tongue,—
Where the gentle maids who fed him,
  And who built his nest when young?

All, alas! are lifeless lying,
  Stretch'd upon their grassy bed;
Nor can all his mournful crying,
  E'er awake the slumbering dead.

Still he calls with voice imploring,
  To a world that heeds him not;
Nought replies but waters roaring—
  No kind soul bewails his lot.

Swift the savage turns his rudder,
  When his eyes the bird behold;
None e'er saw without a shudder
  That Aturian Parrot old!

---

# THE NOCTURNAL LIFE OF ANIMALS

IN THE

# PRIMEVAL FOREST.

---

If the faculty of appreciating nature, in different races of man, and if the character of the countries they now inhabit, or have traversed in their earlier migrations, have more or less enriched the respective languages by appropriate terms, expressive of the forms of mountains, the state of vegetation, the appearances of the atmosphere, and the contour and grouping of the clouds, it must be admitted that by long use and literary caprice many of these designations have been diverted from the sense they originally bore. Words have gradually been regarded as synonymous, which ought to have remained distinct; and languages have thus lost a portion of the expressiveness and force which might else have imparted a physiognomical character to descriptions of natural scenery. As an evidence of the extent to which a communion with nature, and the requirements of a laborious nomadic life, may enrich language, I would recall the abundance of characteristic denominations employed in Arabic and Persian, to distinguish plains, steppes, and deserts (1), according as they are entirely bare, covered with sand, or intersected by tabular masses of rock; or as they are diversified by spots of pasture land and extended tracts of social plants. The old Castilian dialects are no less remarkable (2) for the copiousness of their terms descriptive of the physiognomy of mountains, especially in reference to those features which recur in all regions of the earth, and which proclaim afar off the nature of the rock. As the declivities of the Andes and the mountainous parts of the Canaries, the Antilles, and the Philippines, are all inhabited by races of Spanish descent; and as the nature of the soil has there influenced the mode of life of the inhabitants to a greater degree than in other parts of the world, excepting perhaps in the Himalaya and the Thibetian Highlands; so also the designations expressive of the forms of mountains in trachytic, basaltic, and porphyritic districts, as well as in schistose, calcareous, and sandstone formations, have been happily preserved in daily use. Under such circumstances, newly formed words become incorporated with the common stock. Speech acquires life from everything which bears the true impress of nature, whether it be by the definition of sensuous impressions received from the external world, or by the expression of thoughts and feelings that emanate from our inner being.

In descriptions of natural phenomena, as well as in the choice of the expressions employed, this truth to nature should be especially kept in view. The object will be the best attained by simplicity in the narration of whatever we have ourselves observed and experienced, and by closely examining the locality with which the subject-matter is connected. Generalisation of physical views, and the enumeration of results, belong principally to the study of the Cosmos, which, indeed, must still be regarded as an inductive science; but the vivid delineation of organic forms (animals and plants,) in their picturesque and local relations to the multiform surface of the earth, although limited to a small section of terrestrial life, still affords materials for this study. It acts as a stimulus to the mind wherever it is capable of appreciating the great phenomena of nature in an æsthetic point of view.

To these phenomena belongs especially the boundless forest district which, in the torrid zone of South America, con-

connects the river basins of the Orinoco and the Amazon. This region deserves, in the strictest sense of the word, to be called a *primeval* forest—a term that has, in recent times, been so frequently misapplied. Primeval (or primitive), as applied to a forest, a nation, or a period of time, is a word of rather indefinite signification, and generally but of relative import. If every wild forest, densely covered with trees, on which man has never laid his destroying hand, is to be regarded as a primitive forest, then the phenomenon is common to many parts both of the temperate and the frigid zones; if, however, this character consists in impenetrability, through which it is impossible to clear with the axe, between trees measuring from 8 to 12 feet in diameter, a path of any length, primitive forests belong exclusively to tropical regions. This impenetrability is by no means, as is often erroneously supposed in Europe, always occasioned by the interlaced climbing "lianes," or creeping plants, for these often constitute but a very small portion of the underwood. The chief obstacles are the shrub-like plants which fill up every space between the trees, in a zone where all vegetable forms have a tendency to become arborescent. If travellers, the moment they set foot in a tropical region, and even while on islands, in the vicinity of the sea-coast, imagine that they are within the precincts of a primeval forest, the misconception must be ascribed to their ardent desire of realizing a long-cherished wish. Every tropical forest is not *primeval* forest. I have scarcely ever used the latter term in the narrative of my travels; although, I believe, that of all investigators of nature now living, Bonpland, Martius, Pöppig, Robert and Richard Schomburgk, and myself, have spent the longest period of time in primeval forests in the interior of a great continent.

Notwithstanding the striking richness of the Spanish language in designations, (descriptive of natural objects, of which I have already spoken), yet one and the same word *monte* is employed for a mountain and a forest, for *cerro* (*montaña*), and for *selva*. In a work on the true breadth and the greatest extension of the chain of the Andes towards the east, I have shown how this two-fold signification of the word *monte* has led to the error, in a fine and extensively circulated English map of South America, of marking ranges of high mountains in districts occupied only by plains. Where the Spanish map of La Cruz Olmedilla, which formed the basis of so many others, indicated Cacao Woods, *Montes de Cacao* (3), Cordilleras were supposed to exist, although the Cacao-tree affects only the hottest of the low lands.

If we comprehend, in one general view, the woody region which embraces the whole of South America, between the grassy plains of Venezuela (*los Llanos de Caracas*) and the Pampas of Buenos Ayres, lying between 8° north and 19° south latitude, we perceive that this connected *Hylæa* of the tropical zone is unequalled in extent by any other on the surface of the earth. Its area is about twelve times that of Germany. Traversed in all directions by rivers, some of whose direct and indirect tributary streams (as well those of the second as of the first order) surpass the Danube and Rhine in the abundance of their waters, it owes the wonderful luxuriance of its vegetation to the two-fold influence of great humidity and high temperature. In the temperate zone, particularly in Europe and Northern Asia, forests may be named from particular genera of trees which grow together as social plants (*plantæ sociales*), and form separate woods. In the Oak, Pine, and Birch forests of the northern regions, and in the Linden or Lime Woods of the eastern, there usually predominates only one species of Amentaceæ, Coniferæ or Tiliaceæ; while sometimes a single species of Piniferæ is intermixed with trees of deciduous foliage. Such uniformity of association is unknown in tropical forests. The excessive variety of their rich sylvan flora renders it vain to ask, of what do the primeval forests consist. Numberless families of plants are here crowded together; and even in small spaces, plants of the same species are rarely associated. Every day, and with every change of place, new forms present themselves to the traveller's attention; often flowers, beyond his reach, although the shape of the leaf and the ramifications of the plant excite his curiosity.

The rivers, with their innumerable branches, are the only means of traversing the country. Astronomical observations, or in the absence of these, determinations by compass of the bends of the rivers, between the Orinoco, the Cassiquiare, and the Rio Negro, have shewn that two lonely mission-stations might be situated only a few miles apart, and yet the monks thereof, in visiting each other would require a day and a half to make the passage in their hollow-tree canoes, along the windings of small streams. The most striking evidence of the impenetrability of some portions of these forests, is afforded by a trait in the habits of the American tiger, or panther-like Jaguar. While the introduction of European horned cattle, horses, and mules, has yielded so abundant a supply of food to the beasts of prey in the extensive grassy and treeless plains of Varinas, Meta, and Buenos Ayres; that these animals, (owing to the unequal contest between them and their prey,) have considerably increased since the discovery of America; other individuals of the same species lead a toilsome life in the dense forests contiguous to the sources of the Orinoco. The distressing loss of a large mastiff, the faithful companion of our travels, while we were bivouacking near the junction of the Cassiquiare with the Orinoco, induced us on our return from the insect-swarming Esmeralda, to pass another night on the same spot (uncertain whether he was devoured by a tiger) where we had already long sought him in vain. We again heard in the immediate neighbourhood the cries of the Jaguar, probably the very same animal to which we owed our loss. As the cloudy state of the sky rendered it impossible to conduct astronomical observations, we made our interpreter (*lenguaraz*) repeat to us what the natives, our boatmen, related of the tigers of the country.

The so called black Jaguar is, as we learnt, not unfrequently found among them. It is the largest and most blood-thirsty variety, and has a dark brown skin marked with scarcely distinguishable black spots. It lives at the foot of the mountain ranges of Maraguaca and Unturan. "The love of wandering, and the rapacity of the Jaguars," said our Indian narrator, one of the Durimond tribe; "often lead them into such impenetrable thickets of the forest, that they can no longer hunt on the ground, and then live for a long time in the trees—the terror of the families of monkeys, and of the prehensile-tailed viverra. (*Cercoleptes.*)"

The journal which I wrote at the time in German, and from which I borrow these extracts, was not entirely exhausted in the narrative of my travels (published in French). It contains a circumstantial description of the nocturnal life of animals; I might say, of their nocturnal voices in the tropical forests. And this sketch seems to me to be especially adapted to constitute one of the chapters of the *Views of Nature.* That which is written down on the spot, or soon after the impression of the phenomena has been received, may at least claim to possess more freshness than what is produced by the recollection of long passed events.

We reached the bed of the Orinoco by descending from west to east along the Rio Apure, whose inundations I have noticed in the sketch of the Deserts and Steppes. It was the period of low water, and the average breadth of the Apure was only a little more than 1200 feet; while the Orinoco, at its confluence with the Apure (near the granite rocks of Curiquima, where I was able to measure a base-line), was still upwards of 12,180 feet. Yet this point (the rock of Curiquima,) is 400 miles in a straight line from the sea and from the delta of the Orinoco. Some of the plains, watered by the Apure and the Payara, are inhabited by Yaruros and Achaguas, who are called savages in the mission-villages established by the monks, because they will not relinquish their independence. In reference to social culture, they however occupy about the same scale as those Indians, who, although baptized and living "under the bell" (*baxo la campana*), have remained strangers to every form of instruction and cultivation.

On leaving the Island del Diamante, where the Zambos, who speak Spanish, cultivate the sugar-cane, we entered into a grand and wild domain of nature. The air was filled with countless flamingoes (*Phœnicopterus*) and other water-fowl, which seemed to stand forth from the blue sky like a dark cloud in ever-varying outlines. The bed of the river had here contracted to less than 1000 feet, and formed a perfectly straight canal, which was inclosed on both sides by thick woods. The margin of the forest presents a singular spectacle. In front of the almost impenetrable wall of colossal trunks of Cæsalpinia, Cedrela, and Desmanthus, there rises with the greatest regularity on the sandy bank of the river, a low hedge of Sauso, only four feet high; it consists of a small shrub, *Hermesia castanifolia*, which forms a new genus (4) of the family of Euphorbiaceæ. A few slender, thorny palms, called by the Spaniards Piritu and Corozo (perhaps species of *Martinezia* or *Bactris*) stand close alongside; the whole resembling a trimmed garden hedge, with gate-like openings at considerable distances from each other, formed undoubtedly by the large four-footed animals of the forests, for convenient access to the river. At sunset, and more particularly at break of day, the American Tiger, the Tapir, and the Peccary (*Pecari*, *Dicotyles*) may be seen coming forth from these openings accompanied by their young, to give them drink. When they are disturbed by a passing Indian canoe, and are about to retreat into the forest, they do not attempt to rush violently through these hedges of Sauso, but proceed deliberately along the bank, between the hedge and river, affording the traveller the gratification of watching their motions for sometimes four or five hundred paces, until they disappear through the nearest opening. During a seventy-four days' almost uninterrupted river navigation of 1520 miles up the Orinoco, to the neighbourhood of its sources, and along the Cassiquiare, and the Rio Negro—during the whole of which time we were confined to a narrow canoe—the same spectacle presented itself to our view at many different points, and, I may add, always with renewed excitement. There came to drink, bathe, or fish, groups of creatures belonging to the most opposite species of animals; the larger mammalia with many-coloured herons, palamedeas with the proudly-strutting curassow (*Crax Alector*, *C. Pauxi*). "It is here as in Paradise" (*es como en el Paradiso*), remarked with pious air our steersman, an old Indian, who had been brought up in the house of an ecclesiastic. But the gentle peace of the primitive golden age does not reign in the paradise of these American animals, they stand apart, watch, and avoid each other. The Capybara, a cavy (or river-hog) three or four feet long (a colossal repetition of the common Brazilian cavy, (*Cavia Aguti*), is devoured in the river by the crocodile, and on the shore by the tiger. They run so badly, that we were frequently able to overtake and capture several from among the numerous herds.

Below the mission of Santa Barbara de Arichuna we passed the night as usual in the open air, on a sandy flat, on the bank of the Apure, skirted by the impenetrable forest. We had some difficulty in finding dry wood to kindle the fires with which it is here customary to surround the bivouac, as a safeguard against the attacks of the Jaguar. The air was bland and soft, and the moon shone brightly. Several crocodiles approached the bank; and I have observed that fire attracts these creatures as it does our crabs and many other aquatic animals. The oars of our boats were fixed upright in the ground, to support our hammocks. Deep stillness prevailed, only broken at intervals by the blowing of the fresh-water dolphins (5), which are peculiar to the river net-work of the Orinoco (as, according to Colebrooke, they are also to the Ganges, as high up the river as Benares); they followed each other in long tracks.

After eleven o'clock, such a noise began in the contiguous forest, that for the remainder of the night all sleep was impossible. The wild cries of animals rung through the woods. Among the many voices which resounded together, the Indians could only recognise those which, after short pauses, were heard singly. There was the monotonous, plaintive, cry of the Aluates (howling monkeys), the whining, flute-like notes of the small sapajous, the grunting murmur of the striped nocturnal ape (6) (*Nyctipithecus trivirgatus*, which I was the first to describe), the fitful roar of the great tiger, the Cuguar or maneless American lion, the peccary, the sloth, and a host of parrots, parraquas (*Ortalides*), and other pheasant-like birds. Whenever the tigers approached the edge of the forest, our dog, who before had barked incessantly, came howling to seek protection under the hammocks. Sometimes the cry of the tiger resounded from the branches of a tree, and was then always accompanied by the plaintive piping tones of the apes, who were endeavouring to escape from the unwonted pursuit.

If one asks the Indians why such a continuous noise is heard on certain nights, they answer, with a smile, that "the animals are rejoicing in the beautiful moonlight, and celebrating the return of the full moon." To me the scene appeared rather to be owing to an accidental, long-continued, and gradually increasing conflict among the animals. Thus, for instance, the jaguar will pursue the peccaries and the tapirs, which, densely crowded together, burst through the barrier of tree-like shrubs which opposes their flight. Terrified at the confusion, the monkeys on the tops of the trees join their cries with those of the larger animals. This arouses the tribes of birds who build their nests in communities, and suddenly the whole animal world is in a state of commotion. Further experience taught us, that it was by no means always the festival of moonlight that disturbed the stillness of the forest; for we observed that the voices were loudest during violent storms of rain, or when the thunder echoed and the lightning flashed through the depths of the woods. The good-natured Franciscan monk who (notwithstanding the fever from which he had been suffering for many months), accompanied us through the cataracts of Atures and Maypures to San Carlos, on the Rio Negro, and to the Brazilian coast, used to say, when apprehensive of a storm at night, "May Heaven grant a quiet night both to us and to the wild beasts of the forest!"

A singular contrast to the scenes I have here described, and which I had repeated opportunities of witnessing, is presented by the stillness which reigns within the tropics at the noontide of a day unusually sultry. I borrow from the same journal the description of a scene at the Narrows of Baraguan. Here the Orinoco forms for itself a passage through the western part of the mountains of the Parime. That which is called at this remarkable pass a Narrow (*Angostura del Baraguan*), is, however, a basin almost 5700 feet in breadth. With the exception of an old withered stem of Aubletia (*Apeiba Tiburbu*), and a new Apocinea (*Allamanda Salicifolia*), the barren rocks were only covered with a few silvery croton shruhs. A thermometer observed in the shade, but brought within a few inches of the lofty mass of granite rock, rose to more than 122° Fahr. All distant objects had wavy undulating outlines, the optical effect of the *mirage*. Not a breath of air moved the dust-like sand. The sun stood in the zenith; and the effulgence of light poured upon the river, and which, owing to a gentle ripple of the waters, was brilliantly reflected, gave additional distinctness to the red haze which veiled the distance. All the rocky mounds and naked boulders were covered with large, thick-scaled Iguanas, Gecko-lizards, and spotted Salamanders. Motionless, with uplifted heads and widely extended mouths, they seemed to inhale the heated air with ecstasy. The larger animals at such times take refuge in the deep recesses of the forest, the birds nestle beneath the foliage of the trees, or in the clefts of the rocks; but if in this apparent stillness of nature we listen closely for the faintest tones, we detect, a dull, muffled sound, a buzzing and humming of insects close to the earth, in the lower strata of the atmosphere. Everything proclaims a world of active organic forces. In every shrub, in the cracked bark of trees, in the perforated ground inhabited by hymenopterous insects, life is everywhere audibly manifest. It is one of the many voices of nature revealed to the pious and susceptible spirit of man.

---

## ILLUSTRATIONS AND ADDITIONS.

### (1) p. 191.—" *Characteristic denominations in Arabic and Persian.*"

More than twenty words might be cited by which the Arabs distinguish between a Steppe (tanufah), according as it may be a Desert without water, entirely bare, or covered with siliceous sand, and interspersed with spots of pasture land (Sahara, Kafr, Mikfar, Tih, Mehme). Sahl is a depressed plain; Dakkah a desolate elevated plateau. In Persian Beyaban is an arid sandy waste (as the Mongolian Gobi and the Chinese Han-hai and Scha-mo); Yaila is a Steppe covered with grass rather than with low-growing plants (like the Mongolian Küdah, the Turkish Tala or Tschol, and the Chinese Huang). Deschti-reft is a naked elevated plateau.*

### (2) p. 191.—" *The old Castilian dialects.*"

Pico, picacho, mogote, cucurucho, espigon, loma tendida, mesa, panecillo, farallon, tablon, peña, peñon, peñasco, peñoleria, roca partida, laxa, cerro, sierra, serrania, cordillera, monte, montaña, montañuela, cadena de montes, los altos, malpais, reventazon, bufa, &c.

### (3) p. 194.—" *Where the map had indicated Montes de Cacao.*"

On the range of hills from which the lofty Andes de Cuchao have originated, see my *Relation historique*, t. iii. p. 238.

### (4) p. 197.—" *Hermesia.*"

The genus Hermesia, the Sauso, has been described by Bonpland, and is delineated in our *Plantes équinoxiales*, t. i. p. 162, tab. xlvi.

### (5) p. 199.—" *The fresh-water dolphin.*"

These are not sea dolphins, which, like some species of Pleuronectes (flat fish which invariably have both eyes on one side of the body), ascend the rivers to a great distance, as,

* Humboldt, *Relation historique*, t. ii. p. 158.

for instance, the Limande (*Pleuronectes Limanda*), which is found as far inland as Orleans. Some forms of sea fish, as the dolphin and skate (*Raia*), are met with in the great rivers of both continents. The fresh-water dolphin of the Apure and the Orinoco differs specifically from the *Delphinus gangeticus* as well as from all sea dolphins.*

(6) p. 199.—" *The striped nocturnal monkey.*"

This is the Douroucouli or Cusi-cusi of the Cassiquiare which I have elsewhere described as the *Simia trivirgata*,† from a drawing made by myself of the living animal. We have since seen the nocturnal monkey living in the menagerie of the Jardin des Plantes at Paris.‡ Spix also met with this remarkable little animal on the Amazon River and called it *Nyctipithecus vociferans*.

* See my *Relation historique*, t. ii. pp. 223, 239, 406–413.

† *Recueil d'Observations de Zoologie et d'Anatomie comparée*, t. i. pp. 306–311, tab. xxviii.

‡ *Op. cit.*, t. ii. p. 340.

POTSDAM, JUNE 1849.

## HYPSOMETRIC ADDENDA.

I AM indebted to Mr. Pentland, whose scientific labours have thrown so much light on the geology and geography of Bolivia, for the following determinations of position, which he communicated to me in a letter from Paris (October 1848), subsequent to the publication of his great map.

| Nevado of Sorata, or Ancohuma. | South Latitude. | Longitude. | Height. |
|---|---|---|---|
| South Peak . . | 15° 51′ 33″ | 68° 33′ 55″ | 21,286 |
| North Peak . . | 15° 49′ 18″ | 68° 33′ 52″ | 21,043 |
| Illimani. | | | |
| South Peak . . | 16° 38′ 52″ | 67° 49′ 18″ | 21,145 |
| Middle Peak . . | 16° 38′ 26″ | 67° 49′ 17″ | 21,094 |
| North Peak . . | 16° 37′ 50″ | 67° 49′ 39″ | 21,060 |

The numbers representing the heights are, with the exception of the unimportant difference of a few feet in the South Peak of Illimani, the same as those in the map of the Lake of Titicaca. A sketch of the Illimani, as it appears in all its majesty from La Paz, was given at an earlier date by Mr. Pentland in the Journal of the Royal Geographical Society.* But this was five years after the publication of the first measurements in the *Annuaire du Bureau des Longitudes* for 1830, p. 323, which results I myself hastened to disseminate in Germany.† The Nevado de Sorata lies to the east of the village of Sorata or Esquibel, and is called in the Ymarra language, according to Pentland, Ancomani, Itampu, and Illhampu. In *Illimani* we recognize the Ymarra word *illi*, snow.

If, however, in the *eastern* chain of Bolivia the Sorata was long assumed to be 3962 feet, and the Illimani 2851 feet too high, there are in the western chain of Bolivia, according to Pentland's map of Titicaca (1848), four peaks east of Arica between the latitudes 18° 7′ and 18° 25′, all of which exceed Chimborazo in height, which itself is 21,422 feet.

These four peaks are:—

| | English feet. | French feet. |
|---|---|---|
| Pomarape . . | 21,700 . . | 20,360 |
| Gualateiri . . | 21,960 . . | 20,604 |
| Parinacota . . | 22,030 . . | 20,670 |
| Sahama . . . | 22,350 . . | 20,971 |

* Vol. v. (1835), p. 77.

† Hertha, *Zeitschrift für Erd und Völkerkunde*, von Berghaus, bd. xiii. 1829, s. 3–29.

Berghaus has applied to the chains of the Andes in Bolivia, the investigation which I published* regarding the proportion, which varies extremely in different mountain-chains, of the mountain ridge (the mean height of the passes), to the highest summits (or the culminating points). He finds,† according to Pentland's map, that the mean height of the passes in the eastern chain is 13,505, and in the western chain 14,496 feet. The culminating points are 21,285 and 22,350 feet; consequently the ratio of the height of the ridge to that of the highest summit is, in the eastern chain, as 1 : 1·57, and in the western chain as 1 : 1.54. This ratio, which is, as it were, the measure of the subterranean upheaving force, is very similar to that in the Pyrenees, but very different from the plastic form of the Alps, the mean height of whose passes is far less in comparison with the height of Mont Blanc. In the Pyrenees these ratios are as 1 : 1·43, and in the Alps as 1 : 2·09.

But, according to Fitzroy and Darwin, the height of the Sahama is still surpassed by 848 feet by that of the volcano Aconcagua (south lat. 32° 39′), in the north-east of Valparaiso in Chili. The officers of the expedition of the Adventure and Beagle found, in August 1835, that the Aconcagua was between 23,000 and 23,400 feet in height. If we reckon it at 23,200 feet it is 1776 feet, higher than Chimborazo.‡ According to more recent calculations,§ Aconcagua is determined to be 23,906 feet.

Our knowledge regarding the systems of mountains, which, north of the parallels of 30° and 31°, are distinguished as the *Rocky Mountains* and the *Sierra Nevada of California*, has been vastly augmented during the last few years in the astronomico-geographical, hypsometric, geognostic, and botanical departments, by the excellent works of Charles Frémont,‖ of Dr. Wislizenus,¶ and of Lieutenants Abert and Peck.** There

* *Annales des Sciences Naturelles*, t. iv. 1825, pp. 225–253.

† Berghaus, *Zeitschrift für Erdkunde*, band. ix. s. 322–326.

‡ Fitzroy, *Voyages of the Adventure and Beagle*, 1839, vol. ii. p. 481; Darwin, *Journal of Researches*, 1845, pp. 253 and 291.

§ Mary Somerville, *Physical Geogr.*, 1849, vol. ii. 425.

‖ *Geographical Memoir upon Upper California, an illustration of his Map of Oregon and California*, 1848.

¶ *Memoir of a Tour in Northern Mexico, connected with Col. Doniphan's Expedition*, 1848.

** *Expedition on the Upper Arkansas*, 1845, and *Examination of New Mexico in* 1846 *and* 1847.

prevails throughout these North American works a scientific spirit deserving of the warmest acknowledgment. The remarkable plateau, referred to in p. 34, between the Rocky Mountains and the Sierra Nevada of California, which rises uninterruptedly from 4000 to 5000 French (4260 to 5330 English) feet high, and is termed the Great Basin, presents an interior closed river-system, thermal springs, and salt lakes. None of its rivers, Bear River, Carson River, and Humboldt River, find a passage to the sea. That which, by a process of induction and combination, I represented in my great map of Mexico, executed in 1804, as the Lake of Timpanogos, is the *Great Salt Lake* of Frémont's map. It is 60 miles long from north to south, and 40 miles broad, and it communicates with the fresh-water Lake of Utah, which lies at a higher level, and into which the Timpanogos or Timpanaozu River enters from the eastward, in lat. 40° 13. The fact of the Lake of Timpanogos not having been placed in my map sufficiently to the north and west, arose from the entire absence, at that period, of all astronomical determinations of position of Santa Fé in New Mexico. For the western margin of the lake the error amounts to almost fifty minutes, a difference of absolute longitude which will appear less striking when it is remembered that my itinerary map of Guanaxuato could only be based for an extent of 15° of latitude on determinations made by the compass (magnetic surveys), instituted by Don Pedro de Rivera.§ These determinations gave my talented and prematurely lost fellow-labourer, Herr Friesen, 105° 36′ as the longitude of Santa Fé, while, by other combinations, I calculated it at 104° 51′. According to actual astronomical determinations the true longitude appears to be 106°. The relative position of the strata of rock salt found in thick strata of red clay, south-east of the Great Salt Lake (Laguna de Timpanogos), with its many islands, and near the present Fort Mormon and the Utah Lake, is accurately given in my large map of Mexico. I may refer to the most recent evidence of the traveller who made the first trustworthy determinations of position in this region. "The mineral or rock salt, of which a specimen is placed in Congress Library, was found in the place marked by Humboldt in his map of New Spain (north-

* Humboldt, *Essai polit. sur la Nouvelle Espagne*, t. i. pp. 127–136.

ern half), as derived from the journal of the Missionary Father Escalante, who attempted (1777) to penetrate the unknown country from Santa Fé of New Mexico to Monterey of the Pacific Ocean. South-east of the Lake Timpanogos is the chain of the Wha-satch Mountains; and in this, at the place where Humboldt has written *Montagnes de sel gemme*, this mineral is found."*

A great historical interest is attached to this part of the highland, especially to the neighbourhood of the Lake of Timpanogos, which is probably identical with the Lake of Teguayo, the ancestral seat of the Aztecs. This people, in their migration from Aztlan to Tula, and to the valley of Tenochtitlan in Mexico, made three stations at which the ruins of *Casas grandes* are still to be seen. The first halting-place of the Aztecs was at the Lake of Teguayo, south of Quivira, the second on the Rio Gila, and the third not far from the Presidio de Llanos. Lieutenant Abert found on the banks of the Rio Gila the same immense quantity of elegantly painted fragments of delf and pottery scattered over a large surface of country, which, at the same place, had excited so much astonishment in the missionaries Francisco Garces and Pedro Fonte. From these products of the hand of man, it may be inferred that there was a time when a higher human civilization existed in this now desolate region. Repetitions of the singular architectural style of the Aztecs, and of their houses of seven stories, are at the present time to be found far to the east of the Rio Grande del Norte; as, for instance, at Taos.† The Sierra Nevada of California is parallel to the coast of the Pacific; but between the latitudes of 34° and 41°, between San Buenaventura and the Bay of Trinidad, there runs, west of the Sierra Nevada, a small coast chain whose culminating point, *Monte del Diablo*, is 3674 feet high. In the narrow valley, between this coast chain and the great Sierra Nevada, flow from the south the Rio de San Joaquin, and from the north the Rio del Sacramento. It is in the alluvial soil on the banks of the latter river that the rich gold-

* Frémont, *Geogr. Mem. of Upper California*, 1848, pp. 8 and 67; see also Humboldt, *Essai politique*, t. ii. p. 261.

† Compare Abert's *Examination of New Mexico*, in the *Documents of Congress*, No. 41, pp. 489 and 581-605, with my *Essai pol.*, t. ii. pp. 241-244.

washings occur, which are now proceeding with so much activity.

Besides the hypsometric levelling and the barometric measurements to which I have already referred (see page 33), between the mouth of the Kanzas River in the Missouri and the coast of the Pacific, throughout the immense expanse of 28° of longitude, Dr. Wislizenus has successfully prosecuted the levelling commenced by myself in the equinoctial zone of Mexico, to the north as far as to lat. 35° 38′, and consequently to Santa Fé del Nuevo Mexico. We learn with astonishment that the plateau which forms the broad crest of the Mexican Andes by no means sinks down to an inconsiderable height, as was long supposed to be the case. I give here, for the first time, according to recent measurements, the line of levelling from the city of Mexico to Santa Fé, which is within 16 miles from the Rio del Norte.

| | French feet. | English feet. | |
|---|---|---|---|
| Mexico | 7008 | 7469 | Ht. |
| Tula | 6318 | 6733 | Ht. |
| San Juan del Rio | 6090 | 6490 | Ht. |
| Queretaro | 5970 | 6362 | Ht. |
| Celaya | 5646 | 6017 | Ht. |
| Salamanca | 5496 | 5761 | Ht. |
| Guanaxuato | 6414 | 6836 | Ht. |
| Silao | 5546 | 5911 | Br. |
| Villa de Leon | 5755 | 6133 | Br. |
| Lagos | 5983 | 6376 | Br. |
| Aguas Calientes | 5875 | 6261 | Br. |
| San Luis Potosi | 5714 | 6090 | Br. |
| Zacatecas | 7544 | 8038 | Br. |
| Fresnillo | 6797 | 7244 | Br. |
| Durango | 6426 | 6848 | (Oteiza) |
| Parras | 4678 | 4985 | Ws. |
| Saltillo | 4917 | 5240 | Ws. |
| El Bolson de Mapimi | from 3600 to 4200 | 3836 4476 | Ws. |
| Chihuahua | 4352 | 4638 | Ws. |
| Cosiquiriachi | 5886 | 6273 | Ws. |
| Passo del Norte (on the Rio Grande del Norte) | 3577 | 3810 | Ws. |
| Santa Fé del Nuevo Mexico | 6612 | 7047 | Ws. |

The attached letters Ws., Br., and Ht., indicate the barometric measurements of Dr. Wislizenus, Obergrath Burkart, and myself. To the valuable memoir of Dr. Wislizenus there are appended three profile delineations of the country; one from Santa Fé to Chihuahua over Passo del Norte; one from Chihuahua over Parras to Reynosa; and one from Fort Independence (a little to the east of the confluence of the Missouri and the Kanzas River) to Santa Fé. The calculation is based on daily corresponding observations of the barometer, made by Engelmann at St. Louis, and by Lilly in New Orleans. If we consider that in the north and south direction the difference of latitude between Santa Fé and Mexico is more than 16°, and that, consequently, the distance in a direct meridian direction, independently of curvatures on the road, is more than 960 miles; we are led to ask whether, in the whole world, there exists any similar formation of equal extent and height (between 5000 and 7500 feet above the level of the sea). Four-wheeled waggons can travel from Mexico to Santa Fé. The plateau, whose levelling I have here described, is formed solely by the broad, undulating, flattened crest of the chain of the Mexican Andes; it is not the swelling of a valley between two mountain-chains, such as the "Great Basin" between the Rocky Mountains and the Sierra Nevada of California, in the Northern Hemisphere, or the elevated plateau of the Lake of Titicaca, between the eastern and western chains of Bolivia, or the plateau of Thibet, between the Himalaya and the Kuenlün, in the Southern Hemisphere.

---

# IDEAS

### FOR A

## PHYSIOGNOMY OF PLANTS.

When the active spirit of man is directed to the investigation of nature, or when in imagination he scans the vast fields of organic creation, among the varied emotions excited in his mind there is none more profound or vivid than that awakened by the universal profusion of life. Everywhere—even near the ice-bound poles,—the air resounds with the song of birds and with the busy hum of insects. Not only the lower strata, in which the denser vapours float, but also the higher and ethereal regions of the air, teem with animal life. Whenever the lofty crests of the Peruvian Cordilleras, or the summit of Mont Blanc, south of Lake Leman, have been ascended, living creatures have been found even in these solitudes. On the Chimborazo (1), which is upwards of eight thousand feet higher than Mount Etna, we saw butterflies and other winged insects. Even if they are strangers carried by ascending currents of air to those lofty regions, whither a restless spirit of inquiry leads the toilsome steps of man, their presence nevertheless proves that the more pliant organization of animals may subsist far beyond the limits of the vegetable world. The Condor (2), that giant among the vultures, often soared above us at a greater altitude than the summits of the Andes, and even higher than would be the Peak of Teneriffe were it piled upon the snow-crowned summits of the Pyrenees. Rapacity and the pursuit of the soft-woolled Vicuñas, which herd, like the chamois, on the snow-covered pastures, allure this powerful bird to these regions.

But if the unassisted eye shows that life is diffused throughout the whole atmosphere, the microscope reveals yet greater wonders. Wheel-animalcules, *brachioni*, and a host of microscopic insects are lifted by the winds from the evaporating waters below. Motionless and to all appearance dead, they float on the breeze, until the dew bears them back to the nourishing earth, and bursting the tissue which incloses their transparent rotating (3) bodies, instils new life and motion into all their organs, probably by the action of the vital principle inherent in water. The yellow meteoric sand or mist (dust nebulæ) often observed to fall on the Atlantic near the Cape de Verde Islands, and not unfrequently borne in an easterly direction as far as Northern Africa, Italy, and Central Europe, consists, according to Ehrenberg's brilliant discovery, of agglomerations of siliceous-shelled microscopic organisms. Many of these perhaps float for years in the highest strata of the atmosphere, until they are carried down by the Etesian winds or by descending currents of air, in the full capacity of life, and actually engaged in organic increase by spontaneous self-division.

Together with these developed creatures, the atmosphere contains countless germs of future formations; eggs of insects, and seeds of plants, which, by means of hairy or feathery crowns, are borne forward on their long autumnal journey. Even the vivifying pollen scattered abroad by the male blossoms, is carried by winds and winged insects over sea and land, to the distant and solitary female plant (4). Thus, wheresoever the naturalist turns his eye, life or the germ of life lies spread before him.

But if the moving sea of air in which we are immersed, and above whose surface we are unable to raise ourselves, yields to many organic beings their most essential nourishment, they still require therewith a more substantial species of food, which is provided for them only at the bottom of this gaseous ocean. This bottom is of a twofold kind: the smaller portion constituting the dry earth, in immediate contact with the surrounding atmosphere; the larger portion consisting of water,—formed, perhaps, thousands of years ago from gaseous matters fused by electric fire, and now incessantly undergoing decomposition in the laboratory of the clouds and in the pulsating vessels of animals and plants. Organic forms descend deep into the womb of the earth, wherever the meteoric rain-waters can penetrate into natural cavities, or into artificial excavations and mines. The domain of the subterranean cryptogamic flora was early an object of my scientific researches. Thermal springs of the highest temperature nourish small Hydropores, Confervæ and Oscillatoriæ. Not far from the Arctic circle, at Bear Lake, in the New Continent, Richardson saw flowering plants on the ground which, even in summer, remains frozen to the depth of twenty inches.

It is still undetermined where life is most abundant: whether on the earth or in the fathomless depths of the ocean. Ehrenberg's admirable work on the relative condition of animalcular life in the tropical ocean and the floating and solid ice of the Antarctic circle, has spread the sphere and horizon of organic life before our eyes. Siliceous-shelled Polygastrica and even Coscinodiscæ, alive, with their green ovaries, have been found enveloped in masses within twelve degrees of the Pole; even as the small black glacier flea, *Desoria Glacialis*, and Podurellæ, inhabit the narrow tubules of ice of the Swiss glaciers, as proved by the researches of Agassiz. Ehrenberg has shown that on some microscopic infusorial animalcules (*Synedra* and *Cocconeis*), other species live parasitically; and that in the Gallionellæ the extraordinary powers of division and development of bulk are so great, that an animalcule invisible to the naked eye can in four days form two cubic feet of the Bilin polishing slate.

In the ocean, gelatinous sea-worms, living and dead, shine like luminous stars (5). converting by their phosphorescent light the green surface of the ocean into one vast sheet of fire. Indelible is the impression left on my mind by those calm tropical nights of the Pacific, where the constellation of Argo in its zenith, and the setting Southern Cross, pour their mild planetary light through the ethereal azure of the sky, while dolphins mark the foaming waves with their luminous furrows.

But not alone the depths of ocean, the waters, too, of our own swamps and marshes, conceal innumerable worms of wonderful form. Almost indistinguishable by the eye are the Cyclidiæ, the Euglenes, and the host of Naiads divisible by branches like the *Lemna* (Duckweed), whose leafy shade they seek. Surrounded by differently composed atmospheres, and deprived of light, the spotted Ascaris breathes in the skin of the earth-worm, the silvery and bright Leucophra exists in the body of the shore Nais, and a Pentastoma in the large pulmonary cells of the tropical rattle-snake (6). There are animalcules in the blood of frogs and salmon, and even, according to Nordmann, in the fluid of the eyes of fishes, and in the gills of the bream. Thus are even the most hidden recesses of creation replete with life. We purpose in the following pages to consider the different families of plants, since on their existence entirely depends that of the animal creation. Incessantly are they occupied in organizing the raw material of the earth, assimilating by vital forces those elements which after a thousand metamorphoses become ennobled into active nervous tissue. The glance which we direct to the dissemination of vegetable forms, reveals to us the fulness of that animal life which they sustain and preserve.

The verdant carpet which a luxuriant Flora spreads over the surface of the earth is not woven equally in all parts; for while it is most rich and full where, under an ever-cloudless sky, the sun attains its greatest height, it is thin and scanty near the torpid poles, where the quickly-recurring frosts too speedily blight the opening bud or destroy the ripening fruit.

Yet everywhere man rejoices in the presence of nourishing plants. Even where from the depths of the sea, a volcano bursting through the boiling flood, upheaves a scoriaceous rock, (as once happened in the Greek Islands); or, to instance a more gradual phenomenon, where the united labours of the coral animal (Lithophytes) (7) have piled up their cellular dwellings, on the crests of submarine mountains, until after toiling for thousands of years their edifice reaches the level of the ocean, when its architects perish, and leave a coral island. Thus are organic forces ever ready to animate with living forms the naked rock. How seeds are so suddenly transported to these rocks, whether by birds, or by winds, or by the waves of ocean, is a question that cannot be decided, owing to the great distance of these islands from the coasts. But no sooner has the air greeted the naked rock, than, in our northern countries, it gradually acquires a covering of velvet-like fibres, which appear to the eye to be coloured spots. Some of these are bordered by single and others by double rows, while others again are traversed by furrows and divided into compartments. As they increase in age their colour darkens. The bright glittering yellow becomes brown, and gradually the bluish-grey mass of the Leprariæ changes to a dusty black. As the outlines of this vegetable surface merge into each other with increasing age, the dark ground acquires a new covering of fresh circular spots of dazzling whiteness. Thus one organic tissue rises, like strata, over the other; and as the human race in its development must pass through definite stages of civilization, so also is the gradual distribution of plants dependent on definite physical laws. In spots where lofty forest trees now rear their towering summits, the sole covering of the barren rock was once the tender lichen; the long and immeasurable interval was filled up by the growth of grasses, herbaceous plants, and shrubs. The place occupied in northern regions by mosses and lichens is supplied in the tropics by Portulacas, Gomphrenas, and other low and oleaginous marine plants. The history of the vegetable covering and of its gradual extension over the barren surface of the earth, has its epochs, as well as that of the migratory animal world.

But although life is everywhere diffused, and although the organic forces are incessantly at work in combining into new forms those elements which have been liberated by death; yet this fulness of life and its renovation differ according to difference of climate. Nature undergoes a periodic stagnation in the frigid zones; for fluidity is essential to life. Animals and plants, excepting indeed mosses and other Cryptogamia, here remain many months buried in a winter sleep. Over a great portion of the earth, therefore, only those organic forms are capable of full development, which have the property of resisting any considerable abstraction of heat, or those which, destitute of leaf-organs, can sustain a protracted interruption of their vital functions. Thus, the nearer we approach the tropics, the greater the increase in variety of structure, grace of form, and mixture of colours, as also in perpetual youth and vigour of organic life.

This increase may readily be doubted by those who have never quitted our own hemisphere, or who have neglected the study of physical geography. When in passing from our thickly foliated forests of oak, we cross the Alps or the Pyrenees and enter Italy or Spain, or when the traveller first directs his eye to some of the African coasts of the Mediterranean, he may easily be led to adopt the erroneous inference that absence of trees is a characteristic of hot climates. But they forget that Southern Europe wore a different aspect, when it was first colonised by Pelasgian or Carthaginian settlers; they forget too that an earlier civilization of the human race sets bounds to the increase of forests, and that nations, in their change-loving spirit, gradually destroy the decorations which rejoice our eye in the North, and which, more than the records of history, attest the youthfulness of our civilization. The great catastrophe by which the Mediterranean was formed, when the swollen waters of an inland sea burst their way through the Dardanelles and the Pillars of Hercules, appears to have stripped the contiguous lands of a large portion of their alluvial soil. The records of the Samothracian traditions (8) preserved by Greek writers seem to indicate the recent date of this great convulsion of nature. Moreover, in all the lands bathed by the Mediterranean, and which are characterised by the tertiary and cretaceous formations (Nummulites and Neocomian rocks), a great portion of the earth's surface is naked rock. The picturesque beauty of Italian scenery depends mainly on the pleasing contrast between the bare and desolate rock and the luxuriant vegetation which, island-like, is scattered over its surface. Where the rock is less intersected by fissures, so that the water rests longer on its surface, and where it is covered with earth (as on the enchanting banks of Lake Albano), there even Italy has her oak-forests, as shady and verdant as could be desired by an inhabitant of the North.

The boundless plains or steppes of South America, and the deserts beyond the Atlas range of mountains, can only be regarded as mere local phenomena. The former are found to be covered, at least in the rainy season, with grasses and low almost herbaceous Mimosæ; while the latter are seas of sand in the interior of the Old Continent,—vast arid tracts surrounded by borders of evergreen forests. Here and there only a few isolated fan-palms remind the wanderer that these dreary solitudes are a portion of animated nature. Amid the optical delusions occasioned by the radiation of heat, we see the bases of these trees at one moment hovering in the air, at the next their inverted image reflected in the undulating strata of the atmosphere. To the west of the Peruvian Andes, on the shores of the Pacific, I have passed weeks in traversing these waterless deserts.

The origin of this absence of plants over large tracts of

land, in regions characterised on every side by the most exuberant vegetation, is a geological phenomenon which has hitherto received but little attention; it undoubtedly arises from former revolutions of nature, such as inundations, or from volcanic convulsions of the earth's surface. When once a region loses its vegetable covering, if the sand is loose and devoid of springs, and if vertically ascending currents of heated air prevent the precipitation of vapour (9), thousands of years may elapse before organic life can penetrate from the green shores to the interior of the dreary waste.

Those who are capable of surveying nature with a comprehensive glance, and abstract their attention from local phenomena, cannot fail to observe that organic development and abundance of vitality gradually increase from the poles towards the equator, in proportion to the increase of animating heat. But in this distribution every different climate has allotted to it some beauty peculiar to itself: to the Tropics belong variety and magnitude in vegetable forms; to the North the aspect of its meadows and the periodical renovation of nature at the first genial breath of spring. Every zone, besides its own peculiar advantages, has its own distinctive character. The primeval force of organization, notwithstanding a certain independence in the abnormal development of individual parts, binds all animal and vegetable structures to fixed ever-recurring types. For as in some individual organic beings we recognise a definite physiognomy, and as descriptive botany and zoology are, strictly speaking, analyses of animal and vegetable forms, so also there is a certain natural physiognomy peculiar to every region of the earth.

That which the painter designates by the expressions "Swiss scenery" or "Italian sky" is based on a vague feeling of the local natural character. The azure of the sky, the effects of light and shade, the haze floating on the distant horizon, the forms of animals, the succulence of plants, the bright glossy surface of the leaves, the outlines of mountains, all combine to produce the elements on which depends the impression of any one region. It must be admitted, however, that in all latitudes the same kind of rocks, as trachyte, basalt, porphyritic schist, and dolomite, form mountain groups of exactly similar physiognomy. Thus the greenstone cliffs of South America and Mexico resemble those of the Fichtel mountains of Germany, in like manner as among animals, the form of the Allco, or the original canine race of the New Continent, is analogous to that of the European race. The inorganic crust of the earth is as it were independent of climatic influences; perhaps, because diversity of climate arising from difference of latitude is of more recent date than the formations of the earth, or that the hardening crust, in solidifying and discharging its caloric, acquired its temperature from internal and not from external causes (10). All formations are, therefore, common to every quarter of the globe and assume the like forms. Everywhere basalt rises in twin mountains and truncated cones; everywhere trap-porphyry presents itself to the eye under the form of grotesquely-shaped masses of rock, while granite terminates in gently rounded summits. Thus, too, similar vegetable forms, as pines and oaks, alike crown the mountain declivities of Sweden and those of the most southern portion of Mexico (11). But notwithstanding all this coincidence of form, and resemblance of the outlines of individual portions, the grouping of the mass, as a whole, presents the greatest diversity of character.

As the oryctognostic knowledge of minerals differs from geology, so also does the general study of the physiognomy of nature differ from the individual branches of the natural sciences. The character of certain portions of the earth's surface has been described with inimitable truthfulness by George Forster in his travels and smaller works, by Goethe in the descriptive passages which so frequently occur in his immortal writings, by Buffon, Bernardin de St. Pierre, and Chateaubriand. Such descriptions are not only calculated to yield an enjoyment of the noblest kind, but the knowledge of the character of nature in different regions is also most intimately associated with the history of the human race and its mental culture. For although the dawn of this culture cannot have been determined solely by physical influence, climatic relations have at any rate to a great extent influenced its direction, as well as the character of nations, and the degree of gloom or cheerfulness in the dispositions of men. How powerfully did the skies of Greece act on its inhabitants! Was it not among the nations who settled in the beautiful and happy region between the Euphrates, the Halys, and the Ægean Sea, that social polish and gentler feelings were first awakened? and was it not from these genial climes that our forefathers, when religious enthusiasm had suddenly opened to them the Holy Lands of the East, brought back to Europe, then relapsing into barbarism, the seeds of a gentler civilization? The poetical works of the Greeks and the ruder songs of the primitive northern races owe much of their peculiar character to the forms of plants and animals, to the mountain-valleys in which their poets dwelt, and to the air which surrounded them. To revert to more familiar objects, who is there that does not feel himself differently affected beneath the embowering shade of the beechen grove, or on hills crowned with a few scattered pines, or in the flowering meadow where the breeze murmurs through the trembling foliage of the birch? A feeling of melancholy, or solemnity, or of light buoyant animation is in turn awakened by the contemplation of our native trees. This influence of the physical on the moral world—this mysterious reaction of the sensuous on the ideal, gives to the study of nature, when considered from a higher point of view, a peculiar charm which has not hitherto been sufficiently recognised.

However much the character of different regions of the earth may depend upon a combination of all these external phenomena, and however much the total impression may be influenced by the outline of mountains and hills, the physiognomy of plants and animals, the azure of the sky, the form of the clouds, and the transparency of the atmosphere, still it cannot be denied that it is the vegetable covering of the earth's surface which chiefly conduces to the effect. The animal organism is deficient in mass, while the mobility of its individual members and often their diminutiveness remove them from the sphere of our observation. Vegetable forms, on the other hand, act on the imagination by their enduring magnitude—for here massive size is indicative of age, and in the vegetable kingdom alone are age and the manifestation of an ever-renewed vigour linked together. The colossal Dragon Tree (12), which I saw in the Canary Isles, and which measured more than sixteen feet in diameter, still bears, as it then did, the blossoms and fruit of perpetual youth. When the French adventurers, the Béthencourts, conquered these Fortunate Isles in the beginning of the fifteenth century, the Dragon Tree of Orotava, regarded by the natives with a veneration equal to that bestowed on the olive tree of the Acropolis at Athens, or the elm at Ephesus, was of the same colossal magnitude as at present. In the tropics a grove of Hymeneæ and Cesalpiniæ is probably a memorial of more than a thousand years.

On taking one general view of the different phanerogamic species which have already been collected into our herbariums (13), and which may now be estimated at considerably more than 80,000, we find that this prodigious quantity presents some few forms to which most of the others may be referred. In determining those forms, on whose individual beauty, distribution, and grouping, the physiognomy of a country's vegetation depends, we must not ground our opinion (as from other causes is necessarily the case in botanical systems) on the smaller organs of propagation, that is, the blossoms and fruit; but must be guided solely by those elements of magnitude and mass from which the total impression of a district receives its character of individuality. Among the principal forms of vegetation there are, indeed, some which constitute entire families, according to the so-called "natural system" of botanists. Bananas and Palms, Casuarineæ and Coniferæ, form distinct species in this mode of arrangement. The systematising botanist, however, separates into different groups many plants which the student of the physiognomy of nature is compelled to associate together. Where vegetable forms occur in large masses, the outlines and distribution of the leaves, and the form of the stems and branches lose their individuality and become blended together. The painter—and here his delicate artistical appreciation of nature comes especially into play—distinguishes between pines or palms and beeches in the background of a landscape, but not between forests of beech and other thickly foliated trees.

The physiognomy of nature is principally determined by sixteen forms of plants. I merely enumerate such as I have observed in my travels through the old and new world during many years' study of the vegetation of different latitudes, between the parallels of 60° north and 12° south. The number of these forms will no doubt be considerably increased by travellers penetrating further into the interior of continents, and discovering new genera of plants. We are still wholly ignorant of the vegetation of the south-east of Asia, the interior of Africa and New Holland, and of South America from the Amazon to the province of Chiquitos. Might not a region be some day discovered in which ligneous fungi, *Cenomyce rangiferina*, or mosses, form high trees? *Neckera dendroïdes*, a German species of moss, is in fact arborescent, and the sight of a wood of lofty mosses could hardly afford greater astonishment to its discoverers than that experienced by Europeans at the aspect of arborescent grasses (bamboos) and the tree-ferns of the tropics, which are often equal in height to our lindens and alders. The maximum size and degree of development attainable by organic forms of any genus, whether of animals or plants, are determined by laws with which we are still unacquainted. In each of the great divisions of the animal kingdom, as insects, reptiles, crustacea, birds, fishes, or mammalia, the dimensions of the body oscillate between certain extreme limits. But these limits, based on the observations hitherto contributed to science, may be enlarged by new discoveries of species with which we are at present unacquainted.

In land animals a high degree of temperature, depending on latitude, appears to have exercised a favourable influence on the genetic development of organization. Thus the small and slender form of our lizards expands in the south into the colossal, unwieldy, and mail-clad body of the formidable crocodile. In the huge cats of Africa and America, the tiger, lion, and jaguar, we find, repeated on a larger scale, the form of one of the smallest of our domestic animals. But if we penetrate into the recesses of the earth, and search the tombs of plants and animals, the fossil remains thus brought to light not only manifest a distribution of forms at variance with the present climates, but they also reveal colossal structures, which exhibit as marked a contrast with the small types that now surround us, as does the simple yet dignified heroism of the ancient Greeks, when compared with what is recognized at the present day as "greatness of character." If the temperature of the earth has undergone considerable, perhaps periodically recurring changes, and, if even the relations between sea and land, and the height and pressure of the atmospheric ocean (14), have not always been the same, then the physiognomy of nature, and the magnitude and forms of organic bodies, must also have been subject to many variations. Enormous Pachydermata, elephantine Mastodons, Owen's Mylodon robustus, and the Colossochelys,* a land tortoise upwards of six feet in height, once inhabited forests of colossal Lepidodendra, cactus-like Stigmariæ, and

* Fossil remains of this gigantic antediluvian tortoise are now in the British Museum.—Ed.

numerous genera of Cycadeæ. Unable accurately to delineate the physiognomy of our aging and altering planet according to its present features, I will only attempt to bring prominently forward those characteristics which specially appertain to each individual group of plants. Notwithstanding all the richness and adaptability of our language, the attempt to designate in words, that which, in fact, appertains only to the imitative art of the painter, is always fraught with difficulty. I would also wish to avoid that wearying effect which is almost unavoidably inseparable from a long enumeration of individual forms.

We will begin with Palms (15), the loftiest and most stately of all vegetable forms. To these, above all other trees, the prize of beauty has always been awarded by every nation; and it was from the Asiatic palm-world, or the adjacent countries, that human civilization sent forth the first rays of its early dawn. Marked with rings, and not unfrequently armed with thorns, the tall and slender shaft of this graceful tree rears on high its crown of shining, fan-like, or pinnated leaves, which are often curled like those of some gramineæ. Smooth stems of the palm, which I carefully measured, rose to a height of 190 feet. The palm diminishes in size and beauty as it recedes from the equatorial towards the temperate zones. Europe owns amongst its indigenous trees only one representative of this form of vegetation, the dwarfish coast palm (*Chamærops*), which, in Spain and Italy, is found as far north as 44° lat. The true palm climate has a mean annual temperature of 78° to 81°.5 Fahr., but the date-palm, which has been brought to us from Africa, and is less beautiful than other species of this family, vegetates in the south of Europe in districts whose mean temperature is only from 59° to 62° 4′ Fahr. Stems of palms and skeletons of elephants are found buried in the interior of the earth in Northern Europe; their position renders it probable that they were not drifted from the tropics towards the north, but that, in the great revolutions of our planet, climates, and the physiognomy of nature which is regulated by climate, have been, in many respects, altered.

In all regions of the earth the palm is found associated with the plantain or banana; the *Scitamineæ* and *Musaceæ* of botanists, *Heliconia*, *Amomum*, and *Strelitzia*. This form has a low, succulent, and almost herbaceous stem, the summit of which is crowned with delicately striped, silky, shining leaves of a thin and loose texture. Groves of bananas form the ornament of humid regions; and on their fruit the natives of the torrid zone chiefly depend for subsistence. Like the farinaceous cereals or corn-yielding plants of the north, the banana has accompanied man from the earliest infancy of his civilization (16). By some Semitic traditions the primitive seat of these nutritious tropical plants has been placed on the shores of the Euphrates, and by others, with greater probability, in India, at the foot of the Himalaya mountains. Greek legends cite the plains of Enna as the home of the cereals. Whilst, however, the cereals, spread by culture over the northern regions, in monotonous and far extending tracts, add but little to the beauty of the landscape; the inhabitant of the tropics, on the other hand, is enabled, by the propagation of the banana, to multiply one of the noblest and most lovely of vegetable productions.

The form of the Malvaceæ (17) and Bombaceæ, represented by *Ceiba*, *Cavanillesia*, and the Mexican hand tree (*Cheirostemon*), has immensely thick stems, with lanuginous, large, cordate, or indented leaves, and magnificent flowers, frequently of a purple-red. To this group belongs the Bahobab, or monkey bread-tree, *Adansonia digitata*, which, with a moderate height, has occasionally a diameter of 32 feet,* and may probably be regarded as at once the largest and most ancient organic memorial of our planet. The Malvaceæ already begin to impart to the vegetation of Italy a peculiarly southern character.

* The weight of the lower branches bends them to the ground, so that a single tree forms a hemispherical mass of verdure sometimes 150 feet in diameter.—Ed.

The temperate zone in our old continent unfortunately is wholly devoid of the delicately pinnate Mimosas (18), whose predominating forms are *Acacia*, *Desmanthus*, *Gleditschia*, *Porleria*, and *Tamarindus*. This beautiful form occurs in the United States of North America, where, under equal parallels of latitude, vegetation is more varied and luxuriant than in Europe. The Mimosas are generally characterised, like the Italian pine, by an umbellate expansion of their branches. An extremely picturesque effect is produced by the deep blue of a tropical sky gleaming through the delicate tracery of their foliage.

Heaths (19), which more especially belong to an African group of plants, include, according to physiognomic character and general appearance, the Epacrideæ and Diosmeæ, many Proteaceæ, and the Australian Acacias, which have no leaves but mere flattened petioles (phyllodia). This group bears some resemblance to acicular-leaved forms, with which it contrasts the more gracefully by the abundance of its campanulate blossoms. The arborescent heaths, like some few other African plants, extend as far as the northern shores of the Mediterranean. They adorn the plains of Italy, and the Cistus groves of southern Spain, but I have nowhere seen them growing more luxuriantly than on the declivities of the Peak of Teyde at Teneriffe. In the countries bordering on the Baltic, and further northward, the appearance of this form of plants is regarded with apprehension, as the precursor of drought and barrenness. Our heaths, *Erica* (*Calluna*) *vulgaris*, and *Erica tetralix*, *E. carnea* and *E. cinerea*, are social plants, against whose extension agricultural nations have contended for centuries, with but little success. It is singular that the principal representative of this family should be peculiar to one side of our planet alone. There is only one of the three hundred known species of *Erica* to be met with in the new continent, from Pennsylvania and Labrador to Nootka Sound and Alaschka.

The *Cactus* form (20), on the other hand, is almost peculiar to the new continent; it is sometimes globular, sometimes articulated, sometimes rising in tall polygonal columns not unlike organ-pipes. This group forms the most striking contrast with the Lily and Banana families, and belongs to that class of plants which Bernardin de St. Pierre felicitously terms vegetable fountains of the Desert. In the parched arid plains of South America, the thirsting animals eagerly seek the *Melon-cactus*, a globular plant half-buried in the dry sand, whose succulent interior is concealed by formidable prickles. The stems of the columnar cactus attain a height of more than 30 feet; their candelabra-like ramifications, frequently covered with lichens, reminding the traveller, by some analogy in their physiognomy, of certain of the African Euphorbias.

While these plants form green Oases in the barren desert, the Orchideæ (21) shed beauty over the most desolate rocky clefts, and the seared and blackened stems of those tropical trees which have been discoloured by the action of light. The *Vanilla* form is distinguished by its light green succulent leaves, and by its variegated and singularly shaped blossoms. Some of the orchideous flowers resemble in shape winged insects, while others look like birds, attracted by the fragrance of the honey vessels. An entire life would not suffice to enable an artist, although limiting himself to the specimens afforded by one circumscribed region, to depict the splendid Orchideæ which embellish the deep alpine valleys of the Peruvian Andes.

The form of the Casuarineæ (22), leafless, like almost all the species of Cactus, comprises a group of trees having branches resembling the Equisetum, and is peculiar to the islands of the Pacific and to the East Indies. Traces of this type, which is certainly more singular than beautiful, may however be found in other regions of the earth. Plumier's *Equisetum altissimum*, Forskäl's *Ephedra aphylla* of North

Africa, the Peruvian *Colletia*, and the Siberian *Calligonum Pallasia*, are nearly allied to the form of the Casuarinas.

While the Banana form presents us with the greatest degree of expansion, the Casuarinas and the acicular-leaved (23) trees exhibit the greatest contraction of the leaf-vessels. Pines, Thujas, and Cypresses constitute a northern form but rarely met with in the tropics and in some coniferæ (*Dammara Salisburia*), the leaves are both broad and acicular. Their evergreen foliage enlivens the gloom of the dreary winter landscape, while it proclaims to the natives of the polar regions that, although snow and ice cover the surface, the inner life of plants, like the Promethean fire, is never wholly extinct on our planet.

Besides the Orchideæ, the Pothos tribe of plants (24) also yields a graceful covering to the aged stems of forest trees in the tropical world, like the parasitic mosses and lichens of our own climes. Their succulent herbaceous stalks are furnished with large leaves, arrow-shaped, digitate, or elongated, and invariably furnished with thick veins. The blossoms of the Aroideæ are inclosed in spathes, by which their vital heat is increased; they are stemless, and send forth aërial roots. Pothos, Dracontium, Caladium, and Arum are all kindred forms; and the last-named extends as far as the coasts of the Mediterranean, contributing, together with succulent Tussilago (Coltsfoot), high thistles, and the Acanthus, to give a luxuriant southern character to the vegetation of Spain and Italy.

This Arum form is associated, in the torrid regions of South America, with the tropical *Lianes* or creeping plants (25), which exhibit the utmost luxuriance of vegetation in Paullinias, Banisterias, Bignonias, and Passion-flowers. Our tendrilled hops and vines remind us of this tropical form. On the Orinoco the leafless branches of the Bauhinia are often upwards of 40 feet in length, sometimes hanging perpendicularly from the summit of lofty Swieteniæ, (Mahogany trees), sometimes stretched obliquely like ropes from a mast; along these the tiger-cat may be seen climbing to and fro with wonderful agility.

The self-sustaining form of the bluish-flowered Aloe tribe (26) presents a marked contrast to the pliant climbing lianes with their fresh and brilliant verdure. When there is a stem it is almost branchless, closely marked with spiral rings, and surrounded by a crown of succulent, fleshy, long-pointed leaves, which radiate from a centre. The lofty-stemmed aloe does not grow in clusters like other social plants, but stands isolated in the midst of dreary solitudes, imparting to the tropical landscape a peculiar melancholy (one might almost say African) character.

To this aloe form belong, in reference to physiognomic resemblance and the impression they produce on the landscape: the Pitcairnias, from the family of the Bromeliaceæ, which in the chain of the Andes grow out of clefts in the rock; the great *Pournetia pyramidata* (the *Atschupalla* of the elevated plateaux of New Grenada); the American aloe (Agave), *Bromelia Ananas* and *B. Karatas;* those rare species of the family of the Euphorbiaceæ, which have thick, short, candelabra-like divided stems; the African aloe, and the Dragon tree, *Dracæna Draco,* of the family of the Asphodeleæ; and lastly the tall flowering Yucca, allied to the Liliaceæ.

While the Aloe form is characterised by an air of solemn repose and immobility, the grass form (27), especially as regards the physiognomy of the arborescent grasses, is expressive of buoyant lightness and flexible slenderness. In both the Indies, bamboo groves form arched and shady walks.

The smooth and often inclined and waving stem of the tropical grasses exceeds in height our alders and oaks. As far north as Italy, this form already begins, in the *Arundo Donax,* to raise itself from the ground, and to determine, by height as well as mass, the natural character of the country.

The form of Ferns (28), like that of grasses, also assumes nobler dimensions in the torrid regions of the earth, and the arborescent ferns, which frequently attain the height of above forty feet, have a palm-like appearance, although their stem is thicker, shorter, and more rough and scaly, than that of the palm. The leaf is more delicate, of a loose and more transparent texture, and sharply serrated on the margins. These colossal ferns belong almost exclusively to the tropics, but there they prefer the temperate localities. As in these latitudes diminution of heat is merely the consequence of an increase of elevation, we may regard mountains that rise 2000 or 3000 feet above the level of the sea as the principal seat of these plants. Arborescent ferns grow in South America, side by side with that beneficent tree whose stem yields the febrifuge bark, and both forms of vegetation are indicative of the happy region where reigns the genial mildness of perpetual spring.

I have now to mention the form of the Liliaceous plants (29), *Amaryllis*, *Ixia*, *Gladiolus*, and *Pancratium*, with their flag-like leaves and splendid blossoms, the principal home of which is Southern Africa; also the Willow form (30), which is indigenous in all latitudes, and is represented in the plateaux of Quito, not by the shape of its leaves, but in the form of its ramification, in Schinus Molle; also the Myrtle-form (31) (*Metrosideros*, *Eucalyptus*, *Escallonia myrtelloides*); Melastomaceæ (32); and the Laurel form (33).

It would be an undertaking worthy of a great artist to study the character of all these vegetable groups, not in hot-houses, or from the descriptions of botanists, but on the grand theatre of tropical nature. How interesting and instructive to the landscape painter (34) would be a work that should present to the eye accurate delineations of the sixteen principal forms enumerated, both individually and in collective contrast. What can be more picturesque than the arborescent Ferns. which spread their tender foliage above the Mexican laurel-oak! what more charming than the aspect of banana-groves, shaded by those lofty grasses, the Guadua and Bamboo! It is peculiarly the privilege of the artist to separate these into groups, and thus the beautiful images of nature, if we may be permitted the simile, resolve themselves beneath his touch, like the written works of man, into a few simple elements.

It is beneath the glowing rays of a tropical sun, that the noblest forms of vegetation are developed. In the cold North the bark of trees is covered only with dry lichens and mosses, while beneath the tropics the Cymbidium and the fragrant Vanilla adorn the trunks of the Anacardias and the gigantic Fig-tree. The fresh green of the Pothos leaves and of the Dracontias contrast with the many coloured blossoms of the Orchideæ; climbing Bauhinias, Passion-flowers and golden flowered Banisterias encircle every tree of the forest. Delicate blossoms unfold themselves from the roots of the *Theobroma*, and from the thick and rough bark of the *Crescentia* and *Gustavia* (35). Amid this luxuriant abundance of flowers and foliage, amid this exuberance and tangled web of creeping plants, it is often difficult for the naturalist to recognise the stems to which the various leaves and blossoms belong. A single tree, adorned with Paullinias, Bignonias, and Dendrobias, forms a group of plants, which, separated from each other, would cover a considerable space of ground.

In the tropics, plants are more succulent, of a fresher green, and have larger and more glossy leaves, than in the northern regions. Social plants, which give such a character of uniformity to European vegetation, are almost wholly absent in the equatorial zone. Trees, almost twice as high as our oaks, there bloom with flowers as large and splendid as our lilies. On the shady banks of the Magdalena River, in South America, grows a climbing *Aristolochia*, whose blossoms, measuring four feet in circumference, the Indian children sportively draw on their heads as caps (36). In the South Indian Archipelago, the flower of the Rafflesia is nearly three feet in diameter, and weighs above fourteen pounds.

The extraordinary height to which not only individual mountains but even whole districts rise in tropical regions, and the consequent cold of such elevations, affords the inhabitant of the tropics a singular spectacle. For besides his own palms and bananas, he is surrounded by those vegetable forms which would seem to belong solely to northern latitudes. Cypresses, pines, and oaks, barberry shrubs and alders (nearly allied to our own species) cover the mountain plains of Southern Mexico and the chain of the Andes at the equator. Thus nature has permitted the native of the torrid zone to behold all the vegetable forms of the earth without quitting his own clime, even as are revealed to him the luminous worlds which spangle the firmament from pole to pole (37).

These and many other of the enjoyments which nature affords are denied to the nations of the North. Many constellations and many vegetable forms, including more especially the most beautiful productions of the earth (palms, tree-ferns, bananas, arborescent grasses, and delicately feathered mimosas), remain for ever unknown to them; for the puny plants pent up in our hothouses, give but a faint idea of the majestic vegetation of the tropics. But the rich development of our language, the glowing fancy of the poet, and the imitative art of the painter, afford us abundant compensation; and enable the imagination to depict in vivid colours the images of an exotic Nature. In the frigid North, amid barren heaths, the solitary student may appropriate all that has been discovered in the most remote regions of the earth, and thus create within himself a world as free and imperishable as the spirit from which it emanates.

## ILLUSTRATIONS AND ADDITIONS.

(1) p. 210—"*On the Chimborazo, upwards of eight thousand feet higher than Etna.*"

SMALL singing birds, and even butterflies, (as I have myself witnessed in the Pacific,) are often met with at great distances from the shore, during storms blowing off land. In a similar manner insects are involuntarily carried into the higher regions of the atmosphere, to an elevation of 17,000 to 19,000 feet above the plains. The light bodies of these insects are borne upwards by the vertically ascending currents of air caused by the heated condition of the earth's surface. M. Boussingault, an admirable chemist, who ascended the Gneiss Mountains of Caracas, while holding the appointment of Professor in the newly established Mining Academy at Santa Fé de Bogotá, witnessed, during his ascent to the summit of the Silla, a phenomenon which confirmed in a most remarkable manner this vertical ascent of air. He and his companion, Don Mariano de Rivero, observed at noon a number of luminous whitish bodies rise from the valley of Caracas to the summit of the Silla, an elevation of 5755 feet, and then sink towards the adjacent sea coast. This phenomenon was uninterruptedly prolonged for a whole hour, when it was discovered that the bodies, at first mistaken for a flock of small birds, were a number of minute balls of grass-haums. Boussingault sent me some of this grass, which was immediately recognised by Professor Kunth as a species of Vilfa, a genus of grass which together with Agrostis is of frequent occurrence in the provinces of Caracas and Cumana. It was the *Vilfa tenacissima* of our *Synopsis Plantarum æquinoctialium Orbis Novi*, t. i. p. 205. Saussure found butterflies on Mont Blanc, and Ramond observed them in the solitudes around the summit of Mont Perdu. When MM. Bonpland, Carlos Montufar, and myself, on the 23rd of June, 1802, ascended the eastern declivity of Mount Chimborazo, to a height of 19,286 feet, and where the barometer had fallen to 14·84 inches, we found winged insects buzzing around us. We recognised them to be Diptera, resembling flies, but it was impossible to catch these insects standing on the rocky ledges (*cuchilla*), often less than a foot in breadth, and between masses of snow precipitated from above. The elevation at which we observed these insects was almost the same as that in which the naked trachytic rock, which projected from the eternal snows around, exhibited the last traces of vegetation in Lecidea geographica. These insects were flying at an elevation of 18,225 feet, or nearly 2660 feet higher than the summit of Mont Blanc: and somewhat below this height, at an elevation of 16,626 feet, and therefore also above the region of snow, M. Bonpland saw yellow butterflies flying close to the ground. The mammalia which live nearest to the region of perpetual snow, are, in the Swiss Alps, the hybernating marmot, and a very small field-mouse, (Hypudæus nivalis,) described by Martius, which on the Faulhorn lays up, almost under the snow, a store of the roots of phanerogamic alpine plants.* The opinion prevalent in Europe, that the beautiful rodent, the Chinchilla, whose soft and glossy fur is so much esteemed, is found in the highest mountain regions of Chili, is an error. The Chinchilla laniger (Gray) lives only in a mild lower zone, and does not advance further south than the parallel of 35°.†

Whilst among our European Alps, Lecideas, Parmelias, and Umbilicarias but scantily clothe with a few coloured patches those rocks that are not wholly covered with snow, we found in the Andes, at elevations of 13,700 to nearly 15,000 feet, some phanerogamic plants which we were the first to describe; as for instance, the woolly species of Fraylejon, (Culcitium nivale, C. rufescens, and C. reflexum, Espeletia grandiflora, and E. argentea) Sida pichinchensis, Ranunculus nubigenus, R. Gusmanni with red or orange-coloured flowers, the small moss-like umbelliferous plant, Myrrhis andicola, and Fragosa arctioides. On the declivity of the Chimborazo, the Saxifraga Boussingaulti, described by Adolph Brongniart, grows beyond the limits of perpetual snow on loose blocks of stone at an elevation of 15,770 feet above the level of the sea, and not at 17,000 as has been stated in two admirable English

* *Actes de la Société Helvétique*, 1843, p. 324.

† Claudio Gay, *Historia fisica y politica de Chile, Zoologia*, 1844, p. 91.

journals.* This Saxifrage, discovered by Boussingault, must therefore be regarded as the highest growing phanerogamic plant in the world.

The vertical height of Chimborazo is, according to my measurement, 21,422 feet.† This result is a mean between those which have been given by the French and Spanish Academicians. The principal differences do not here depend on different assumptions for the refraction, but on a difference in reducing the measured line to the level of the sea. This reduction can only be made in the Andes by the barometer, and hence every so-called trigonometric measurement must also necessarily be a barometric one, whose result will vary according to the different formulæ employed. Owing to the enormous mass of the mountain chain, we can only obtain very small angles of altitude, when the greater portion of the whole height has to be measured trigonometrically, and the observation is made at some low and distant point near the plain or the level of the sea. It is on the other hand extremely difficult to obtain a convenient base line, as the space that is to be determined barometrically increases with every step we advance towards the mountain. These obstacles have to be encountered by every traveller who on the high table-lands, which surround the summit of the Andes, selects a spot for performing a geodetic operation. On the pumice-covered plain of Tapia, to the west of the Rio Chambo, at a height of 9477 feet, barometrically determined, I measured the Chimborazo. The Llanos de Luisa, and more especially the plain of Sisgun, whose elevation is 12,150 feet, would yield greater angles of altitude. I had on one occasion made every preparation necessary for the measurement of Mount Chimborazo, from the plain of Sisgun, when the summit of the mountain was suddenly shrouded in a dense cloud.

Some hypothetical suggestions, regarding the probable derivation of the name of the far-famed "Chimborazo," may not be wholly unwelcome to etymologists. The district in which the mountain is situated is called Chimbo, a word which La

* Compare my *Asie centrale,* t. iii. p. 262, with Hooker, *Journal of Botany,* vol. i. 1834, p. 327, and the *Edinburgh New Philosophical Journal,* vol. xvii. 1834, p. 380.

† *Recueil d'Observ. astron.,* t. i. Intr. p. lxxii.

Condamine* derives from *chimpani*, to cross a river. "Chimboraço" means, according to him, "the snow of the opposite bank," from the fact of a brook being crossed at the village of Chimbo, in sight of the huge snow-covered mountain. (In the Qquichua language *chimpa* signifies the opposite bank or side; *chimpani* to cross a river, bridge, &c.) Several natives of the province of Quito assured me that Chimborazo meant simply the snow of Chimbo. In Carguai-razo we meet with the same termination, and it would appear that "razo" is a provincial word. The Jesuit Holguin, whose excellent vocabulary† I possess, is not acquainted with the word razo. The genuine term for snow is ritti. On the other hand, my friend, Professor Buschmann, an admirable linguist, remarks that in the Chinchaysuyo dialect, (employed north of Cuzco as far as Quito and Pasto) raju, the *j* being apparently guttural, signifies snow.‡ As chimpa and chimpani do not well suit on account of the *a*, we may seek a definite meaning for the first portion of the name of the mountain and of the village Chimbo, in the Qquichua word "chimpu," which is used to express a coloured thread or fringe (señal de lana, hilo ó borlilla de colores); the redness of the sky (arreboles), and the halo round the sun and moon. The name of the mountain might be thus derived from this word, without reference to the district or village. At all events, whatever may be the etymology of the word Chimborazo, it should be written in the Peruvian manner Chimporazo, as the Peruvians have no *b* in their alphabet.

May not the name of this colossal mountain be wholly independent of the Inca language, and have come down from a bygone age? The Inca or Qquichua language had not been introduced long prior to the Spanish invasion into the kingdom of Quito, where the now wholly extinct Puruay language had been previously used. The names of other mountains, as Pichincha, Ilinissa, and Cotopaxi, are wholly devoid of meaning in the language of the Incas, and are therefore undoubt-

* *Voyage à l'Equateur,* 1751, p. 184.

† *Vocabulario de la Lengua general de todo el Peru llamada Lengua Qquichua ó del Inca*, Lima, 1608.

‡ See the word in Juan de Figueredo's vocabulary of Chinchaysuyo words appended to Diego de Torres Rubio, *Arte, y Vocabulario de la Lengua Quichua,* reimpr. en Lima, 1751, fol. 222, b.

edly of higher antiquity than the introduction of the worship of the sun, and of the court-language of the rulers of Cuzco The names of mountains and rivers belong in all regions of the earth to the most ancient and authentic relics of languages; and my brother, Wilhelm von Humboldt, in his investigations into the former distribution of the Iberian races, has made ingenious use of these names. A singular and unexpected statement has recently been made,* "that the Incas, Tupac Yupanqui, and Huayna Capac, were astonished on their first conquest of Quito, to find a dialect of their Qquichua language in use among the natives." Prescott, however, seems to regard this as a very bold assertion.†

If we could suppose the pass of St. Gothard, Mount Athos, or the Rigi, piled on the summit of the Chimborazo, we should have the elevation which is at present ascribed to the Dhawalagiri in the Himalaya. The geologist who regards the interior of our planet from a more general point of view, and to whom not the directions, but the relative heights of the rocky projections, which we designate mountain chains, appear but as phenomena of little importance, will not be astonished if at some future period mountain summits should be discovered between the Himalaya and the Altai, which should surpass in height those of Dhawalagiri and Djewahir as much as these exceed that of Chimborazo.‡ The great height to which the snow-line recedes *in summer* on the northern declivity of the Himalaya, owing to the heat radiated from the elevated plateaux in Central Asia, renders the mountain, notwithstanding that it is situated in 29 to 30½° north lat., as accessible as are the Peruvian Andes in the region of the tropics. Captain Gerard has moreover recently ascended the Tarhigang as high, if not 117 feet higher,§ than I ascended the Chimborazo. Unfortunately, as I have elsewhere more fully shown, these mountain ascents, beyond the line of perpetual snow, however they may engage the curiosity of the public, are of very little scientific utility.

* Velasco, *Historia de Quito,* t. i. p. 185.

† *Hist. of the Conquest of Peru,* vol. i. p. 125.

‡ See my *Vues des Cordillères et Monumens des peuples indigènes de l'Amérique,* t. i. p. 116; and the Memoir entitled *Ueber zwei Versuche den Chimborazo zu besteigen* 1802 and 1831, in Schumacher's *Jahrbuch für* 1837, S. 176.

§ *Critical Researches on Philology and Geography,* 1824, p. 144.

(2) p. 210—"*The Condor, that giant among vultures.*"

I have elsewhere* given the natural history of the Condor, which before my travels had been variously misstated. The name is properly *Cuntur* in the Inca language; *Mañque* among the Araucanes in Chili; *Sarcoramphus Condor* according to Duméril. I sketched the head of this bird from life, of the natural size, and had my drawing engraved. Next to the Condor, the Lämmergeier of Switzerland, and the *Falco destructor* (Daud.), probably Linnæus' *Falco Harpyia*, are the largest of all *flying* birds.

The region which may be regarded as the common resort of the Condor, begins at the elevation of Mount Etna. It embraces atmospheric strata which are from 10,000 to 19,000 feet above the level of the sea. Humming birds also, which in their summer flights advance as far as 61° north lat. on the western coast of America, and are on the other hand found in the Archipelago of the Tierra del Fuego, were seen by Von Tschudi in Puna at an elevation of 14,600 feet.† There is a pleasure in comparing the largest and the smallest of the feathered inhabitants of the air. The largest among the Condors found in the Cordilleras, near Quito, measure nearly 15 feet across the expanded wings, and the smaller ones 8½ feet. This size, and the visual angle at which the birds are seen vertically above one's head, afford an idea of the enormous height to which the Condor soars in a clear sky. A visual angle of four minutes, for instance, would give a vertical elevation of 7330 feet. The cavern (Mackay) of Antisana, opposite the mountain of Chussulongo, and where we measured the birds soaring over the chain of the Andes, lies at an elevation of nearly 16,000 feet above the surface of the Pacific; the absolute height which the Condor reached must therefore be 23,273 feet, a height at which the barometer scarcely stands at 12·7 inches; but which, however, does not exceed that of the loftiest summit of the Himalaya. It is a remarkable physiological phenomenon that the same bird, which wheels for hours together through these highly rarefied regions, should be able suddenly, as for instance on the western declivity of the volcano of Pichincha, to descend to

* See my *Recueil d'Observations de Zoologie et d'Anatomie comparée*, vol. i. p. 26—45.

† *Fauna Peruana, Ornithol.* p. 12.

the sea-shore, and thus in the course of a few hours traverse, as it were, all climates. At heights of 23,000 feet and upwards the membranous air-sacs of the Condor must undergo a remarkable degree of inflation after being filled in lower regions of the atmosphere.

Ulloa, more than a hundred years ago, expressed his astonishment that the Vulture of the Andes could soar at heights where the pressure of the atmosphere was less than fifteen inches.* An opinion was at that time entertained, from the analogy of experiments made with the air-pump, that no animal could exist under this slight amount of atmospheric pressure. I have myself, as has already been mentioned, seen the barometer fall to 14·85 inches on the Chimborazo; and my friend, M. Gay-Lussac, breathed for a quarter of an hour an atmosphere in which the pressure was only 12·9 inches. It must be admitted that man, when wearied by muscular exertion, finds himself in a state of painful exhaustion at such elevations; but in the Condor, the respiratory process seems to be performed with equal facility under a pressure of 30 or of 13 inches. This bird probably raises itself *voluntarily* to a greater height from the surface of our earth than any other living creature. I use the expression "voluntarily," since small insects and siliceous-shelled infusoria are frequently borne to greater elevations by a rising current of air. It is probable that the Condor flies even higher than the above calculations would appear to show. I remember observing near the Cotopaxi, in the pumice plain of Suniguaicu, at an elevation of 14,471 feet above the level of the sea, this bird soaring at such a height above my head that it appeared like a black speck. But what is the smallest angle under which faintly illumined objects can be distinguished? Their form (linear extension) exercises a great influence on the minimum of this angle. The transparency of the mountain air is so great under the equator, that in the province of Quito, as I have elsewhere stated, the white cloak (*poncho*) of a horseman may be distinguished with the naked eye at a horizontal distance of 89,664 feet, and therefore under an angle of thirteen seconds. It was my friend Bonpland whom we observed, from the pleasant country-seat of the Marques de Selvalegre,

* *Voyage de l'Amérique méridionale,* t. ii. p. 2. 1752; *Observations astronomiques et physiques,* p. 110.

moving along a black rocky precipice on the volcano of Pichincha. Lightning conductors, being thin elongated objects, are visible, as Arago has observed, from the greatest distances and under the smallest angles.

The account I have given in my Monograph of the Condor (*Zoologie*, pp. 26—45) of the habits of this powerful bird in the mountain districts of Quito and Peru has been confirmed by a more recent traveller, Gay, who has explored the whole of Chili, and described it in his admirable work, *Historia fisica y politica de Chile.* This bird which, singularly enough, like the Lamas, Vicuñas, Alpacas and Guanacos, is not found beyond the equator in New Granada, penetrates as far south as the Straits of Magellan. In Chili, as in the elevated plateaux of Quito, the Condors, which usually live in pairs, or even alone, congregate in flocks for the purpose of attacking lambs and calves, or seizing on young Guanacos (Guanacillos). The havoc annually committed by the Condor among the herds of sheep, goats and cattle, as well as among the wild vicuñas, alpacas and guanacos of the chain of the Andes is very considerable. The Chilians assert that this bird when in captivity can endure hunger for forty days; when in a free state, however, its voracity is excessive, and it then, like the vulture, feeds by preference on carrion.

The mode of catching these birds, by an inclosure of palisades such as I have already described, is as successful in Chili as in Peru, for the bird after being rendered heavy from excess of food is obliged to run a short distance with half-extended wings before it can take flight. A dead ox which is already in an incipient state of decomposition, is strongly inclosed with palisades, within which narrow space the Condors throng together; being unable, as already observed, to fly on account of the excess of food which they have devoured, and impeded in their run by the palisades, these birds are either killed by the natives with clubs, or are caught alive by the lasso. The Condor was represented as a symbol of strength on the coinage of Chili immediately after the first declaration of political independence.*

The different species of Gallinazos, which are much more considerable in point of numbers than the Condors, are also

* Claudio Gay, *Historia fisica y politica de Chile,* publicada bajo los auspicios del Supremo Gobierno; Zoologia, pp. 194—198.

far more useful than the latter in the great economy of Nature for destroying and removing animal substances that are becoming decomposed, and thus purifying the atmosphere in the neighbourhood of human dwellings. In tropical America, I have sometimes seen seventy or eighty of these creatures collected round a dead ox; and I am able, as an eye-witness, to confirm the fact that has of late erroneously been called in question by ornithologists, that the appearance of one single king-vulture (who is not larger than the Gallinazos) is sufficient to put a whole assemblage of these birds to flight. No contest ever takes place; but the Gallinazos (two species of which, (Cathartes urubu and C. aura,) have been confounded together by an unfortunately fluctuating nomenclature) are intimidated by the sudden appearance and the courageous demeanour of the richly coloured "*Sarcoramphus Papa.*" As the ancient Egyptians protected the Percnopteri, which purified the atmosphere, so also the wanton destruction of Gallinazos is punished in Peru by a fine (*multa*), which, according to Gay, amounts in some cities to 300 piastres for every bird. It is a remarkable fact, that this species of vulture, as was already testified by Don Felix de Azara, if trained early, will so accustom themselves to the person who has reared them, that they will follow him on a journey for many miles, flying after his carriage across the Pampa.

(3) p. 211—"*Encloses their rotating bodies.*"

Fontana, in his admirable treatise "on the poison of the viper," vol. i. p. 62, mentions that he succeeded in restoring to animation, after two hours' immersion in a drop of water, a wheel-animalcule which had lain in a dried and motionless condition for the space of two years and a half.*

The so-called reanimation of Rotifera has very recently again been made a subject of lively discussion, since observations have been conducted with more exactness and subjected to a stricter criticism. Baker affirmed that in 1771, he had revived paste-eels which Needham had given him in the year 1744! Franz Bauer saw his *Vibrio tritici*, which had lain four years in a dry state, move on being moistened. The

* On the action of water, see my *Versuche über die gereizte Muskel- und Nervenfaser*, Bd. ii. S. 250.

remarkably careful and experienced observer, Doyère,* draws the following conclusions from his beautiful experiments: that Rotifera revive, i. e. pass from a motionless state to one of motion, after being exposed to a cold of 11°.2 Fahr., or to a heat of 113° Fahr.; that they preserve the property of reviving in dry sand up to a temperature of 159° Fahr.; but that they lose this property and remain immoveable if warmed in *moist sand* to 131° Fahr. only;† and that the possibility of this so-called revivification is not prevented by their being exposed to desiccation for twenty-eight days in barometric tubes, in vacuo, even should chloride of lime or sulphuric acid be employed.‡

Doyère has also seen Rotifera slowly revive after being dried without sand, (desséchés à nu,) a fact which Spallanzani denies.§ "Desiccation conducted in an ordinary temperature might be open to many objections which are not perhaps wholly obviated by the employment of a dry vacuum; but when we observe that the *Tardigrades* irrevocably perish in a temperature of 131° Fahr. if their tissues are permeated with water, whereas they can, when dried, support a temperature that may be estimated at 248° Fahr., we are disposed to admit that the sole condition required for *animal* revivification is the perfect integrity of organic structure and continuity."

In like manner, the sporules, or germinating cells of cryptogamic plants, which Kunth compares to the propagation of certain phanerogamic plants by buds (bulbillæ), retain their power of germination in the highest temperature. According to the most recent experiments of Payen, the sporules of a small fungus (Oïdium aurantiacum), which invests the crumb of bread with a reddish feathery coating, do not even lose their vegetative powers by being exposed in closed tubes for half an hour to a temperature of 183° to 208° Fahr. before being strewn on fresh, unspoilt dough. May not the newly discovered and wonderful monad (Monas prodigiosa), which causes blood-like spots in mealy substances, have been mixed with this fungus?

* See his *Mémoire sur les Tardigrades et sur leur propriété de revenir à la vie* (1842).

† Doyère, *Op. cit.* p. 119.

‡ Doyère, *Op. cit.* pp. 130—133.

§ Doyère, *Op. cit.* pp. 117 and 129.

Ehrenberg, in his great work on Infusoria (p. 492—496), has given the most complete history of all the observations instituted on the so-called revivification of Rotifera. He believes, that notwithstanding all the means of desiccation employed, the organization-fluid still remains in the apparently dead animal. He contests the hypothesis of "latent life"; for death, he says, "is not life in a torpid state, but the absence of life."

The hybernation or winter-sleep of both warm and cold-blooded animals, as dormice, marmots, sand-martins (*Hirundo riparia*, according to Cuvier)*, and of frogs and toads, affords us evidence of the diminution, if not of the complete suspension, of the organic functions. Frogs awakened from their winter-sleep by warmth, can remain eight times longer under water, without drowning, than frogs in the breeding season. It seems as if the respiratory functions of the lungs require a less degree of activity after the long suspension of their excitability. The circumstance of the sand-martin burying itself during the winter in marshes, is a phenomenon which, while it scarcely admits of a doubt, is the more remarkable, because in birds, the function of respiration is so extremely energetic, that, according to Lavoisier's experiments, two sparrows in an ordinary condition will, in the same time, decompose as much atmospheric air as a Guinea-pig.† Winter-sleep is not supposed to be general to the whole species of these sand-martins, but only to some few individuals.‡

As in the frigid zone deprivation of warmth produces winter-sleep in some animals, so in the torrid regions, within the tropics, an analogous phenomenon is manifested that has not hitherto been sufficiently regarded, and to which I have applied the term *summer-sleep*.§ Drought and a continuous high temperature act like the cold of winter in reducing excitability. Madagascar, excepting a very small portion of its southern extremity, lies within the tropics, and here, as was already observed by Bruguière, the hedgehog-like Tenrecs (*Centeres*, Illiger), one species of which (*C. ecau-*

* *Règne animal*, 1829, t. i. p. 396.

† Lavoisier, *Mémoires de Chimie*, t. i. p. 119.

‡ Milne Edwards, *Eléments de Zoologie*, 1834, p. 543.

§ *Relat. hist.*, t. ii. pp. 192, 626.

*datus*) was introduced into the Isle of France (20° 9′, latitude), sleep during excessive heat. The objection advanced by Desjardins, that the time of their sleep falls within the season of winter in the southern hemisphere, can scarcely be regarded as applicable in reference to a country, where the mean temperature of the coldest month is nearly 7° Fahr. above that of the hottest month in Paris; and this circumstance cannot therefore change the three months' summer-sleep of the Tenrec in Madagascar and Port Louis (Isle of France) into actual hybernation.

In a similar manner, the Crocodile in the Llanos of Venezuela, the land and water Tortoises on the Orinoco, and the colossal Boa, and many of the smaller species of serpents, lie torpid and motionless in the hardened ground, throughout the hot and dry season of the year. The missionary Gilij relates, that the natives, in seeking the dormant Terekai (land-tortoises), which lie buried in dry mud to the depth of 16 or 17 inches, are often bitten by serpents suddenly awakened, and which had buried themselves with the tortoises. An admirable observer, Dr. Peters, who has only just returned from the eastern coast of Africa, writes to me as follows: "I could not obtain any certain information regarding the Tenrec during my short stay in Madagascar, but I am, on the other hand, well aware, that in the portion of eastern Africa where I spent several years, different species of tortoises (Pentonyx and Trionices) remain enclosed for months together, without food, in the parched and indurated ground, during the dry season of this tropical country. The *Lepidosiren* also remains motionless and coiled up in the hardened earth, from May to December, wherever the swamps have been dried up."

We thus meet with an enfeeblement of certain vital functions in numerous and very different classes of animals, and, what is peculiarly striking, without the same phenomenon presenting itself in organisms nearly allied, and belonging to one and the same family. The northern glutton (Gulo), allied to the badger (Meles), does not, like the latter, sleep during the winter; whilst, according to Cuvier, "a Myoxus (Dormouse of Senegal, Myoxus Coupeii) which had probably never experienced a winter-sleep in its tropical home, fell into a state of hybernation at the beginning of winter, the first year it was brought to Europe." This enfeeblement of the vital functions and vital activity passes through several gradations, according as it extends to the processes of nutrition, respiration and muscular movement, or induces a depression of the cerebral and nervous systems. The winter-sleep of the solitary bear and of the badger is not attended with rigidity, and hence the awakening of these animals is easy, and, as I frequently heard in Siberia, very dangerous to the hunters and country people. The recognition of the gradation and connection of these phenomena leads us to the so-called *vita minima* of the microscopic organisms, which occasionally fall in the Atlantic in showers of meteoric dust, and some of which have green ovaries and are engaged in a self-generating process. The apparent revivification of the Rotifera and of the siliceous-shelled Infusoria is only the renewal of long enfeebled vital functions—a condition of vitality never entirely extinguished, but merely revived by excitation. Physiological phenomena can only be comprehended by being traced through the entire series of analogous modifications.

(4) p. 211—" *Winged Insects.*"

The fructification of diœcious plants was at one time principally ascribed to the agency of the wind. It has been shown by Kölreuter, and also with much ingenuity by Sprengel, that bees, wasps and numerous small winged insects, are the main agents in this process. I use the phrase "main agents", since I cannot regard it as consonant to nature that fructification should be impossible without the intervention of these insects, as Willdenow has also fully shewn.* On the other hand dichogamy, sap-marks, (*maculæ indicantes*), coloured spots indicating the presence of honey-vessels, and fructification by insects, appear to be almost inseparable from one another.†

The statement often repeated since Spallanzani, that the diœcious common hemp (*Cannabis sativa*), which was introduced into Europe from Persia, bears ripe seeds without being in the neighbourhood of pollen-tubes, has been entirely refuted by more recent investigations. When seeds have been obtained, anthers in a rudimentary state have been found near the ovarium, and these may have been capable of yield-

* *Grundriss der Kräuterkunde,* 4te Aufl. Berl. 1805. s. 405—412.

† Auguste de St. Hilaire, *Lecons de Botanique,* 1840, pp. 565— 571.

ing some grains of fructifying pollen. Such hermaphrodism is frequent in the whole family of *Urticeæ*, but a singular and hitherto unexplained phenomenon is manifested in the forcing-houses at Kew by a small New Holland shrub, the Cœlebogyne of Smith. This phanerogamic plant brings forth seeds in England without exhibiting any trace of male organs, and without the bastard introduction of the pollen of any other plant. "A species of Euphorbiaceæ," (?) writes the distinguished botanist, Jussieu, "the *Cœlebogyne*, which, although but recently described, has been cultivated for many years in English conservatories, has several times borne seeds, which were evidently perfect, since the well-formed embryos they contained have produced similar plants. The most careful observations have hitherto failed in discovering the slightest trace of anthers or even pollen in the flowers, which are diœcious. No male plants of this kind are known to exist in England. The embryo cannot therefore have come from the pollen, which is wholly deficient, but must have been formed entirely in the ovule."*

In order to obtain a fresh and confirmatory explanation of this important and isolated physiological phenomenon, I lately addressed myself to my young friend, Dr. Joseph Hooker, who after having accompanied Sir James Ross in his Antarctic voyage, has now joined the great Thibeto-Himalayan expedition. Dr. Hooker wrote to me as follows from Alexandria, at the close of December, 1847, prior to his embarkation at Suez: "Our Cœlebogyne still flowers with my father at Kew, as well as in the Gardens of the Horticultural Society. It ripens its seeds regularly. I have repeatedly examined it with care, but have never been able to discover a penetration of pollen utricles into the stigma, nor any traces of their presence in the latter or in the style. In my herbarium the male blossoms are in small catkins."

(5) p. 212—"*Like luminous stars.*"

The phosphorescence of the ocean is one of those splendid phenomena of nature which excite our admiration, even when we behold its recurrence every night for months together. The ocean is phosphorescent in all zones of the earth, but he who has not witnessed the phenomenon in the tropics, and

* Adrien de Jussien, *Cours élémentaire de Botanique,* 1840, p. 463.

especially in the Pacific, can form but a very imperfect idea of the majesty of this brilliant spectacle. The traveller on board a man-of-war, when ploughing the foaming waves before a fresh breeze, feels that he can scarcely satisfy himself with gazing on the spectacle presented by the circling waves. Wherever the ship's side rises above the waves, bluish or reddish flames seem to flash lightning-like upwards from the keel. The appearance presented in the tropical seas on a dark night is indescribably glorious, when shoals of dolphins are seen sporting around, and cutting the foaming waves in long and circling lines, gleaming with bright and sparkling light. In the Gulf of Cariaco, between Cumana and the Peninsula of Maniquarez, I have spent hours in enjoying this spectacle.

Le Gentil and the elder Forster ascribed these flames to the electrical friction of the water on the vessel as it glides forward—an explanation that must, in the present condition of our physical knowledge, be regarded as untenable.*

There are probably few subjects of natural investigation which have excited so many and such long-continued contentions as the phosphorescence of sea-water. All that is known with certainty regarding this much disputed question may be reduced to the following simple facts. There are many luminous mollusca which possess the property when alive of emitting at will a faint phosphoric light; which is of a bluish tinge in *Nereis noctiluca*, *Medusa pelagica var.* β,† and in the pipe-like *Monophora noctiluca*, discovered in Baudin's expedition.‡ The luminosity of sea-water is in part owing to living light-bearing animals, and in part to the organic fibres and membranes of the same, when in a state of decomposition. The first-named of these causes of the phosphorescence of the ocean is undoubtedly the most common and the most widely diffused. The more actively and the more efficiently that travellers engaged in the study

* Joh. Reinh. Forster, *Bemerkungen auf seiner Reise um die Welt*, 1783, s. 57; Le Gentil, *Voyage dans les Mers de l'Inde*, 1772, t. i. pp. 685—698.

† Forskaal, *Fauna ægyptiaco-arabica, s. Descriptiones animalium quæ in itinere orientali observavit*, 1775, p. 109.

‡ Bory de St.-Vincent, *Voyage dans les Iles des Mers d'Afrique*, 1804 t. i. p. 107, pl. vi.

of nature have learnt to employ powerful microscopes, the more our zoological systems have been enriched by new groups of mollusca and infusoria, whose property of emitting light either at will or from external stimulus has been recognised.

The luminosity of the sea, as far as it depends on living organisms, is principally owing, among zoophytes, to the Acalephæ (the families of Medusæ and Cyaneæ), to some Mollusca, and to an innumerable host of Infusoria. Among the small Acalephæ (Sea-nettles), the *Mammaria scintillans* presents us, as it were, with the glorious image of the starry firmament reflected in the surface of the sea. When full-grown this little creature scarcely equals in size the head of a pin. The existence of siliceous-shelled luminous infusoria was first shown by Michaelis at Kiel. He observed the coruscation of the Peridinium, (a ciliated animalcule,) of the Cuirass-monad (*Prorocentrum micans*), and of a rotifer, which he named Synchata baltica,* the same that Focke subsequently found in the lagoons of Venice. My distinguished friend and fellow traveller in Siberia, Ehrenberg, succeeded in keeping two luminous Infusoria of the Baltic alive for nearly two months at Berlin. I examined them with him in 1832; and saw them coruscate in a drop of sea-water on the darkened field of the microscope. When these luminous Infusoria (the largest of which was only $\frac{1}{8}$ and the smallest from $\frac{1}{48}$ to $\frac{1}{96}$ of a Parisian line in length) were exhausted, and ceased to emit sparks, they would renew their flashing on being stimulated by the addition of acids or by the application of a little alcohol to the sea-water.

By repeatedly filtering fresh sea-water, Ehrenberg succeeded in procuring a fluid in which a large number of these light-emitting animalcules were accumulated.† This acute observer has found in the organs of the Photocharis which give off flashes of light (either voluntarily or when stimulated), a cellular structure of a gelatinous character in the interior, and which manifests some similarity with the electric organ of the Gymnotus and the Torpedo. "When the Photocharis is irritated, in each cirrus a kindling and a gleaming of separate sparks may be observed, which gradually increase and at length illuminate the

* Michaelis, *Ueber das Leuchten der Ostsee bei Kiel*, 1830, s. 17.

† *Abhandlungen der Akad. der Wiss. zu Berlin aus dem J.* 1833, s. 307, 1834, s. 537—575, 1838, s. 45, 258.

whole cirrus; until the living flame runs also over the back of this nereid-like animalcule, making it appear under the microscope like a burning thread of sulphur with a greenish-yellow light. In the *Oceania* (*Thaumanthias*) *hemisphærica*, the number and position of the sparks correspond accurately, at the thickened base, with the larger cirri or organs which alternate with them, a circumstance that merits special attention. The manifestation of this wreath of fire is an act of vitality, and the whole development of light an organic vital process, which exhibits itself in Infusorial animals as a momentary spark of light, and is repeated after short intervals of rest."*

The luminous animals of the ocean appear, from these conjectures, to prove the existence of a magneto-electric light-generating vital process in other classes of animals besides fishes, insects, mollusca, and acalephæ. Is the secretion of the luminous fluid which is effused in some animalcules, and which continues to shine for a long period *without further influence of the living organism* (as, for instance, in Lampyrides and Elaterides, in the German and Italian glow-worms, and in the South American Cucuyo of the sugar-cane), merely the consequence of the first electric discharge, or is it simply dependent on chemical composition? The luminosity of insects surrounded by air assuredly depends on physiological causes different from those which give rise to a luminous condition in aquatic animals, fishes, Medusæ, and Infusoria. The small Infusoria of the ocean, being surrounded by strata of salt-water which constitutes a powerful conducting medium, must be capable of an enormous electric tension of their flashing organs to enable them to shine so vividly in the water. They strike like the Torpedo, the Gymnotus, and the Electric Silurus of the Nile, through the stratum of water: whilst electric fishes which, in connection with the galvanic circuit, are capable of decomposing water, and of imparting magnetic power to steel needles, (as I showed more than half a century ago,† and as John Davy has more recently confirmed,‡) yield

* Ehrenberg, *Ueber das Leuchten des Meeres*, 1836, s. 110, 158, 160, 163.

† *Versuche über die gereizte Muskel- und Nervenfaser*, bd. i. s. 438—441; see also *Obs. de Zoologie et d'Anatomie comparée*, vol. i. p. 84.

‡ *Philosophical Transactions for the year* 1834, part ii. pp. 545—547.

no indications of electricity through the smallest intervening stratum of flame.

The considerations which we have here developed render it probable that one and the same process operates, alike in the smallest living organisms invisible to the naked eye, in the contests of the serpent-like Gymnoti, in the flashing luminous Infusoria which impart such glorious brilliancy to the phosphorescence of the sea, in the thunder-cloud and in the terrestrial or polar light (the silent magnetic flashes), which, caused by an increased tension of the interior of the earth, are announced, for some hours previously, by the sudden variations of the magnetic needle.*

Sometimes one cannot, even with high magnifying powers, discover any animalcules in the luminous water; and yet, wherever a wave breaks in foam against a hard body, and, indeed, wherever water is violently agitated, flashes of light become visible. The cause of this phenomenon depends probably on the decomposing fibres of dead Mollusca, which are diffused in the greatest abundance throughout the water. If this luminous water be filtered through finely woven cloths, the fibres and membranes appear like separate luminous points. When we bathed at Cumana, in the gulf of Cariaco, and walked naked on the solitary beach in the beautiful evening air, parts of our bodies remained luminous from the bright fibres and organic membranes which adhered to the skin, nor did they lose this light for some minutes. If we consider the enormous quantity of Mollusca which animate all tropical seas, we can hardly wonder that sea-water should be luminous, even where no fibres can be visibly separated from it. From the endless subdivision of the masses of dead *Dagysæ* and *Medusæ* the whole ocean may, in fact, be regarded as a fluid containing gelatine, and, as such, luminous and of a nauseous taste; unfit for the use of man, but capable of affording nourishment to many species of fish. On rubbing a board with a portion of the *Medusa hysocella*, the surface thus rubbed recovers its phosphorescence when friction is applied by means of the dry finger. During my voyage to South America I occasionally placed a Medusa on a tin plate, and I then observed that if I struck the plate with another metallic

* See my letter to the editor of the *Annalen der Physik und Chemie*, bd. xxxvii. 1836, s. 242—244.

substance the slightest vibrations of the tin were sufficient to cause the animal to emit light. How do the blow and the vibrations here act? Is the temperature momentarily augmented, or are new surfaces presented? or, again, does some gaseous matter such as phosphuretted hydrogen, exude in consequence of this impulse, and burn when it comes in contact with the oxygen of the atmosphere, or with that dissolved in the sea-water, and by which the respiration of the Mollusca is maintained? This light-exciting effect of the blow is most remarkable in a cross or sugar-loaf sea, (*mer clapoteuse,*) where the waves, clashing from opposite directions, rise in a conical form.

I have seen the ocean, in the tropics, luminous in the most opposite kinds of weather, but most strongly so before a storm, or in a sultry and hazy atmosphere with thick clouds. Heat and cold appear to exercise but little influence on this phenomenon, for, on the Bank of Newfoundland, the phosphorescence is frequently very brilliant in the severest winter. Occasionally, too, the sea will be highly luminous one night, and not at all so on the following, notwithstanding an apparent identity of external conditions. Does the atmosphere favour this development of light? or do all the differences observed during this phenomenon depend on the accidental circumstance of the sea being more or less impregnated, in some parts, with the gelatinous portions of mollusca? Perhaps these phosphorescent social animalcules only rise to the surface under certain conditions of the atmosphere. It has been asked, why our fresh-water swamps which are filled with polyps are not phosphorescent. It would appear that, both in animals and plants, a peculiar mixture of organic particles favours this development of light; thus, for instance, the wood of the willow is more frequently found to be luminous than that of the oak. In England, salt-water has been rendered luminous by mixing herring-brine with it; indeed, it will be easy for any one to convince himself by galvanic experiments, that the luminosity of living animals depends on nervous irritation. I have observed strong phosphorescence emitted from a dying *Elater noctilucus*, on touching the ganglion of its fore leg with zinc and silver. Medusæ also occasionally emit a stronger light at the moment the galvanic circuit is completed.*

* Humboldt, *Relat. hist.*, t. i. pp. 79, 533. Respecting the wonder-

(6) p. 213—" *Which inhabits the lungs of the Rattlesnake of the tropics.*"

The animal which I formerly named an *Echinorhynchus*, and to which I even applied the term *Porocephalus*, appears, on a closer inspection, according to Rudolphi's better grounded opinion, to belong to the division of *Pentastoma*.* It is found in the abdominal cavity and the wide-celled lungs of a species of *Crotalus*, which, in Cumana, occasionally infests even the interior of houses, and preys on mice. The *Ascaris lumbrici*† lives beneath the skin of the common earth-worm, and is the smallest of all the species of Ascaris. *Leucophra nodulata*, Gleichen's pearl animalcule, has been observed by Otto Friedrich Müller in the interior of the reddish *Nais littoralis*.‡ It is probable that these microscopic animals are. in their turn, inhabited by others. All are surrounded by air, deficient in oxygen, and copiously charged with hydrogen and carbonic acid. It is extremely doubtful whether any animal could exist in *pure nitrogen*, although such an opinion did, formerly indeed, seem warranted with reference to Fischer's *Cistidicola farionis*, since, according to Fourcroy's experiments, the swimming-bladder of fish was presumed to contain air wholly devoid of oxygen. But the experiments made by Erman, and confirmed by myself, prove that the swimming-bladder of fresh-water fish never contains pure nitrogen § In sea fish as much as 0·80 parts of oxygen have been found, while, according to Biot's views, the purity of the air depends on the depth at which the fishes live.‖

(7) p. 214—" *The united Lithophytes.*"

According to Linnæus and Ellis the calcareous Zoophytes, (among which Madrepores, Meandrinæ, Astrææ, and Pocil-

ful development of mass and power of increase in the Infusorial animalcules, see Ehrenberg, *Infus.*, s. xiii. 291 and 512. " The galaxy of the smallest organisms," he says, "passes through the genera Monas (where they are often only $\frac{1}{3000}$ of a line), Vibrio, and Bacterium," (s. xix. 244.)

* Rudolphi, *Entozoorum Synopsis*, pp. 124, 434.

† See Gözen's *Eingeweidewürmer*, tab. iv. fig. 10.

‡ Müller, *Zoologia danica*, Fasc. ii. tab. lxxx. a—e.

§ Humboldt et Provençal, *Sur la respiration des Poissons*, in *Rec. d'Obs. de Zoologie*, vol. ii. pp. 194—216.

‖ *Mémoires de Physique et de Chimie de la Société d'Arcueil*, t. i. 1807, pp. 252—281.

loporæ especially produce mural coral-reefs,) are inhabited and invested by animalcules, which were long supposed to be allied to the Nereids belonging to Cuvier's Annelida (jointed worms). The anatomy of these gelatinous animalcules has been made known by the acute and comprehensive researches of Cavolini, Savigny, and Ehrenberg. We have learned that, in order to understand the whole organism of the (so-called) rock-building animals, we must not consider the scaffolding which remains after their death, namely, the layers of lime formed into delicate lamellæ by a vital function of secretion, as foreign to the soft membranes of the food-receiving animal.

Besides our increased knowledge of the wonderful formation of the living coral-stocks, a more correct view has gradually gained ground respecting the extensive influence which the coral world has exercised on the appearance of low island groups above the level of the sea, on the migration of land-plants, and the successive extension of the domain of the Floras, and, indeed, in some parts of the ocean, on the distribution of the human race and of languages.

As minute social organisms the corals play an important part in the general economy of nature, although they do not, as people began to believe after Capt. Cook's voyages of discovery, build up islands or enlarge continents from almost unfathomable depths of the ocean. They excite the liveliest interest, whether regarded as physiological objects, and as illustrating the various gradations of animal form, or in connection with the geography of plants, and the geognostic relations of the earth's crust. According to the comprehensive views of Leopold von Buch, the whole Jura-formation consists of "large elevated coral-banks of the ancient world, surrounding at a certain distance the old mountain chains."

According to Ehrenberg's classification,* coral-animals, (in English works often incorrectly termed coral-insects,) are separable into the monostomous *Anthozoa,* which are either free and with the power of detaching themselves, as *Animal-corals;* or are attached in the manner of plants, as *Phyto-corals.* To the first order (Zoocorallia) belong the Hydras or Armpolyps of Trembley, the Actiniæ, radiant with the most

* *Abhandlungen der Akad. der Wiss. zu Berlin aus dem J.* 1832, s. 393—432.

splendid colours, and the mushroom-corals; and to the second order belong the Madrepores, the Astrææ, and the Ocellinæ. The Polyps of the second order are those which from their cellular, wave-resisting, wall-works are the principal subject of this illustration. The wall-work is composed of the aggregate of the coral-trunks, which, however, do not suddenly lose their combined vitality, like a dead forest tree.

Every coral-trunk arises by a process of gemmation in accordance with certain laws, and forms one complete structure, each portion being formed by a great number of organically distinct individual animals. In the group of Phyto-corals these cannot separate themselves spontaneously, but remain united with one another by lamellæ of carbonate of lime. Hence each coral-trunk by no means possesses a central point of common vitality.* The propagation of coral-animals, according to the difference of the orders, is by eggs, spontaneous division or gemmation. This last kind of propagation presents the greatest variety of forms in the development of individuals.

The Coral-reefs (or, as Dioscorides designates them, sea-plants, a forest of stony-trees, Lithodendra), are of three kinds; namely, *Coast-reefs*, (shore-reefs, fringing-reefs), which are directly connected with continental or insular coasts, as on the north-east coast of New Holland, between Sandy Cape and the dreaded Torres Straits, and almost all the coral-banks of the Red Sea examined for eighteen months by Ehrenberg and Hemprich; *Island-surrounding reefs* (barrier-reefs, encircling-reefs), as at Vanikoro in the small archipelago of Santa Cruz, north of the New Hebrides, and at Puynipete, one of the Carolinas; and *Coral-banks surrounding lagoons* (Atolls or Lagoon-islands). This very natural division and nomenclature have been introduced by Charles Darwin, and are most intimately connected with the very ingenious explanation which this intellectual naturalist has given of the gradual origin of these wonderful forms. While, on the one hand, Cavolini, Ehrenberg, and Savigny have completed the scientific anatomical knowledge of the organization of coral-animals, on the other, the geographical and geological relations of coral-islands have been investigated, first by Reinhold and George Forster in Cook's second voyage, and then, after a long

* Ehrenberg, *Op. cit.*, s. 419.

interval, by Chamisso, Péron, Quoy and Gaimard, Flinders, Lütke, Beechey, Darwin, d'Urville, and Lottin.

The coral-animals and their stony cellular scaffoldings belong, for the most part, to the warm tropical seas; and the reefs occur most frequently in the Southern Hemisphere. Thus we find the Atolls or Lagoon Islands crowded together in the so-called coral-sea between the north-east coast of New Holland, New Caledonia, Solomon's Islands, and the Louisiade Archipelago; in the group of the Low Islands (Low Archipelago), eighty in number; in the Fidji, Ellice, and Gilbert Islands; and in the Indian Ocean, north-east of Madagascar, under the name of the Atoll group of Saya de Malha.

The great Chagos Bank, whose structure and dead coral-trunks have been thoroughly investigated by Captains Moresby and Powell, is the more interesting to us, because we may regard it as a prolongation of the more northern Laccadive and Maldive Islands. I have previously directed attention in another work* to the importance of the order of succession of the Atolls, which are exactly in the direction of a meridian as far as 7° south lat., in reference to the general mountain system, and the form of the earth's surface, in Central Asia. The meridian-chains, which mark the intersection of many mountain-systems running from east to west at the great bend of the Thibetian river Tzang-bo, correspond with the great meridian mountain rampart of the Ghauts and of the more northern Bolor in further or trans-Gangetic India. Here lie the parallel chains of Cochin China, Siam, and Malacca, as well as those of Ava and Arracan, which, after courses of unequal length, all terminate in the gulfs of Siam, Martaban, and Bengal. The bay of Bengal appears like an arrested effort of nature to produce an inland sea. A deep inbreak of the waters, between the simple western system of the Ghauts, and the very complex eastern trans-Gangetic system, has swallowed up a great part of the eastern lowlands, but met with an impediment not so easily overcome in the early existing and extensive table-land of Mysore.

An oceanic inbreak of this nature has given rise to two almost pyramidal peninsulas of very different length and narrowness; and the prolongation of two opposing meridian systems, the mountain system of Malacca in the east, and the

* *Asie centrale*, t. i. p. 218.

Ghauts of Malabar in the west, manifests itself in submarine, symmetrical series of islands, on the one side in the Andaman and Nicobar Islands, which are poor in corals, and on the other in three long-extended archipelagos of Atolls—the Laccadives, the Maldives, and Chagos. The last, called by mariners the Chagos Bank, forms a lagoon, belted by a narrow, and already much broken coral-reef. The length of this lagoon is 88, and its breadth 72 miles. Whilst the enclosed lagoon is only from 17 to 40 fathoms deep, bottom was scarcely found at a depth of 210 fathoms at a small distance from the outer margin of the coral wall, which appears to be now sinking.* At the coral-lagoon, known as Keeling-Atoll, south of Sumatra, Captain Fitz-Roy states, that at only 2000 yards from the reef, no soundings were found with 7200 feet of line.

"The forms of coral, which in the Red Sea rise in thick wall-like masses, are Mæandrinæ, Astrææ, Favia, Madrepores (Porites), Pocillopora (Hemprichii), Millepores, and Heteropores. The latter are among the most massive, although they are branched. The deepest coral trunks, which magnified by the refraction of light, appear to the eye to resemble the dome of a cathedral, belong, as far as could be determined, to Mæandrinæ and Astrææ."† A distinction must be made between single and in part free polyp-trunks, and those which form wall-like rocks.

If the accumulation of building polyp-trunks in some regions is so striking, it is no less astonishing to observe the perfect absence of these structures in other and often adjacent regions. Their presence or absence must be determined by certain, still uninvestigated, relations of currents, by the partial temperature of the water, and by the abundance or deficiency of nutriment. That certain delicate-branched corals, with less calcareous deposition on the side opposite to the mouth, prefer the stillness of the interior lagoons, is not to be denied; but this preference for still water must not, as has too often happened,‡ be regarded as a peculiarity of the whole class of these animals. According to the experiences of Ehrenberg and Chamisso in the Red Sea and in

* Darwin, *Structure of Coral Reefs*, pp. 39, 111 and 183.

† Ehrenberg's *Manuscript Notes.*

‡ *Annales des Sciences naturelles,* t. vi., 1825, p. 277.

the Marshall Islands, which abound in Atolls and lie east of the Caroline Islands, and according to the observations of Captains Bird Allen and Moresby in the West Indies and in the Maldives, we find that living Madrepores, Millepores, Astræas, and Mæandrinas, can support "a tremendous surf;"* and indeed seem to prefer localities the most exposed to the action of storms. The vital forces of the organism regulating the cellular structure, which with age acquires a rocky hardness, resist most triumphantly the mechanical forces,—the shock of moving waters.

In the South Pacific there is a perfect absence of coral-reefs at the Galapagos and along the whole of the west coast of the New Continent, notwithstanding their vicinity to the numerous Atolls of the Low Islands, and the Archipelago of Mendaña or the Marquesas. It is true that the current of the South Pacific, which washes the coasts of Chili and Peru, (and whose low temperature I observed in the year 1802,) is only 60°.1 Fahr., while the undisturbed water at the sides of the cold current is from 81°.5 to 83°.7 Fahr. at Punta Parima, where it deflects to the west. Moreover at the Galapagos there are small currents between the islands, having a temperature of only 58°.3 Fahr. But this lower temperature does not prevail further northwards along the coasts of the Pacific from Guayaquil to Guatimala and Mexico, neither does it prevail in the Cape de Verd Islands, on the whole west coast of Africa, or at the small islands of St. Paul, St. Helena, Ascension, and San Fernando Noronha; yet in none of these are there coral-reefs.

If this absence of reefs characterises the *western* coasts of America, Africa, and New Holland, they are, on the other hand, of frequent occurrence on the *eastern* coasts of tropical America, on the African coast of Zanzibar, and on the southern coast of New South Wales. The best opportunities I have enjoyed for personally examining coral banks have been in the Gulf of Mexico, and south of the Island of Cuba, in the so-called "Gardens of the King and Queen" (*Jardines y Jardinillos del Rey y de la Reyna*). It was Christopher Columbus himself who, on his second voyage, in May, 1494, gave this name to this little group of islands, because from the pleasant association of the silver-leaved arborescent Tour-

* Darwin, *Coral Reefs*, p. 63—65.

nefortia gnapholoides, of flowering species of Dolichos, of Avicennia nitida, and mangrove-thickets (Rhizophora), the coral-islands formed as it were an archipelago of floating gardens. "*Son Cayos verdes y graciosos llenos de arboledas,*" says the admiral. On my voyage from Batabano to Trinidad de Cuba, I remained for several days in these gardens, which lie to the east of the great Isle of Pines, abounding in mahogany, for the purpose of determining the longitude of the different *Cayos.*

The *Cayos Flamenco, Bonito, de Diego Perez,* and *de Piedras,* are coral islands, rising only from 8 to 15 inches above the level of the sea. The upper edge of the reef does not consist merely of dead polyp-trunks, but is rather formed of a true conglomerate, in which angular pieces of coral, lying in various directions, are embedded in a cement composed of granules of quartz. In Cayo de Piedras I saw such embedded masses of coral, some of them measuring upwards of three cubic feet. Several of the West Indian smaller coral islands have fresh water, a phenomenon which merits a careful investigation wherever it occurs (as for instance near Radak in the South Sea),* since it has sometimes been ascribed to hydrostatic pressure, acting from a distant coast (as in Venice, and in the Bay of Xagua, east of Batabano), and sometimes to the filtration of rain-water.†

The living gelatinous covering of the calcareous fabric of the coral-trunks attracts fishes and even turtles in search of food. In the time of Columbus the now desolate district of the Jardines del Rey was animated by a singular branch of industry pursued by the inhabitants of the sea-coasts of Cuba, who availed themselves of a little fish, the Remora, or sucking-fish (the so-called Ship-holder), probably the Echeneis naucrates, for catching turtles. A long and strong line, made of the fibres of the palm, was attached to the tail of the fish. The Remora (called in Spanish *Reves*, or reversed, because at first sight the back and abdomen might easily be mistaken for each other), attaches itself by suction to the turtle through the indented and moveable cartilaginous plates of the upper shell that covers

* Chamisso, in *Kotzebue's Entdeckungsreise,* bd. iii., s. 108.

† See my *Essai Politique sur l'Ile de Cuba,* t. ii. p. 137.

the head. The Remora, says Columbus, would rather let itself be torn to pieces than relinquish its prey, and the little fish and the turtle are thus drawn out of the water together. "Nostrates," says Martin Anghiera, the learned secretary of Charles V, "piscem Reversum appellant, quod versus venatur. Non aliter ac nos canibus gallicis per æquora campi lepores insectamur, illi (incolæ Cubæ insulæ) venatorio pisce pisces alios capiebant."* We learn from Dampier and Commerson, that this artifice of employing a sucking-fish to catch other fishes is very common on the eastern coasts of Africa, near Cape Natal and Mozambique, as well as on the island of Madagascar.† An acquaintance with the habits of animals, and the same necessities, lead to similar artifices and modes of capture amongst tribes having no connection with one another.

Although, as we have already remarked, the actual seat of the Lithophytes who build calcareous walls, lies within a zone extending from 22 to 24 degrees on either side of the equator, yet coral-reefs, favoured, it is supposed, by the warm Gulf Stream, are met with around the Bermudas in 32° 23′ lat., and these have been admirably described by Lieutenant Nelson.‡ In the southern hemisphere corals (Millepores and Cellepores) are found singly as far as Chiloe and even to the Chonos-Archipelago and Tierra del Fuego, in 53° lat., while Retepores have even been found as far as 72½° lat.

Since Captain Cook's second voyage, the hypothesis advanced by him as well as by Reinhold and George Forster, that the flat coral islands of the South Pacific have been built up by living agents from the depths of the sea's bottom, has found numerous advocates. The distinguished naturalists Quoy and Gaimard, who accompanied Captain Freycinet on his voyage of circumnavigation in the frigate "Uranie," were the first who expressed themselves, in 1823, with much freedom against the views advanced by the two Forsters (father and son), by Flinders, and Péron.§ "In directing the

* Petr. Martyr, *Oceanica*, 1532, Dec. 1, p. 9; Gomara, *Hist. de las Indias*, 1553, fol. xiv.

† Lacépède, *Hist. nat. des Poissons*, t. i. p. 55.

‡ *Transactions of the Geological Soc.*, 2nd Ser. vol. v. P. 1, 1837, p. 103.

§ *Annales des Sciences naturelles*, t. vi., 1825, p. 273.

attention of naturalists to coral-animalcules," they say, "we hope to be able to prove that all which has been hitherto affirmed or believed up to the present time, regarding the immense structures they are capable of raising, is for the most part inexact, and in all cases very greatly exaggerated. We are rather of opinion that coral-animalcules, instead of rearing perpendicular walls from the depths of the Ocean, only form strata or incrustrations of some few toises in thickness." Quoy and Gaimard (p. 289) have also expressed an opinion, that Atolls (coral walls inclosing a lagoon) owe their origin to submarine volcanic craters. They have undoubtedly underrated the depth at which animals who construct coral-reefs (as for example the Astræa) can exist, as they place the extreme limits at from 26 to 32 feet below the level of the sea. Charles Darwin, a naturalist, who has known how to enhance the value of his own observations by a comparison with those of others in many parts of the world, places the region of living coral-animals at a depth of 20 or 30 fathoms,* which corresponds with that in which Professor Edward Forbes found the greatest number of corals in the Ægean Sea. This is Professor Forbes's fourth region of marine-animals, as given in his ingenious memoir on the *Provinces of Depth*, and the geographical distribution of Mollusca at perpendicular distances from the surface.† It would appear, however, that the depth at which corals live is very different in the different species, especially in the more delicate ones which do not form such considerable structures.

Sir James Ross, in his Antarctic expedition, brought up corals from a great depth with the lead; and these he remitted for accurate examination to Mr. Stokes and Professor Forbes. Westward of Victoria Land, in the neighbourhood of the Coulman Island, in 72° 31′ south lat., and at a depth of 270 fathoms, Retepora cellulosa, a Hornera, and Prymnoa Rossii, (the latter very similar to a species common to the coasts of Norway,) were found alive and in a perfectly

* See *Darwin's Journal,* 1845, p. 467, also his *Structure of Coral Reefs,* pp. 84—87; and Sir Robert Schomburgk, *Hist. of Barbadoes,* 1848, p. 636.

† *Report on Ægean Invertebrata* in the *Report of the Thirteenth Meeting of the British Association, held at Cork in* 1843, pp. 151, 161.

fresh condition.* In the far north too, the Greenland *Umbellaria Grœnlandica* has been brought up alive by whale fishers from a depth of 236 fathoms.† The same relation between species and locality is met with among sponges, which however are now regarded as belonging more to plants than to zoophytes. On the shores of Asia Minor, the common marine sponge is brought up from depths varying from 5 to 30 fathoms, although one very small species of the same genus is only found at a depth of at least 180 fathoms.‡ It is difficult to divine what hinders the Astræas, Madrepores, Mæandrinas, and the whole group of tropical phyto-corals, which are capable of constructing large cellular calcareous walls, from living in very deep strata of water. The decrease of temperature is very gradual, the diminution of light nearly the same, and the existence of numerous Infusoria at great depths of the Ocean proves that there cannot here be any deficiency of food for polyps.

In opposition to the hitherto generally adopted opinion respecting the absence of all organisms and living creatures in the Dead Sea, it is worthy of notice that my friend and fellow-labourer, M. Valenciennes, has received, through the Marquis Charles de l'Escalopier, and through the French Consul Botta, beautiful specimens of Porites elongata from the Dead Sea. This fact is the more interesting, because this species is not found in the Mediterranean, but only in the Red Sea, which, according to Valenciennes, has but few organisms in common with the Mediterranean. As a sea-fish, a species of Pleuronectes, advances far into the interior of France, and accustoms itself to gill-respiration in fresh water, so also does a remarkable flexibility of organization exist in the above-mentioned coral-animal (Porites elongata of Lamarck), as the same species lives both in the Dead Sea, which is super-saturated with salt, and in the open ocean near the Séchelles Islands.§

According to the most recent chemical analyses of the younger

* See Ross, *Voyage of Discovery in the Southern and Antarctic Regions*, vol. i. pp. 334, 337.

† Ehrenberg, in the *Abhandl. der Berl. Akad. aus dem J.* 1832, s. 430.

‡ Forbes and Spratt, *Travels in Lycia*, 1847, vol. ii. p. 124.

§ See my *Asie centrale*, t. ii. p. 517.

Silliman, the genus Porites, like many other cellular coral-trunks (Madrepores, Astræas, and Mæandrinas of Ceylon and the Bermudas), contains besides from 92 to 95 per cent. of carbonate of lime and magnesia, a portion of fluorine and phosphoric acid.* The presence of fluorine in the hard skeleton of the polyps reminds us of the fluoride of calcium found in fish bones according to Morechini's and Gay-Lussac's experiments at Rome. Silex is mixed only in very small quantities, with the fluoride of calcium and phosphate of lime found in the coral-trunks; but one coral animal allied to the Horn corals (Gray's *Hyalonema*, Glass thread) has an axis of fibres of pure silex, resembling a hanging tuft of hair. Professor Forchhammer, who has recently been engaged in a thorough analysis of sea-water in the most opposite parts of the earth's surface, finds the quantity of lime in the Caribbean Sea remarkably small, it being only $\frac{247}{10000}$, whilst in the Cattegat it amounts to $\frac{371}{10000}$. He is disposed to ascribe this difference to the numerous coral-banks near the West India Islands, which appropriate the lime to themselves, and thus exhaust the sea-water.†

Charles Darwin has with great ingenuity developed the genetic connection between shore-reefs, island-encircling reefs, and lagoon islands, *i. e.*, narrow, annular coral banks which surround inner lagoons. According to his views, these three kinds of structure depend upon the oscillating condition of the bottom of the sea, or on periodical elevations and subsidences. The often-advanced hypothesis, according to which the lagoon-islands, or atolls, mark by their circularly enclosed coral-reefs, the outline of a submarine crater, raised on a volcanic crater-margin, is opposed by the great extent of their diameters, which are in some instances upwards of 30, 40, or even 60 miles. Our fire-emitting mountains have no such craters, and if we would compare the lagoon, with its submerged mural surface and narrow encircling reef, with one of the annular lunar mountains, we must not forget that these annular mountains are not volcanoes, but tracts of land

* Compare James Dana (geologist in the United States' Exploring Expedition under the command of Captain Wilkes), *On the Structure and Classification of Zoophytes*, 1846, pp. 124—131.

† *Report of the Sixteenth Meeting of the British Association for the Advancement of Science, held in* 1846, p. 91.

enclosed by walls. According to Darwin, the following is the process of formation. An island mountain closely encircled by a coral reef subsides, while the *fringing* reef that had sunk with it, is constantly recovering its level owing to the tendency of the coral animals to regain the surface by renewed perpendicular structures; these constitute first a reef encircling the island at a distance, and subsequently, when the inclosed island has wholly subsided, an *atoll.* According to this view, which regards islands as the most prominent parts, or the culminating points of the submarine land, the relative position of the coral islands would disclose to us what we could scarcely hope to discover by the sounding line, viz., the former configuration and articulation of the land. This attractive subject (to the connection of which with the migrations of plants and the distribution of the races of men we drew attention at the beginning of this note), can only be fully elucidated when we shall succeed in acquiring further knowledge of the depth and nature of the different rocks which serve as a foundation for the lower strata of the dead polyp-trunks.

(8) p. 216—" *Of the Samothracian Traditions.*"

Diodorus has preserved to us these remarkable traditions, the probability of which has invested them with almost historical certainty in the eyes of geologists. The island of Samothrace, once also named Ethiopea, Dardania, and Leucania or Leucosia in the Scholiast of Apollonius Rhodius, the seat of the ancient mysteries of the Cabiri, was inhabited by the remnant of an aboriginal people, several words of whose vernacular language were preserved in later times in sacrificial ceremonies. The position of Samothrace, opposite to the Thracian Hebrus, and near the Dardanelles, explains why a more circumstantial tradition of the great catastrophe of an outburst of the waters of the Pontus (Euxine) should have been especially preserved in this island. Sacred rites were here performed at altars erected on the supposed limits of this inundation; and among the Samothracians, as well as the Bœotians, a belief in the periodical destruction of the human race (a belief which also prevailed among the Mexicans in their myth of the four destructions of the world) was associated with historical recollections of individual inundations.* According to Diodorus, the Samothracians related that the Black Sea had been an inland lake, which, swelled by the influx of rivers (long prior to the inundations which had occurred among other nations) had burst, first through the straits of the Bosphorus, and subsequently through those of the Hellespont.† These ancient revolutions of nature have been considered in a special treatise, by Dureau de la Malle, and all the facts known regarding them collected by Carl von Hoff, in an important work on the subject.‡ The Samothracian traditions seem reflected as it were in the Sluice-theory of Strato of Lampsacus, according to which the swelling of the waters in the Euxine first formed the passage of the Dardanelles, and next the opening through the Pillars of Hercules. Strabo, in the first book of his Geography, has preserved among the critical extracts from the works of Eratosthenes, a remarkable fragment of the lost work of Strato, which presents views that embrace almost the whole circumference of the Mediterranean.

"Strato of Lampsacus," says Strabo,§ "enters more fully than the Lydian Xanthus (who has described the impressions of shells far from the sea) into a consideration of the causes of these phenomena. He maintains, that the Euxine had formerly no outlet at Byzantium, but that the pressure of the swollen mass of waters caused by the influx of rivers had opened a passage, whereupon the water rushed into the Propontis and the Hellespont. The same thing also happened to *our* sea (the Mediterranean), for here too a passage was opened through the isthmus at the Pillars of Hercules, in consequence of the filling of the sea by currents, which in flowing off left the former swampy banks uncovered and dry. In proof of this, Strato affirms, first, that the outer and inner bottoms of the sea are different; then that there is still a bank running under the sea from Europe to Lybia, which shows that the inner and outer sea were formerly not united; next that the Euxine is extremely shallow, while the Cretan,

* Otfr. Müller, *Geschichten Hellenischer Stämme und Städte,* bd. i. s. 65, 119.

† Diodor. Sicul. lib. v. cap. 47, p. 369. Wesseling.

‡ *Geschichte der natüralichen Veränderungen der Erdoberfläche,* Th. i. 1822, s. 105—162, and Creuzer's *Symbolik,* 2te Aufl. th. ii. s. 285, 318, 361.

§ Lib. i. p. 49, 50. Casaub.

the Sicilian and the Sardinian seas are, on the contrary, very deep; the cause of this being that the former is filled with mud from the numerous large rivers flowing into it from the north. Hence too the Euxine is the freshest, and the streams flowing from it are directed towards the parts where the bottom is deepest. It would also appear that if these rivers continue to flow into the Euxine, it will some day be completely choked with mud, for even now, its left side is becoming marshy in the direction of Salmydessus (the Thracian Apollonia), at the part called by mariners 'The Breasts,' before the mouth of the Ister and the desert of Scythia. Perhaps, therefore, the Lybian Temple of Ammon may also have once stood on the sea-shore, its present position in the interior of the country being in consequence of such off-flowings of rivers. Strato also conjectures that the fame and celebrity of the Oracle (of Ammon) is more easily accounted for, on the supposition that the temple was on the sea-shore, since its great distance from the coast would otherwise make its present distinction and fame inexplicable. Egypt also was in ancient times overflowed by the sea as far as the marshes of Pelusium, Mount Casius, and Lake Serbonis; for whenever in digging it happened that salt-water was met with, the borings passed through strata of sea-sand and shells, as if the country had been inundated, and the whole district around Mount Casius and Gerrha had been a marshy sea, continuous with the Gulf of the Red Sea. When the sea (the Mediterranean) retreated, the country was uncovered, leaving, however, the present Lake Serbonis. Subsequently the waters of this lake also flowed off, converting its bed into a swamp. In like manner the banks of Lake Mœris resemble more the shores of a sea than those of a river." An erroneous reading introduced as an emendation by Grosskurd, in consequence of a passage in Strabo,* gives in place of Mœris, "the Lake Halmyris," but the latter was situated near the southern mouth of the Danube.

The Sluice-theory of Strato led Eratosthenes of Cyrene (the most celebrated in the series of the librarians of Alexandria) to investigate the problem of the uniformity of level in all external seas flowing round continents, although with less success than Archimedes in his treatise on floating

* Lib. xvii. p. 809. Casaub.

bodies.* The articulation of the northern coasts of the Mediterranean as well as the form of its peninsulas and islands had given origin to the geognostic myth of the ancient land of Lyctonia. The origin of the lesser Syrtis, of the Triton Lake,† and of the whole of Western Atlas,‡ had been embodied in an imaginary scheme of fire-eruptions and earthquakes.§ I have recently entered more fully into this question,|| in a passage with which I would be allowed to close this note:

"The northern shore of the Mediterranean possesses the advantage of being more richly and variously articulated than the southern or Lybian shore, and this was, according to Strabo, already noticed by Eratosthenes. Here we find three peninsulas, the Iberian, the Italian, and the Hellenic, which, owing to their various and deeply indented contour, form, together with the neighbouring islands and the opposite coasts, many straits and isthmuses. Such a configuration of continents and of islands that have been partly severed and partly upheaved by volcanic agency in rows, as if over far-extending fissures, early led to geognostic views regarding eruptions, terrestrial revolutions, and outpourings of the swollen higher seas into those below them. The Euxine, the Dardanelles, the Straits of Gades, and the Mediterranean with its numerous islands, were well fitted to originate such a system of sluices. The Orphic Argonaut, who probably lived in the Christian era, has interwoven old mythical narrations in his composition. He sings of the division of the ancient Lyctonia into separate islands, 'when the dark-haired Poseidon in anger with Father Kronion struck Lyctonia with the golden trident.' Similar fancies, which may often certainly have sprung from an imperfect knowledge of geographical relations, were frequently elaborated in the erudite Alexandrian school, which was so devoted to everything connected with antiquity. Whether the myth of the breaking up of Atlantis be a vague and western reflection of that of

* Strabo, lib. i. p. 51—56, lib. ii. p. 104. Casaub.

† Diod. iii. 53—55.

‡ Maximus Tyrius, viii. 7.

§ Compare my *Examen critique de l'hist. de la Géographie,* t. i. p. 179, t. iii. p. 136.

|| *Cosmos,* vol. ii. p. 481. (Bohn's edition).

Lyctonia, as I have elsewhere shown to be probable, or whether, according to Otfried Müller, 'the destruction of Lyctonia (Leuconia) refers to the Samothracian tradition of a great flood, which changed the form of that district,' is a question which it is here unnecessary to decide."

(9) p. 217—"*Precipitation from the clouds.*"

The vertical ascent of currents of air is one of the principal causes of the most important meteorological phenomena. Where a desert or a sandy surface devoid of vegetation is surrounded by a high mountain-chain, the sea-wind may be observed driving a dense cloud over the desert, without any precipitation of vapour taking place before it reaches the crest of the mountains. This phenomenon was formerly very unsatisfactorily referred to an *attraction* supposed to be exercised by the mountain-chain on the clouds. The true cause appears to lie in the ascent from the sandy plain of a column of warm air, which prevents the condensation of the vesicles of vapour. The more barren the surface, and the greater the degree of heat acquired by the sand, the higher will be the ascent of the clouds, and the less readily will the vapour be precipitated. Over the declivities of mountains these causes cease. The play of the vertical column of air is there weaker; the clouds sink, and their disintegration is effected by a cooler stratum of air. Thus *deficiency of rain* and *absence of vegetation in the desert* stand in a reciprocal action to one another. It does not rain because the barren and bare surface of sand becomes more strongly heated and radiates more heat; and the desert is not converted into a steppe or grassy plain because without water no organic development is possible.

(10). p. 218—"*The indurating and heat-emitting mass of the earth.*"

If according to the hypothesis of the Neptunists (now long since obsolete), the so-called primitive rocks were also precipitated from a fluid, the transition of the earth's crust from a condition of fluidity to one of solidity, must have been followed by the liberation of an enormous quantity of caloric, which would have given rise to new evaporation and new precipitations. The more recent these precipitations, the more rapid, the more tumultuous, and the more uncrystalline would they have been. Such a sudden liberation of caloric from the indurating crust of the earth, independent of the latitude, and the position of the earth's axis, might indeed occasion local elevations of temperature in the atmosphere, which would influence the distribution of plants. The same cause might also occasion a kind of porosity which seems to be indicated by many enigmatical geological phenomena in floetz rocks. I have developed my conjectures on this subject in detail in a small memoir on primitive porosity.* According to the views I have more recently adopted, it appears to me that the variously shattered and fissured earth, with its fused interior, may long have continued in the primeval period, to impart to its oxidised surface a high degree of temperature, independent of its position with respect to the sun and to latitude. What an influence would not, for instance, be exercised for ages to come on the climate of Germany by an open fissure a thousand fathoms in depth, extending from the Adriatic Gulf to the northern coast? Although in the present condition of the earth, long-continued radiation has almost entirely restored the stable equilibrium of temperature first calculated by Fourier in his *Théorie analytique de la Chaleur*, and the outer atmosphere is now only brought into direct communication with the molten interior of the earth, by means of the insignificant openings of a few volcanoes; yet in the primitive condition of our planet, this interior emitted hot streams of air into the atmosphere through the various clefts and fissures formed by the frequently recurring foldings (or corrugations) of the mountain strata. This emission was wholly independent of latitude. Every newly formed planet must thus in its earliest condition have regulated its own temperature, which was, however, subsequently changed and determined by its position in relation to the central body, the sun. The moon's surface also exhibits traces of this reaction of the interior upon the crust.

(11) p. 218—" *The mountain-declivities of the most southern parts of Mexico.*"

The spherical greenstone in the mountain district of Gua-

* See my work, *Versuche über die chemische Zersetzung des Luftkreises,* 1799, p. 177; and Moll's *Jahrbücher der Berg- und Hüttenkunde,* 1797, p. 234.

naxuato is perfectly similar to that of the Fichtelberg in Franconia. Both form grotesque domes, which break through and are superimposed on transition argillaceous schists. In the same manner pearl-stone, porphyritic schist, trachyte and pitch-stone porphyry present analogous forms in the Mexican mountains, near Cinapecuaro and Moran, in Hungary, Bohemia, and in Northern Asia.

(12) p. 220—" *The Colossal Dragon-tree of Orotava.*"

This colossal dragon-tree (Dracæna draco) stands in the garden of M. Franqui, in the little town of Orotava, called formerly Taoro, one of the most charming spots in the world. In June, 1799, when we ascended the Peak of Teneriffe, we found that this enormous tree measured 48 feet in circumference. Our measurement was made at several feet above the root. Nearer to the ground Le Dru found it nearly 79 feet. Sir G. Staunton asserts that at an elevation of ten feet from the ground, its diameter is still 12 feet. The height of the tree is not much more than 69 feet. According to tradition it would appear that this tree was venerated by the Guanches (as was the ash-tree of Ephesus by the Greeks, the Plantain of Lydia, which Xerxes decorated with ornaments, also the sacred Banyan-tree of Ceylon), and that in the year 1402, which was the period of Béthencourt's first expedition, it was as large and as hollow as in the present day. When it is remembered that the dragon-tree is everywhere of very slow growth, we may conclude that the one at Orotava is of extreme antiquity. Berthollet says, in his description of Teneriffe, " On comparing the young dragon-trees which grows near this colossal tree, the calculations we are led to make on the age of the latter strike the mind with astonishment."* The Dragon-tree has been cultivated from the most ancient times in the Canary isles, in Madeira, and Porto Santo, and that accurate observer, Leopold von Buch, found it growing wild near Iguesti in Teneriffe. Its original habitat is not therefore the East Indies, as has long been believed; and its appearance does not afford any refutation of the opinion of those who regard the Guanches as a wholly isolated primitive Atlantic race, having no intercourse

* *Nova Acta Acad. Leop. Carol. Naturæ Curiosorum,* t. xiii. 1827, p. 781.

with African or Asiatic nations: The form of the *Dracænæ* is repeated on the southern extremity of Africa, in the Isle of Bourbon, in China, and in New Zealand. In these remotely distant regions we recognise species of the same genus, but none are to be found in the New Continent, where this form is supplied by the Yucca. The *Dracæna borealis* of Aiton is a true *Convallaria*, the nature of both being perfectly identical.*

I have given a representation, in the last plate of the Picturesque Atlas of my American journey,† of the dragon-tree of Orotava, taken from a drawing made in 1776 by F. d'Ozonne, and which I found among the posthumous papers of the celebrated Borda, in the still unprinted journal entrusted to me by the Dépôt de la Marine, and from which I have borrowed important astronomically-determined geographical, data besides many barometrical and trigonometrical notices.‡ The measurement of the dragon-tree in the Villa Franqui was made in Borda's first voyage with Pingré in 1771, and not in the second, made 1776 with Varela. It is asserted, that in the fifteenth century, during the early periods of the Norman and Spanish conquests, mass was performed at a small altar erected in the hollow trunk of this tree. Unfortunately, the Dracæna of Orotava lost one side of its leafy top in the storm of the 21st of July, 1819. There is a fine large English copper-plate engraving, which gives an exceedingly true representation of the present condition of the tree.

The monumental character of these colossal living forms, and the impression of reverence which they have created among all nations, have led, in modern times, to a more careful study of the numerical determination of their age, and of the size of their trunks. The results of such investigations induced the elder Decandolle, (the author of the important treatise, entitled *De la Longévité des Arbres*,) Endlicher, Unger, and other distinguished botanists to conjecture, that the age of many existing vegetable forms may extend to the earliest historical times, if not to the records of the Nile, at least to those of Greece and Italy. In the *Bibliothèque Universelle*

* Humboldt, *Rélat. hist.*, t. i. pp. 118, 639.

† *Vues des Cordillères et Monumens des peuples indigènes de l'Amerique*, pl. lxix.

‡ *Rélat. hist.*, t. i., p. 282.

*de Genève* (t. xlvii. 1831, p. 50) we find the following passage: "Numerous examples seem to confirm the idea, that there still exist, on our planet, trees of a prodigious antiquity—the witnesses, perhaps, of one or more of its latest physical revolutions. If we consider a tree as the combination of as many individual forms as there have been buds developed on its surface, one cannot be surprised if the aggregate resulting from the continual addition of new buds to the older ones, should not necessarily have any fixed termination to its existence." In the same manner, Agardh says: "If in each solar year new parts be formed in the plant, and the older hardened ones be replaced by new parts capable of conducting sap, we have a type of growth limited by external causes alone." He ascribes the short duration of the life of herbaceous plants, "to the preponderance of the production of blossoms and fruit over the formation of leaves." Unfruitfulness in a plant insures a prolongation of its life. Endlicher adduces the instance of an individual plant of Medicago sativa, var. β versicolor, which lived eighty years because it bore no fruit.*

To the dragon-trees, which, notwithstanding the gigantic development of their closed vascular bundles, must be classed, in respect to their floral parts, in the same natural family as Asparagus and the garden onion, belongs the Adansonia, (the monkey bread-tree, *Baobab*), undoubtedly among the largest and most ancient inhabitants of our planet. In the earliest voyages of discovery made by Catalans and Portuguese, the sailors were accustomed to carve their names on these two species of trees; not always from a mere wish of perpetuating their memory, but also as "marcos," or signs of possession, and of the rights which nations assume in virtue of first discovery. The Portuguese mariners often selected for carving on the trees, as a "marco," or mark of possession, the elegant French motto *talent de bien faire*, so frequently employed by the Infante Don Henrique, the Discoverer. Thus Manuel de Faria y Sousa says expressly;† "Era uso de los primeros Navegantes de dexar inscrito el motto del Infante, *talent de bien faire*, en la corteza de los arboles."‡ (It was the custom

* *Grundzüge der Botanik*, 1843, § 1003.

† *Asia Portuguesa*, t. i., cap. 2., pp. 14, 18.

‡ Compare also Barros, *Asia*, dec. i. liv. ii., cap. 2, t. i. (Lisboa, 1778;) p. 148.

of the early navigators to inscribe the motto of the Infante in the bark of the trees.)

The above-named motto, cut on the bark of two trees by Portuguese navigators in the year 1435, and therefore twenty-eight years before the death of the Infante Don Henrique, Duke of Viseo, is singularly connected, in the history of discoveries, with the discussions that have arisen from a comparison of Vespucci's fourth voyage with that of Gonzalo Coelho (1503). Vespucci relates, that the Admiral's ship of Coelho's squadron was wrecked on an island which was sometimes supposed to be that of San Fernando Noronha; sometimes, Peñedo de San Pedro; and sometimes, the problematical island of St. Matthew. The last-named island was discovered on the 15th of October, 1525, by Garcia Jofre de Loaysa in 2½ south lat., in the meridian of Cape Palmas, and almost in the Gulf of Guinea. He remained there eighteen days at anchor, and found crosses, orange-trees that had become wild, and two trunks of trees having inscriptions that bore the date of ninety years back.* I have in another place,† in an inquiry regarding the trustworthiness of Amerigo Vespucci, more fully considered this problem.

The oldest description of the Baobab (Adansonia digitata) is that of the Venetian, Aloysius Cadamosto, (whose real name was Alvise da Ca da Mosto) in 1454. He found at the mouth of the Senegal, (where he joined Antoniotto Usodimare), trunks, whose circumference he estimated at 17 fathoms, or 112 feet.‡ He might have compared them to dragon-trees, which he had already seen. Perrottet says,§ that he had seen monkey-bread fruit trees, which had a diameter of about thirty-two feet, with a height of only from seventy to eighty-five feet. The same dimensions had been given by Adanson in his voyage, 1748. The largest trunks of the monkey bread-fruit trees, which he himself saw, in 1749, some on one of the small Magdalena islands near Cape de Verd, and others at the mouth of the Senegal, were from 26 to nearly 29 feet in diameter, with a height of little more than 70 feet, and a top measuring upwards of 180 feet across.

* Navarrete, t. v, pp. 8, 247, 401.

† *Examen critique de l'Hist. de la Géographie*, t. v. pp. 129–132.

‡ Ramusio, vol. i. p. 109.

§ *Flore de Sénégambie*, p. 76.

Adanson, however, makes the remark that other travellers had found trunks having a diameter of about 32 feet.* French and Dutch sailors had carved their names on the trunks in characters six inches in length. One of these inscriptions was of the fifteenth century,† while all the others were of the sixteenth. From the depth of the cuts, which are covered with new layers of wood,‡ and from a comparison of the thickness of trunks, whose various ages were known, Adanson computed the age of trees having a diameter of 32 feet at 5150 years.§ He however cautiously subjoins the following remarks, in a quaint mode of spelling which I do not alter: "le calcul de l'aje de chake couche n'a pas d'exactitude géometrike." In the village of Grand Galarques, also in Senegambia, the negroes have adorned the entrance of a hollow Baobab with carvings cut out of wood still green. The inner cavity serves as a place of general meeting in which the community debate on their interests. This hall reminds us of the hollow (specus) in the interior of a plantain in Lycia, in which the Roman ex-consul, Lucinius Mutianus, entertained twenty-one guests. Pliny (xii. 3) gives to a cavity of this kind the somewhat ample breadth of eighty Roman feet. The Baobab was seen by René Caillié in the valley of the Niger near Jenne, by Cailliaud in Nubia, and by Wilhelm Peters along the whole eastern coast of Africa, where this tree, which is called *Mulapa, i.e. Nlapa-tree*, or more correctly *muti-nlapa*, advances as far as Lourenzo Marques, almost to 26° south lat. The oldest and thickest trunks seen by Peters "measured from 60 to 75 feet in circumference." Although Cadamosto observed, in the fifteenth century, *eminentia non quadrat magnitudini;* and although Golberry‖ found, in the "Vallée des deux Gagnacks," trunks only

* This tree was formerly called "the Ethiopian sour gourd;" Julius Scaliger, who gave it the name of Guanabanus, instances one, which seventeen men with outstretched arms could not encompass. The wood is very perishable, and the negroes place in the hollow of these trees the corpses of their conjurors, or of such persons who they suppose would enchant or desecrate the ground, if buried in the usual way.—Ed.

† *Familles des Plantes d'Adanson*, 1763, P. I. pp. ccxv—ccxviii. The fourteenth century is here stated, but this is no doubt an error.

‡ Adrien de Jussieu, *Cours de Botanique*, p. 62.

§ *Voyage au Sénégal*, 1757, p. 66.

‖ *Fragmens d'un voyage en Afrique*, t. ii. p. 92.

64 feet in height whose diameter was 36 feet," this disproportion between thickness and height must not be assumed to be general. "Very old trees," says the learned traveller, Peters, "lose their crowns by gradual decay, while they continue to increase in circumference. On the eastern coast of Africa one not unfrequently meets with trees having a diameter of more than 10 feet which reach the height of nearly 70 feet.'

While therefore the bold calculations of Adanson and Perrottet assign to the Adansonias measured by them, an age of 5150 or even 6000 years, which would make them coeval with the builders of the Pyramids, or even with Menes, and would place them in an epoch when the Southern Cross was still visible in Northern Germany;* the more certain estimations yielded by annular rings, and by the relation found to exist between the thickness of the layer of wood and the duration of growth, give us, on the other hand, shorter periods for our temperate northern zone. Decandolle finds that of all European species of trees, the yew attains the greatest age; and according to his calculations, 30 centuries must be assigned as the age of the *Taxus baccata* of Braburn in Kent, from 25 to 26 to the Scotch yew of Fortingal, and 14½ and 12 respectively to those of Crowhurst in Surrey and Ripon (Fountains Abbey) in Yorkshire.† Endlicher remarks that "another yew-tree in the churchyard of Grasford, North Wales, which measures more than 50 feet in girth below the branches, is more than 1400 years old, whilst one in Derbyshire is estimated at 2096 years. In Lithuania linden trees have been felled which measured 87 feet round, and in which 815 annular rings have been counted."‡ In the temperate zone of the southern hemisphere some species of the Eucalyptus attain an enormous girth, and as they at the same time attain a height of nearly 250 feet, they afford a singular contrast to our yew trees, which are colossal only in thickness. Mr. Backhouse found in Emu Bay, on the shore of Van Diemen's Land,

* *Cosmos*, vol. ii. p. 662. (Bohn's Edition.)

† Decandolle, *de la Longévité des Arbres*, p. 65. Fine engravings of the venerable yew at Fortingal, Fountains Abbey, Ankerwyke, &c., will be found in Strutt's magnificent work on forest trees. A very full account of the Yew-tree, with engravings, will also be found in Loudon's *Arboretum Britannicum*.—Ed.

‡ Endlicher, *Grundzüge der Botanik*, s. 399.

Eucalyptus trunks which, with a circumference of 70 feet at the base, measured as much as 50 feet at a little more than 5 feet from the ground.*

It was not Malpighi, as has been generally asserted, but the intellectual Michel Montaigne, who had the merit of first showing, in 1581, in his *Voyage en Italie*, the relation that exists between the annual rings and the age of the tree.† An intelligent artisan, engaged in the preparation of astronomical instruments, first drew Montaigne's attention to the significance of the annual rings, asserting that the part of the trunk directed towards the north had narrower rings. Jean Jacques Rousseau entertained the same opinion; and his Emile, when he loses himself in the forest, is made to direct his course in accordance with the deposition of the layers of wood. Recent phyto-anatomical observations‡ teach us, however, that the acceleration of vegetation as well as the remission of growth, and the varying production of the circles of the ligneous bundles (annual deposits) from the cambium cells, depend on other influences than position with respect to the quarter of the heavens.

Trees which in the case of some examples attain a diameter of more than 20 feet, and an age of many centuries, belong to very different natural families. We may here instance Baobabs, Dragon trees, various species of Eucalyptus, Taxodium distichum, (Rich.,) Pinus Lambertiana, (Douglasii,) Hymenæa Courbaril, Cæsalpinieæ, Bombax, Swietenia Mahagoni, the Banyan tree (*Ficus religiosa*), Liriodendron tulipifera (?), Platanus orientalis, and our Lindens, Oaks, and Yews. The celebrated Taxodium distichon, the Ahuahuete of the Mexicans (*Cupressus disticha*, Linn., *Schubertia disticha*, Mirbel), of Santa Maria del Tule, in the State of Oaxaca, has not a diameter of 60 feet, as stated by Decandolle, but exactly 40½ feet.§ The two beautiful Ahuahuetes which I have frequently seen at Chapoltepec (growing in what was probably once a garden or pleasure ground of Montezuma) measure, according to the instructive account in Burkardt's

* Gould, *Birds of Australia*, vol. i. Introd. p. xv.

† Adrien de Jussieu, *Cours élémentaire de Botanique*, 1840, p. 61.

‡ Kunth, *Lehrbuch der Botanik*, th. i. 1847, s. 146, 164; Lindley, *Introduction to Botany*, 2nd ed. p. 75.

§ Mühlenpfordt, *Versuch einer getreuen Schilderung der Republik Mexico*, bd. i. s. 153.

travels (bd. i. s. 268) only 36 and 38 feet in circumference, and not in diameter, as has often been erroneously maintained. The Buddhists of Ceylon venerate the colossal trunk of the sacred fig-tree of Anurahdepura. The Banyan, which takes root by its branches, often attains a thickness of 30 feet, and forms, as Onesicritus truly expresses himself, a leafy roof resembling a many-pillared tent.* On the Bombax Ceiba see early notices from the time of Columbus in Bembo.†

Among those oak trees which have been very accurately measured, the largest in Europe is undoubtedly the one near Saintes on the road to Cozes, in the Department de la Charente inférieure. This tree, which has an elevation of 64 feet, measures very nearly 30 feet in diameter near the ground, while 5 feet higher up it is nearly 23 feet, and where the main branches begin more than 6 feet. A little room, from 10 feet 8 inches to 12 feet 9 inches in width and 9 feet 7 inches in height, has been cleared in the dead part of the trunk, and a semicircular bench cut within it from the green wood. A window gives light to the interior, and hence the walls of this little room, which is closed by a door, are gracefully clothed with ferns and lichens. From the size of a small piece of wood that had been cut out over the door, and in which two hundred ligneous rings were counted, the age of the oak of Saintes must be estimated at 1800 or 2000 years.‡

With respect to the rose-tree (*Rosa canina*) reputed to be a thousand years old, which grows in the crypt of the Cathedral of Hildesheim, I learn from accurate information, based on authentic records, for which I am indebted to the kindness of the Stadtgerichts-Assessor Römer, that the main stem only has an age of eight hundred years. A legend connects this rose-tree with a vow of the first founder of the cathedral, Louis the Pious; and a document of the eleventh century says, "that when Bishop Hezilo rebuilt the cathedral, which had been burnt down, he enclosed the roots of the rose-tree within a vault still

* Lassen, *Indische Althertlumskunde*, bd. i. s. 260. See an interesting account of the Banyan tree in Forbes' *Oriental Memoirs*, vol. i. pp. 25—28. The tree there described (the famous *Cubbeer-Burr*) comprises 350 large trunks and more than 3000 small ones, and extends over an area of several thousand feet. Milton alludes to the Banyan tree in his *Paradise Lost*, book ix. line 1100, &c.—Ed.

† *Historiæ Venetæ*, 1551, fol. 83.

‡ *Annales de la Société d'Agriculture de la Rochelle*, 1843, p. 380.

remaining, raised on the latter the walls of the crypt, which was re-consecrated in 1061, and spread the branches of the rose-tree over its sides." The stem, still living, is nearly 27 feet in height, and only 2 inches thick, and spreads across a width of 32 feet over the outer wall of the eastern crypt. It is undoubtedly of very considerable antiquity, and well worthy of the renown it has so long enjoyed throughout Germany.

If excessive size, in point of organic development, may in general be regarded as a proof of a long protraction of life, special attention is due, among the thalassophytes of the submarine vegetable world, to a species of fucus, *Macrocystis pyrifera*, Agardh (*Fucus giganteus*). This marine plant attains, according to Captain Cook and George Forster, a length of 360 feet, and exceeds therefore the height of the loftiest Coniferous trees, not excepting *Sequoia gigantea*, Endl. (*Taxodium sempervirens*, Hook, and Arnott) of California.* Captain Fitz-Roy has confirmed this statement.† Macrocystis pyrifera grows from 64° south lat. to 45° north lat., as far as the Bay of San Francisco on the north-west coast of the New Continent; indeed Joseph Hooker believes that this species of Fucus advances as far as Kamtschatka. In the waters of the Antarctic seas it is even seen floating between the pack-ice.‡ The cellular band and thread-like structures of the Macrocystis (which are attached to the bottom of the sea by an adhesive organ resembling a claw) seem to be limited in their length by accidental disturbing causes alone.

### (13) p. 220—" *Phanerogamic plants already recorded in herbariums.*"

Three questions must be carefully distinguished from one another: 1. How many species of plants have been described in printed works? 2. How many of those discovered—that is to say included in herbariums—still remain undescribed? 3. How many species probably exist on the surface of the earth? Murray's edition of the Linnæan system contains, including cryptogamic plants, only 10,042 species. Willdenow, in his edition of the *Species Plantarum* from 1797 to

* Darwin, *Journal of Researches into Nat. Hist.*, 1845, p. 239.

† *Voyages of the Adventure and Beagle*, vol. ii. p. 363.

‡ *Flora Antarctica*, p. vii, 1 and 178; and Camille Montagne, *Botanique cryptogame du Voyage de la Bonite*, 1846, p. 36.

1807, has described as many as 17,457 species of phanerogamia, reckoning from Monandria to Polygamia diœcia. If to these we add 3000 species of cryptogamic plants, we shall bring the number as given by Willdenow to 20,000. More recent investigations have shown how far this estimate of the species described, and of those preserved in herbariums, falls short of the truth. Robert Brown* first enumerated above 37,000 phanerogamia, and I at that time attempted to describe the distribution of 44,000 species of phanerogamic and cryptogamic plants, over the different portions of the world already explored.† Decandolle finds, on comparing Persoon's *Enchiridium* with his *Universal System divided into twelve families*, that more than 56,000 species of plants may be enumerated from the writings of botanists and European herbariums.‡ If we consider how many new species have been described by travellers since that time, (my expedition alone afforded 3600 of the 5800 collected species of equinoctial plants), and if we bear in mind that there are assuredly upwards of 25,000 phanerogamic plants, cultivated in all the different botanical gardens, we shall soon see how much Decandolle's estimate is below the truth. From our complete ignorance of the interior of South America (Mato-Grosso, Paraguay, the eastern declivity of the Andes, Santa-Cruz de la Sierra, and all the countries lying between the Orinoco, the Rio Negro, the Amazon, and Puruz), of Africa, of Madagascar, and Borneo, and of Central and Eastern Asia, the idea involuntarily presents itself to the mind that we are not yet acquainted with one third, or probably even with one fifth part of the plants existing on the earth. Drège has collected 7092 phanerogamic species in Southern Africa alone; and he believes that the flora of that region consists of more than 11,000 phanerogamic species, seeing that in Germany and Switzerland, on an equal area (192,000 square miles,) Koch has described only 3300, and Decandolle only 3645 phanerogamia in France. I would here also instance the new genera, consisting partly of high forest trees, which are still being discovered in the neighbourhood of large commercial towns in the lesser Antilles, although they have been visited by Europeans for the last three hundred years. Such considerations,

* *General Remarks on the Botany of Terra Australis*, p. 4.

† Humboldt, *de distributione geographica Plantarum*, p. 23.

‡ *Essai élémentaire de Géographie botanique*, p. 62.

which I purpose developing more fully at the close of this illustration, seem to verify the ancient myth of the Zend-Avesta, that "the creating primeval force called forth 120,000 vegetable forms from the sacred blood of the bull."

If therefore no direct scientific solution can be afforded to the question, how many vegetable forms—leafless cryptogamia (water algæ, fungi, and lichens), characeæ, liverworts, foliaceous mosses, marsilaceæ, lycopodiaceæ, and ferns—exist on the dry land, and in the wide basin of the sea, in the present condition of the organic terrestrial life of our planet, it only remains for us to employ an approximative method for ascertaining with some degree of probability certain "extreme limits" (numerical data of minima). Since the year 1815, I have, in my arithmetical considerations on the geography of plants, calculated the numbers expressing the ratio which the aggregate of species of different natural families bears to the whole mass of the phanerogamia in those countries where the latter is sufficiently determined. Robert Brown,* the greatest botanist of our age, had, prior to my researches, already determined the numerical proportion of the principal divisions of vegetable forms, as for instance of acotyledons (*Agamæ*, cryptogamic or cellular plants) to cotyledons (*Phanerogamia*, or vascular plants), and of monocotyledons (*Endogenæ*) to dicotyledons (*Exogenæ*). He finds the ratio of monocotyledons to dicotyledons in the tropical zone as in the proportion of 1 to 5, and in the frigid zone, in the parallels of 60° north, and 55° south lat. as 1 to 2½.† The absolute numbers of the species are compared together in the three great divisions of the vegetable kingdom, according to the method developed in Brown's work. I was the first who passed from these principal divisions to the individual families, and considered the number of the species contained in each, in their ratio to the whole mass of phanerogamia belonging to one zone.‡

* Formerly librarian to Sir Joseph Banks, now President of the Linnæan Society.—Ed.

† Robert Brown, *General remarks on the botany of Terra Australis*, in *Flinders' Voyage*, vol. ii. p. 338.

‡ Compare my essay, *De distributione geographica Plantarum secundum cœli temperiem et altitudinem montium*, 1817, pp. 24—44; and see the further development of numerical relations as given by me in the *Dictionnaire des Sciences naturelles*, t. xviii. 1820, pp. 422—436; and in the *Annales de Chimie et de Physique*, t. xvi. 1821, pp. 267—292.

The numerical relations of the forms of plants, and the laws observed in their geographical distribution, admit of being considered from two very different points of view. When we study plants in their arrangement according to natural families, without regard to their geographical distribution, the question arises: What are the fundamental forms or types of organization, in accordance with which the greater number of their species are formed? Are there more Glumaceæ than Compositæ on the earth's surface? Do these two orders of plants combined, constitute one-fourth of the phanerogamia? What numerical relation do monocotyledons bear to dicotyledons? These are questions of general phytology, a science that investigates the organization of plants and their mutual connection, and therefore has reference to the now existing state of vegetation.

If, on the other hand, the species of plants that have been connected together according to their structural analogy, are considered not abstractedly, but in accordance with their climatic relations, and their distribution over the earth's surface, these questions acquire a totally different interest. We then examine what families of plants predominate in the torrid zone more than towards the polar circle over other phanerogamia? We inquire, whether the Compositæ are more numerous in the new than in the old world, under equal geographical latitudes or between equal isothermal lines? Whether the forms which gradually lose their predominance in advancing from the equator to the poles, follow a similar law of decrease in ascending mountains situated in the equatorial region? Whether the relations of the different families to the whole mass of the phanerogamia differ under equal isothermal lines in the temperate zones on either side of the equator? These questions belong to the geography of plants properly so called, and are connected with the most important problems that can be presented by meteorology and terrestrial physics. Thus the predominance of certain families of plants determines the character of a landscape, and whether the aspect of the country is desolate or luxuriant, or smiling and majestic. Grasses, forming extended Savannahs, or the abundance of fruit-yielding palms, or social coniferous trees, have respectively exerted a powerful influence on the material condition, manners, and character of nations, and on the more or less rapid development of their prosperity.

In studying the geographical distribution of forms, we may consider the species, genera, and natural families of plants separately. A single species, especially among social plants, frequently covers an extensive tract of land. Thus we have in the north, Pine or Fir forests, and Heaths (*ericeta*); in Spain, Cistus groves; and in tropical America, collections of one and the same species of Cactus, Croton, Brathys, or Bambusa Guadua. It is interesting to study more closely these relations of individual increase, and of organic development; and here we may inquire, what species produces the greatest number of individuals in one certain zone; or, merely what are the families to which the predominating species belong in different climates. In a very high northern latitude, where the Compositæ and the Ferns stand in the ratios of 1 : 13 and 1 : 25 to the sum of all the phanerogamia (*i. e.*, where these ratios are found by dividing the sum total of all phanerogamia by the number of species included in the family of the Compositæ, or in that of the Ferns); one single species of Fern may, however, cover ten times more space than all the species of the Compositæ taken together. In this case the Ferns predominate over the Compositæ by their mass, and by the number of the individuals belonging to the same species of Pteris, or Polypodium; but they will not be found to predominate, if we only compare the number of the different specific forms of the Filices, and of the Compositæ, with the sum total of all Phanerogamia. As, therefore, multiplication of plants does not follow the same laws in all species, and as all do not produce an equal number of individuals, the quotients obtained by dividing the sum of all phanerogamic plants by the species of one family, do not *alone* determine the leading features impressed on the landscape, or the physiognomy of nature peculiar to different regions of the earth. If the attention of the travelling botanist be arrested by the frequent repetition of the same species, by its mass, and the uniformity of vegetation thus produced, it will be still more forcibly arrested by the infrequency of many other species useful to man. In tropical regions, where the Rubiaceæ, Myrtles, Leguminosæ, or Terebinthaceæ, compose the forests, one is astonished to meet with so few trees of Cinchona, or of certain species of mahogany (*Swietenia*), of Hæmatoxylon, Styrax, or balsamic Myroxylon. I would also here refer to the scanty and detached occurrence of the precious febrifuge-bark trees (species of Cinchona) which I had an opportunity of observing on the declivity of the elevated plains of Bogota and Popayan, and in the neighbourhood of Loxa, in descending towards the unhealthy valley of the Catamayo, and to the river Amazon. The *febrifuge-bark hunters* (Cazadores de Cascarilla), as those Indians and Mestizoes are called at Loxa, who each year collect the most efficacious of all the medicinal barks, the *Cinchona Condaminea*, among the lonely mountains of Caxanuma, Uritusinga, and Rumisitana, undergo considerable danger in climbing to the summits of the highest forest-trees, in order to obtain an extended view, from which they may distinguish the scattered, slender, and aspiring trunks of the Cinchona, by the reddish tint of their large leaves. The mean temperature of this important forest region (between 4° and 4½° south lat.) varies from 60° to 68° Fahr., at an absolute height of from 6400 to 8000 feet above the level of the sea.*

In considering the distribution of species, we may also, independently of individual multiplication and mass, compare together the absolute number which belong to each family. Such a mode of comparison, which was employed by Decandolle,† has been extended by Kunth to more than 3300 of the species of Compositæ with which we are at present acquainted. It does not show what family preponderates by individual mass, or by the number of its species, over other phanerogamic forms, but it simply indicates how many of the species of one and the same family are indigenous in any one country or portion of the earth. The results of this method are, on the whole, more exact, because they are obtained by a careful study of the separate families, without requiring that the whole number of the phanerogamia of every country should be known. Thus, for instance, the most varied forms of Ferns are found in the tropical zone, each genus presenting the greatest number of species in the temperate, humid, and shaded mountainous parts of islands. While these species are less numerous in passing from tropical regions to the temperate zone, their *absolute number* diminishes still more in approaching nearer to the poles. Although the frigid zone, as, for instance, Lapland, supports species of the families which are

* Humboldt et Bonpland, *Plantes équinoxiales*, t. i. p. 33, tab. 10.

† See his work, *Regni Vegetabilis Systema naturale*, t. i. pp. 128, 396, 439, 464, 510.

best able to resist the cold, Ferns predominate more over other phanerogamia in Lapland than either in France or Germany, notwithstanding the absolute inferiority of the gross number of ferns indigenous to the northern zone, when compared with other countries. These relations are, in France and Germany, as $\frac{1}{73}$ and $\frac{1}{71}$, while in Lapland they are as $\frac{1}{25}$. These numerical relations (obtained by dividing the sum total of all the phanerogamia of the different floras by the species of each family) were published by me in 1817, in my *Prolegomena de distributione geographica Plantarum,* and corrected in accordance with the great works of Robert Brown, in my Essay on the Distribution of Plants over the earth's surface, which I subsequently wrote in French. These relations, as we advance from the equator towards the poles, necessarily vary from the ratios obtained by a comparison of the absolute number of the different species belonging to each family. We often see the value of the fractions increase by the decrease of the denominator, whilst the absolute number of the species is reduced. In the fractional method which I have followed as the most applicable to questions relating to the geography of plants, there are two variable quantities; for in passing from one isothermal line to another, we do not find the sum total of the phanerogamia change in the same proportion as the number of the species of one particular family.

In proceeding from the consideration of these species to that of the divisions established in the natural system according to an ideal series of abstractions, we may direct our attention to genera or races, to families, or even to still higher classes of division. There are some genera, and even whole families, which exclusively belong to certain zones; not merely because they can only thrive under a special combination of climatic relations, but also because they first sprang up within very circumscribed localities, and have been checked in their migrations. The larger number of genera and families have, however, their representatives in all regions of the earth, and at all elevations. The earliest inquiries into the distribution of vegetable forms had reference to genera alone, and are to be found in the valuable work of Treviranus.* This method is, however, less appropriate for yielding general results, than that which compares the number of the species of

* *Biologie,* bd. ii. s. 47, 63, 83, 129.

each family, or the great leading divisions (acotyledons, monocotyledons, and dicotyledons), with the sum total of the phanerogamia. In the frigid zone, the variety of forms, or the number of the genera, does not decrease in an equal degree with that of the species, there being in these regions relatively more genera and fewer species.* The case is almost the same on the summits of high mountain-chains, where are sheltered individual members of many different genera which one would be disposed to regard as belonging exclusively to the vegetation of the plain.

I have deemed it expedient to indicate the different points of view from which the laws of the distribution of vegetable forms may be considered. It is only when these points of view are confounded together, that we meet with contradictions, which have been unjustly attributed to uncertainty of observation.† When expressions like the following are employed: "This form, or this family diminishes as it approaches towards the cold zone," or "the true habitat of this form is in such or such a parallel of latitude;" or "this is a southern form," or, again, "it predominates in the temperate zone;" it should be definitely stated whether reference is made to the absolute number of the species, and the proportion of their predominance according to the increase or decrease of latitude; or whether the meaning conveyed is, that a family, when compared with the whole number of the phanerogamia of a flora, predominates over other families of plants. The impression conveyed to the mind of the predominance of forms, depends literally on the conception of relative quantity.

Terrestrial physics have their numerical elements as well as the cosmical system, and it is only by the united labours of botanical travellers that we can hope gradually to arrive at a knowledge of the laws which determine the geographical and climatic distribution of vegetable forms. I have already observed that in the temperate zone of the northern hemisphere, the Compositæ (Synanthereæ) and the Glumaceæ (in which latter division I place the three families of the Gramineæ, the Cyperoideæ, and the Juncaceæ) constitute the fourth part of all phanerogamia. The following numerical

* Decandolle, *Théorie élémentaire de la Botanique,* p. 190; Humboldt, *Nova genera et species Plantarum,* t. i. pp. xvii. l.

† *Jahrbücher der Gewächskunde,* bd. i. Berlin, 1818, s. 18, 21, 30.

relations are the result of my investigations for seven great families of the vegetable kingdom in one and the same temperate zone:

Glumaceæ $\frac{1}{8}$ (Grasses alone $\frac{1}{12}$)
Compositæ $\frac{1}{8}$
Leguminosæ $\frac{1}{18}$
Labiatæ $\frac{1}{24}$
Umbelliferæ $\frac{1}{40}$
Amentaceæ (Cupuliferæ, Betulineæ, and Salicineæ) $\frac{1}{45}$
Cruciferæ $\frac{1}{19}$

The forms of organic beings are reciprocally dependent on one another. Such is the unity of nature, that these forms limit each other in obedience to laws which are probably connected with long periods of time. When we have ascertained the number of the species on any particular part of the earth's surface belonging to one of the great families of the Glumaceæ, the Leguminosæ, or the Compositæ, we may with some degree of probability, form approximative conclusions regarding the number of all the phanerogamia, as well as of the species belonging to the other families of plants growing in the country. The number of the Cyperoideæ determines that of the Compositæ, and the number of the latter determines that of the Leguminosæ; and these estimates, moreover, enable us to ascertain in what classes and orders the Floras of a country are still incomplete, teaching us what harvests may still be reaped in the respective families, if we guard against confounding together very different systems of vegetation.

The comparison of the numerical proportions of families in the different zones which have as yet been well explored, has led me to a knowledge of the laws which determine the numerical increase or decrease of vegetable forms constituting a natural family, in proceeding from the equator to the poles, when compared, for instance, with the whole mass of phanerogamia peculiar to each zone. We must here have regard not only to the direction, but also to the rapidity or measure of the increase. We see the denominator of the fraction, which expresses the ratio, increase or diminish. Thus, for instance, the beautiful family of the Leguminosæ diminishes in proportion as it recedes from the equinoctial zone to the north pole. If we find its ratio for the torrid zone (from 0° to 10° of latitude) $\frac{1}{10}$, we shall have for the part of the temperate zone (lying between 45° and 52°) $\frac{1}{18}$, and for the frigid zone (between 67° and 70° lat.) only $\frac{1}{35}$. The direction followed by the great family of the Leguminosæ (viz., increase towards the equator) is also that of the Rubiaceæ, the Euphorbiaceæ, and especially the Malvaceæ. On the other hand, the Gramineæ and the Juncaceæ (the latter more than the former), the Ericeæ, and Amentaceæ, diminish towards the torrid zone. The Compositæ, Labiatæ, Umbelliferæ, and Cruciferæ, diminish from the temperate zone towards the pole and the equator, and the two latter families most rapidly in the direction of the equatorial region; whilst in the temperate zone the Cruciferæ are three times more abundant in Europe than in the United States of North America. In Greenland the Labiatæ are reduced to only one species, and the Umbelliferæ to two, while the whole number of the phanerogamia still amounts, according to Hornemann, to 315 species.

It must at the same time be observed that the development of plants of different families, and the distribution of their forms, do not depend alone on the geographical, or even on the isothermal latitude; the quotients not being always equal on one and the same isothermal line in the temperate zone, as for instance in the plains of America and in those of the Old Continent. Within the tropics there is a very marked difference between America, the East Indies, and the western coast of Africa. The distribution of organic beings over the surface of the earth does not depend solely on the great complication of thermic and climatic relations, but also on geological causes which continue almost wholly unknown to us, since they have been produced by the original condition of the earth, and by catastrophes which have not affected all parts of our planet simultaneously. The large pachydermata are no longer found in the New Continent, while they still exist under analogous climates in Asia and Africa. These differences, instead of deterring us from the investigation of the laws of nature, should rather stimulate us to study them in all their intricate modifications.

The numerical laws of families, the frequently striking agreement between the ratios, where the species constituting these families are for the most part different, lead us into that mysterious obscurity which envelopes everything connected with the fixing of organic types in the different species of animals and plants, and with all that refers to formation and development. I will take as examples two neighbouring countries—France and Germany—which have both been long since explored. In France many species of Gramineæ, Umbelliferæ, Cruciferæ, Compositæ, Leguminosæ, and Labiatæ are wanting, which are some of the commonest in Germany, and yet the ratios of these six large families are almost identical in both countries. Their relations, which I here give, are as follows:

| Families. | Germany. | France. |
|---|---|---|
| Gramineæ. | $\frac{1}{13}$ | $\frac{1}{13}$ |
| Umbelliferæ. | $\frac{1}{22}$ | $\frac{1}{21}$ |
| Cruciferæ. | $\frac{1}{18}$ | $\frac{1}{19}$ |
| Compositæ. | $\frac{1}{8}$ | $\frac{1}{7}$ |
| Leguminosæ. | $\frac{1}{18}$ | $\frac{1}{16}$ |
| Labiatæ. | $\frac{1}{26}$ | $\frac{1}{24}$ |

This correspondence in the number of species of one family compared to the whole mass of the phanerogamia of Germany and France would not exist, if the absent German species were not replaced in France by other types of the same families. Those who delight in conjectures respecting the gradual transformation of species, and who regard the different parrots, peculiar to islands situated near each other, as merely transformed species, will ascribe the remarkable uniformity presented by the above numerical ratios to a migration of the same species, which having been altered by climatic influences, continuing for thousands of years, appear to replace each other. But why have our common Heath, (Calluna vulgaris,) and our Oaks not penetrated to the east of the Ural Mountains, and passed from Europe to northern Asia? Why is there no species of the genus Rosa in the southern, and scarcely any Calceolaria in the northern hemisphere? These are points that cannot be explained by peculiarities of temperature. The present distribution of forms (fixed forms of organization) is no more explained by thermal relations alone, than by the hypothesis of migrations of plants radiating from certain central points. Thermal relations are scarcely sufficient to explain the phenomenon why certain species have fixed limits beyond which they cannot pass, either in the plains towards the pole, or in vertical elevation on the declivities of mountains. The cycle of vegetation of each species, however different may be its duration, requires a certain minimum of temperature to enable it to arrive at the full stage of its development.* But all the conditions necessary to the existence of a plant, either within its natural sphere of distribution or cultivation—such as geographical distance from the pole, and elevation of the locality—are rendered still more complicated by the difficulty of determining the beginning of the thermic cycle of vegetation; by the influence which the unequal distribution of the same quantity of heat among days and nights succeeding each other in groups, exerts on the irritability, the progressive development, and the whole vital process; and lastly, by the secondary influence of the hygrometric and electric relations of the atmosphere.

My investigations regarding the numerical laws of the distribution of vegetable forms may, perhaps, at some future time, be applied successfully to the different classes of vertebrate animals. The rich collections of the Muséum d'histoire naturelle in the Jardin des Plantes at Paris, contained in 1820, at a rough estimate, above 56,000 species of phanerogamic and cryptogamic plants in the herbariums, 44,000 insects (probably below the actual number, although they were thus given me by Latreille), 2500 species of fishes, 700 reptiles, 4000 birds, and 500 mammalia. Europe possesses about 80 mammalia, 400 birds, and 30 reptiles; there are, therefore, five times as many birds as mammalia in the northern temperate zone, (as there are in Europe five times as many Compositæ as Amentaceæ and Coniferæ, and five times as many Leguminosæ as Orchideæ and Euphorbiaceæ). In the southern temperate zone the ratio of the Mammalia bears a sufficiently striking accord with that of Birds, being as 1 : 4·3. Birds (and rep-

* Playfair, in the *Transactions of the Royal Soc. of Edinb.*, vol. v. 1805, p. 202; Humboldt, on the sum total of the thermometric degrees required for the cycle of vegetation of the Cereals, in *Mém. sur des lignes isothermes*, p. 96; Boussingault, *Economie rurale*, t. ii. p. 659, 663, 667; and Alphonse Decandolle, *Sur les causes qui limitent les espèces végétales*, 1847, p. 8.

tiles even to a greater extent), increase more than mammalia in advancing towards the torrid zone. We might be disposed to believe, from Cuvier's investigations, that this ratio was different in the earlier age of our planet, and that the number of mammalia that perished by convulsions of nature was much greater than that of birds. Latreille has shown the different groups of insects that increase in advancing towards the pole, or towards the equator, and Illiger has indicated the native places of 3800 birds, according to the quarters of the globe; —a far less instructive method than if they had been given according to zones. We may easily comprehend how, on a given area, the individuals of one class of plants or animals may limit each other's numbers, and how, after the long-continued contests and fluctuations engendered by the requirements of nourishment and mode of life, a condition of equilibrium may have been at length established; but the causes which have determined their typical varieties, and have circumscribed the sphere of the distribution of the forms themselves, no less than the number of individuals of each form, are shrouded in that impenetrable obscurity which still conceals from our view all that relates to the beginning of things and the first appearance of organic life.

If, therefore, as I have already observed at the beginning of this illustration, we attempt to give an approximative estimate of the *numerical limit* ("le nombre limite" of the French mathematicians), *below* which we cannot place the sum of all the phanerogamia on the surface of the earth; we shall find that the surest method will be by comparing the known ratios of the families of plants with the number of the species contained in our herbariums, or cultivated in large botanical gardens. As I have just remarked, the herbariums of the Jardin des Plantes at Paris were, in 1820, already estimated at 56,000 species. I will not hazard a conjecture as to the number that may be contained in the herbariums of England, but the great Paris herbarium, which Benjamin Delessert with the noblest disinterestedness has given up to free and general use, was estimated, at the time of his death, to contain 86,000 species, a number almost equal to that which Lindley, even in 1835,* regarded as the probable number of all the species existing "on the whole earth." Few herbariums are numbered with

* *Introduction to Botany*, 2nd ed. p. 504.

care, according to a complete, severe, and methodical separation of the different varieties; while, moreover, we often find no inconsiderable number of plants wanting in the large so-called general herbariums, which are contained in some of the smaller ones. Dr. Klotzsch estimates the whole number of Phanerogamic plants in the Great Royal Herbarium at Schöneberg, near Berlin, of which he is curator, at 74,000 species.

Loudon's useful work (*Hortus britannicus*) gives a general view of the species which now are or recently have been, cultivated in English gardens. The edition of 1832 enumerates, including indigenous plants, exactly 26,660 Phanerogamia. We must not confound with this large number of plants that either have been, or still are, cultivated in Great Britain, "all the living plants which may simultaneously be found in an individual botanic garden." In this last respect the Botanic Garden of Berlin has long been regarded as one of the richest in Europe. The fame of its extraordinary riches rested formerly on a mere approximative estimate of its contents, and, as my old friend and fellow-labourer Professor Kunth, has very correctly remarked,* "it was only by the completion of a systematic catalogue, based on the most careful examination of the species, that an actual enumeration could be undertaken. This enumeration gave somewhat more than 14,060 species; and when we deduct from these 375 cultivated ferns, there remain 13,685 Phanerogamia, among which there are 1600 Compositæ, 1150 Leguminosæ, 428 Labiatæ, 370 Umbelliferæ, 460 Orchideæ, 60 Palms, and 600 Grasses and Cyperaceæ. If we compare with these numbers the number of species given in recent works, as, for instance, Compositæ (according to Decandolle and Walpers), at about 10,000, Leguminosæ 8070, Labiatæ (Bentham) 2190, Umbelliferæ 1620, Grasses 3544, and Cyperaceæ 2000,† we shall perceive that the Botanic Garden at Berlin cultivates only $\frac{1}{7}$, $\frac{1}{8}$, and $\frac{1}{9}$ of the very large families (Compositæ, Leguminosæ, and Grasses), and as many as $\frac{1}{5}$ and $\frac{1}{4}$ of the already described species belonging to the small families (Labiatæ and Umbelliferæ). If we estimate the number of all the different species

* Manuscript notice communicated to the "Gartenbau-Verein" in Dec. 1846.

† Kunth, *Enumeratio Plantarum.*

of Phanerogamia *simultaneously* cultivated in all the botanical gardens of Europe at 20,000, we shall find, as they appear to constitute about the eighth part of those already described and contained in herbariums, that the whole number of Phanerogamia must amount to nearly 160,000. This estimate need not be regarded as too high, since scarcely the hundredth part of many of the larger families, as, for instance, Guttiferæ, Malpighiaceæ, Melastomeæ, Myrtaceæ, and Rubiaceæ, belong to our gardens." If we take the number (26,660 species), given in Loudon's "Hortus Britannicus," as the basis, we shall find, from the well-grounded series of inferences drawn by Professor Kunth, and which I borrow from his manuscript notice above referred to, that the estimate of 160,000 will increase to 213,000 species; and even this is still very moderate, since Heynhold, in his "Nomenclator botanicus hortensis" (1846), estimates the species of Phanerogamia already cultivated at 35,600. On the whole, therefore,—and the conclusion is, at first sight, sufficiently striking,—the number of species of Phanerogamia at present known by cultivation in gardens, by descriptions, and in herbariums, is almost greater than that of known insects. According to the average estimates of several of the most distinguished entomologists, whose opinion I have been able to obtain, the number of insects at present described, or contained in collections without being described, may be stated as between 150,000 and 170,000 species. The rich collection at Berlin contains fully 90,000, among which there are about 32,000 beetles. Travellers have collected an immense quantity of plants in remote regions, without bringing with them the insects living upon them, or in the neighbourhood. If, however, we limit these numerical estimates to a definite portion of the earth's surface that has been the best explored in regard to its plants and insects, as, for instance, Europe, we find the ratio between the vital forms of Phanerogamic plants and those of insects changed to such a degree, that while Europe counts scarcely 7000 or 8000 Phanerogamia, more than three times that number of European insects are at present known. According to the interesting contributions of my friend Dohrn in Stettin, more than 8700 insects have already been collected from the rich fauna of the neighbourhood, and yet there are still many Micro-

Lepidoptera wanting; while the number of Phanerogamia found there scarcely exceeds 1000. The Insect-fauna of Great Britain is estimated at 11,600. Such a preponderance of animal forms will appear less surprising when we remember that several of the large classes of insects live only on animal substances, whilst others subsist on agamic plants (Fungi), and even on those which are subterranean. Bombyx Pini, the Pine Spider, the most destructive of all forest-insects, is infested, according to Ratzeburg, by no less than thirty-five parasitical Ichneumonidæ.

These considerations have led us to the proportion borne by the number of species growing in gardens to the gross number of those already described and preserved in herbariums; it now remains for us to consider the proportion of the latter to the conjectural number of species existing on the whole earth, or, in other words, to test their minimum by the relative numbers of the different families—*i. e.* by variable *multipla*. A test of this kind gives, however, such low results for the *lower* amount, as plainly to show that even in the large families, which appear to have been the most strikingly enriched in recent times by the researches of descriptive botanists, our knowledge is still limited to a very small portion of the treasure actually existing. The *Repertorium* of Walpers which completes Decandolle's *Prodromus* of 1825 to 1846, gives 8068 species of the family of the Leguminosæ. We may assume the mean ratio to be $\frac{1}{21}$; since it is $\frac{1}{10}$ in the tropical zone, $\frac{1}{18}$ in the middle temperate zone, and $\frac{1}{33}$ in the cold northern zone. The *described* Leguminosæ would therefore only lead us to assume that there were 169,400 species of Phanerogamia existing on the earth, whereas the Compositæ, as already shewn, testify to the existence of more than 160,000 known Phanerogamia, *i. e.* such as have been described or are contained in herbariums. This discrepancy is instructive, and will be further elucidated by the following analogous considerations.

The larger number of the Compositæ, of which Linnæus knew only 785 species, and which have now increased to 12,000, appear to belong to the Old Continent. At least Decandolle described only 3590 American, while he estimated the European, Asiatic, and African species at 5093. This abundance of Compositæ in our vegetable systems is however deceptive, and only apparently considerable; for the quotient of this family (which within the tropical zone is $\frac{1}{15}$, in the temperate zone $\frac{1}{7}$, and in the frigid zone $\frac{1}{13}$) shows that more species of Compositæ than of Leguminosæ have hitherto eluded the diligent research of travellers; for even when multiplied by 12 we only obtain the improbably small number of 144,000 for the sum total of the Phanerogamia! The families of the Grasses and of the Cyperaceæ give still lower results, because a proportionally smaller number of species have been described and collected. We need only cast a glance at the map of South America, and remember that the vast extent of country occupied by the grassy plains of Venezuela the Apure and the Meta, as well as to the south of the woody region of the Amazon, in Chaco, in Eastern Tucuman, and in the Pampas of Buenos Ayres and Patagonia, has either been very imperfectly or not at all explored in relation to botany. Northern and Central Asia present an almost equally extensive territory occupied by steppes; but here a larger proportion of dicotyledonous plants is intermixed with the Gramineæ. If we had sufficient grounds for believing that one-half of all the phanerogamic plants existing on the surface of the earth are known, and if we estimate this number at only 160,000 or at 213,000 known species; we must give to the family of grasses, whose general ratio appears to be $\frac{1}{12}$, in the former case at least 26,000, and in the latter 35,000 different species, of which in the first case $\frac{1}{8}$, and in the second $\frac{1}{10}$ are known.

The following considerations oppose the hypothesis that we are already acquainted with half the Phanerogamia on the earth's surface. Several thousand species of Monocotyledons and Dicotyledons, and among them lofty arborescent forms, have recently been discovered (I would remind the reader of my own expedition) in districts of a very large extent, which had already been explored by distinguished botanists. Yet that portion of the great continents which has never been visited by botanical observers far exceeds the extent of the parts even superficially traversed. The greatest variety of phanerogamic vegetation, *i. e.* the greatest number of species on an equal area, is to be met with in the tropical or sub-tropical zones. It is therefore the more important to bear in mind that we are almost wholly unacquainted, north of the equator, in the New Continent, with the floras of Oaxaca, Yucatan, Guatimala, Nicaragua, the Isthmus of Panama, the Choco, Antioquia, and the Province de los Pastos; while south of the equator, we are equally ignorant of the floras of the boundless forest-region between the Ucayale, the Rio de la Madura, and the Toncantin (three mighty tributaries of the Amazon), as well as of those of Paraguay and the Province de las Missiones. In Africa, we know nothing of the vegetation of the whole of the interior, between 15° north and 20° south lat.; and in Asia we are unacquainted with the floras of the south and south-east of Arabia, where the highlands rise to an elevation of 6400 feet; as also with the floras between the Thian-schan, the Kuen-Lün, and the Himalaya; those of Western China; and those of the great portion of the countries beyond the Ganges. Still more unknown to botanists are the interior portions of Borneo and New Guinea, and of some districts of Australia. Further to the south the number of the species decreases in a most remarkable manner, as Joseph Hooker has ably shown, from his own observation, in his *Antarctic Flora.* The three islands which constitute New Zealand extend from 34½° to 47¼° of latitude, and as they have besides snow-crowned mountains more than 8850 feet in height, they must exhibit considerable differences of climate. The most northern island has been explored with tolerable accuracy from the time of Banks and Solander's voyage (with Capt. Cook), to the visits of Lesson, the brothers Cunningham, and Colenso; and yet in more than seventy years, the number of Phanerogamia with which we have become acquainted is below 700.* This paucity of vegetable species corresponds with the paucity of animal forms. Dr. Joseph Hooker has observed that "Iceland, proverbially barren as it is, and upon which no tree, save a few stunted birches, is to be found, possesses five times as many flowering plants as Lord Auckland's group and Campbell's Islands together, although these are situated at from 8° to 10° nearer the equator in the southern hemisphere. The antarctic flora is at once characterised by uniformity and great luxuriance of vegetation, which is attributable to the influence exerted by an uninterruptedly cool and humid climate. In Southern Chili, Patagonia, and Tierra del

* Ernest Dieffenbach, *Travels in New Zealand,* 1843, vol. i. p. 419.

Fuego (from 45° to 56° lat.) this uniformity is strikingly manifested on the mountains and their declivities no less than in the plains. How great is the difference of species when we compare the flora of the south of France, in the same latitude as the Chonos Islands off the coast of Chili, with the Scottish flora of Argyleshire, in the parallel of Cape Horn. In the southern hemisphere the same types of vegetation pass through many degrees of latitude. In the regions near the north pole ten flowering plants have been collected on Walden Island (80½° north lat.), while there is scarcely a solitary grass to be met with in the South Shetland Islands, although situated 63° south latitude."* These considerations on the distribution of plants prove that the great mass of the still unobserved, uncollected, and undescribed phanerogamia belong to the tropical zone, and to the contiguous regions extending from twelve to fifteen degrees from it.

I have deemed it not unimportant to draw attention to the imperfect state of our knowledge in this slightly cultivated department of numerical botany, and to treat such questions in a more definite manner than has hitherto been possible. In all conjectures regarding relative numbers, we must first examine the practicability of obtaining the *lowest limit;* as in the question, of which I have treated elsewhere, regarding the ratio of the gold and silver coined to the quantity of the precious metals existing in a wrought state; or as in the question of how many stars, from the tenth to the twelfth magnitude, are scattered over the heavens, and how many of the smallest telescopic stars may be contained in the Milky Way?† It is an established fact, that if it were possible to ascertain completely by observation the number of species of the large phanerogamic families, we should at the same time obtain an approximate knowledge of the sum-total of all the phanerogamia on the surface of the earth (that is, the numbers included in every family). The more therefore we are enabled, by the progressive exploration of unknown districts, gradually to determine the number of species belonging to any one great family, the higher will be the gradual rise of the lowest limit, and the nearer we shall

* Joseph Hooker, *Flora Antarctica,* pp. 73—75.

† Sir John Herschel, *Results of Astron. Observ. at the Cape of Good Hope,* 1847, p. 381.

arrive at the solution of a great numerical vital problem, since the forms, in accordance with still unexplained laws of universal organism, reciprocally limit each other. But is the number of the organisms a constant number? Do not new vegetable forms spring from the ground after long intervals of time, whilst others become more and more rare, and finally disappear? Geology confirms the latter part of this question by means of the historical memorials of ancient terrestrial life. "In the primitive world," to use the expression of the intellectual Link,* "elements remote from each other blend together in wondrous forms, indicating, as it were, a higher degree of development and articulation in a future period of the world."

(14) p. 222—"*Whether the height of the aërial ocean and its pressure have always been the same.*"

The pressure of the atmosphere has a decided influence on the form and life of plants. This life, owing to the fulness and abundance of the leafy organs provided with interstitial openings, is principally directed *outwards*. Plants mainly live in and through their surfaces, and hence their dependence on the surrounding medium. Animals are more dependant on *internal* stimuli; they generate and maintain their own temperature, deriving from muscular movements their electric currents, and the chemical vital processes which arise from and re-act upon those currents. A kind of cutaneous respiration constitutes an active vital function of plants, and depends, so far as it is an evaporation, inhalation, and exhalation of fluids, on atmospheric pressure. Hence Alpine plants are more aromatic and hirsute than others, and more amply provided with numerous exhalants.† Zoonomic experiments teach us, as I have shown in another work, that organs are more abundant and more perfectly developed in proportion to the facility with which their functional requirements are fulfilled. The disturbance occasioned in the respiration of their external integuments, by increased barometric pressure, renders it, as I have elsewhere shewn, very difficult for Alpine plants to thrive in the plain.

* *Abhandl. der Akad. der Wiss. zu Berlin aus dem J.* 1846, s. 322.

† See my work, *Ueber die gereizte Muskel-und Nervenfaser*, bd. ii. s. 142—145.

Whether the aërial ocean surrounding the earth has always exerted the same mean pressure is a question wholly undecided. We do not even know for certain whether the mean barometric height has remained the same during a hundred years at any one given spot. According to the observations of Poleni and Toaldo, this pressure appeared variable. Doubts were long entertained regarding the accuracy of these views, but the more recent investigations of the astronomer Carlini render it almost probable that in Milan the mean barometric pressure is on the decrease. Perhaps the phenomenon is very local, and dependent on periodic variations in descending currents of air.

(15) p. 223—"*Palms.*"

It is remarkable, that of this majestic form of plants—the Palms—some of which rise to more than twice the height of the Royal Palace at Berlin, and which the Indian, Amarasinha, has very characteristically called "kings among grasses,"—only fifteen species had been described up to the time of the death of Linnæus. The Peruvian travellers, Ruiz and Pavon, added only eight; whilst Bonpland and myself, traversing a greater extent of country, from 12° south lat. to 21° north lat., described twenty new species, and distinguished as many more which we named, without however being able to procure their blossoms in a perfect state.* At present (forty-four years after my return from Mexico) more than 440 species of palms, from both continents, have already been scientifically described, including the East Indian species arranged by Griffith. The "Enumeratio Plantarum" of my friend Kunth, which appeared in 1841, contains no fewer than 356 species.

The very few palms belonging, like our Coniferæ, Quercineæ, and Betulineæ, to social plants, are the Mauritian Palm (*Mauritia flexuosa*), and the two species of Chamærops, of which the Chamærops humilis covers whole tracts of land at the estuary of the Ebro and in Valencia, while the other, Chamærops Mocini, which we discovered on the Mexican shore of the Pacific, is entirely without prickles. In the same manner as there are some species of palms, including Cocos and Chamærops, which are peculiar to sea-coasts, so also is there a certain group of Alpine palms belonging to the region

* Humboldt, *De distributione geographica Plantarum*, pp. 225–233.

of the tropics, which, if I mistake not, was wholly unknown before my South American journey. Almost all these species of the palm family grow in plains and in a mean temperature of 81°.5 and 86° Fahr., seldom advancing higher up the sides of the Andes than to 1900 feet. The beautiful wax palm (*Ceroxylon andicola*), the Palmetto of Azufral at the Pass of Quindiu, (*Oreodoxa frigida*), and the reed-like Kunthia montana (*Caña de la Vibora*) of Pasto, all flourish at elevations varying from 6400 to 9600 feet above the level of the sea, where the thermometer frequently sinks in the night to 42°.8 and 45°.5 Fahr., and the mean temperature is scarcely 57° Fahr. These Alpine palms are interspersed with nut-trees, yew-leaved species of Podocarpus, and oaks, (*Quercus granatensis*). I have determined, by accurate barometric measurements, the upper and lower limits of the wax palm. We began to observe it first on the eastern declivity of the Cordilleras of Quindiu, at an elevation of 7929 feet, from whence it ascended to the Garita del Paramo, and Los Volcancitos, as high as about 9700 feet. The distinguished botanist, Don José Caldas, who was long our companion in the mountains of New Granada, and who fell a victim to Spanish party hatred, found, many years after my departure from the country, three species of palms in the Paramo de Guanacos, in the immediate vicinity of the limit of perpetual snow, and therefore, probably at an elevation of nearly 14,000 feet.* Even beyond the tropical region (in lat. 28°), Chamærops Martiana† rises on the advanced spurs of the Himalaya range to a height of 5000 feet.

When we consider the extreme geographical and, consequently, also the climatic limits of palms at spots which are but little elevated above the level of the sea, we find that some forms (the Date Palm, *Chamærops humilis*, *Ch. palmetto*, and *Areca sapida* of New Zealand,) advance far within the temperate zone of both hemispheres, to districts where the mean annual temperature scarcely reaches from 57° to 60° Fahr. If we form a progressive scale of cultivated plants in accordance with the different degrees of heat they require, and begin with the maximum, we have Cacao, Indigo, Bananas, Coffee, Cotton, Date Palms, Orange and Lemon trees, Olives, Spanish

* *Semanario de Santa Fé de Bogotá*, 1809, No. 21, p. 163.

† Wallich, *Plantæ asiaticæ*, vol. iii. tab. 211.

Chesnuts, and Vines. In Europe, Date Palms, together with Chamærops humilis, grow in the parallels of 43½° and 44°, as, for instance, on the Genoese Rivera del Ponente, near Bordighera, between Monaco and San Stefano, where there is a palm grove, numbering more than 4000 trees; also in Dalmatia, near Spalatro. It is remarkable that the Chamærops humilis is of frequent occurrence in the neighbourhood of Nice and in Sardinia, whilst it is not found in the Island of Corsica, lying between the two. In the New Continent, the Chamærops palmetto, which is sometimes more than 40 feet high, does not advance further north than 34°; a circumstance that may be explained by the inflection of the isothermal lines. In the southern hemisphere, Robert Brown* found that palms, of which there are only very few (six or seven) species, advance as far as 34°in New Holland; while Sir Joseph Banks saw an Areca, in New Zealand, as far as 38°. Africa, which, contrary to the ancient and still extensively diffused opinion, is poor in species of palms, exhibits only one palm (*Hyphæne coriacea*) which advances south of the equator, only as far as Port Natal, in 30° lat. The continent of South America presents almost the same limits. East of the chain of the Andes, in the Pampas of Buenos Ayres, and in the Cis-Plata province, palms extend, according to Auguste de St.-Hilaire,† as far as 34° and 35°. The Coco de Chile, (our Jubæa spectabilis?), the only species of palm indigenous in Chili, advances on the western side of the chain of the Andes, according to Claude Gay,‡ to an equal latitude, viz., to the Rio Maule.

I will here subjoin the aphoristic observations which, in March, 1801, I noted down while on board ship, at the moment we were leaving the palm region surrounding the mouth of the Rio Sinu, west of Darien, and were setting sail for Cartagena de Indias.

"In the space of two years, we have seen as many as 27 different species of palms in South America. How many then must have been observed by Commerson, Thunberg, Banks, Solander, the two Forsters, Adanson, and Sonnerat, on their extensive travels! Yet, at the moment I am writing, our vegetable systems recognise scarcely more than from

* *General remarks on the Botany of Terra Australis*, p. 45.

† *Voyage au Brésil*, p. 60.

‡ Compare also Darwin, *Journal*, Ed. of 1845, pp. 244, 256.

fourteen to eighteen methodically described species of palms. The difficulties of reaching and procuring the blossoms of palms are, in fact, greater than can well be conceived; and, in our own case, we were made peculiarly sensible of this in consequence of our having directed our attention especially to palms, grasses, cyperaceæ, juncaceæ, cryptogamia, and numerous other subjects hitherto much neglected. Most of the palms flower only once a year, and this period near the equator is generally about the months of January and February. How few travellers are likely to be in the region of palms precisely during this season! The period of blossoming of particular trees is often limited to a few days, and the traveller commonly finds, on his arrival in the region of palms, that the blossoms have passed away, and that the trees present only fructified ovaries and no male flowers. In an area of 32,000 square miles, there are often not more than three or four species of palms to be found. Who can possibly, during the brief period of flowering, simultaneously visit the various palm regions near the Missions on the Rio Caroni, in the Morichales at the mouth of the Orinoco, in the valley of Caura and Erevato, on the banks of the Atabapo and the Rio Negro, and on the declivity of the Duida? There is, moreover, great difficulty when the trees grow in thick woods or on swampy shores (as at the Temi and Tuamini), in reaching the blossoms, which are often suspended from stems formidably armed with huge thorns, and rising to a height of between 60 and 70 feet. They who contemplate distant travels from Europe for the purpose of investigating subjects of natural history, picture to themselves visions of efficient shears and curved knives attached to poles, ready for securing anything that comes in their way; and of boys who, obedient to their mandates, are prepared, with a cord attached to their feet, to climb the loftiest trees! Unfortunately, scarcely any of these visions are ever realised; while the flowers are almost unattainable, owing to the great height at which they grow. In the missionary settlements of the river net-work of Guiana, the stranger finds himself amongst Indians, who, rendered rich and independent by their apathy, their poverty, and their barbarism, cannot be induced either by money or presents to deviate three steps from the regular path, supposing one to exist. This stubborn indifference of the natives provokes the European so much the more, from his being continually a witness of the inconceivable agility with which they will climb any height when prompted by their own inclination, as, for instance, in the pursuit of a parrot, an iguana, or a monkey, which, wounded by their arrows, saves itself from falling by its prehensile tail. In the month of January the stems of the *Palma Real*, our *Oreodoxa Regia*, were covered with snow-white blossoms, in all the most frequented thoroughfares of the Havannah, and in the immediate vicinity of the city; but, although we offered, for several days running, a couple of piastres for a single spadix of the hermaphrodite blossoms to every negro boy we met in the streets of Regla and Guanavacoa, it was in vain, for, in the tropics, no free man will ever undertake any labour attended by fatigue unless he is compelled to do so by imperative necessity! The botanists and painters of the Royal Spanish Commission of Natural History under Count Don Jaruco y Mopox (Estevez, Boldo, Guio, Echeveria), confessed to us that, for several years, they had been unable to examine these blossoms, owing to the absolute impossibility of obtaining them.

"After this statement of the difficulties attending their acquisition, the fact of our being only able, in the course of two years, systematically to describe twelve species of palms, although we had discovered twenty species, may be understood; but I confess it would hardly have been credible to me before I left Europe. How interesting a work might be written on palms by a traveller, who could exclusively devote himself to the delineation, in their natural size, of the spathe, spadix, inflorescence and fruits!" (Thus I wrote many years before the Brazilian travels of Martius and Spix, and the appearance of the admirable work on Palms by the former.)

"There is much sameness in the form of the leaves, which are either feathery (pinnata), or fanlike (palmo-digitata); the leaf-stalk (petiolus) is either without thorns or is sharply serrated (*serrato-spinosus*). The leaf-form of *Caryota urens* and *Martinezia caryotifolia*, which we saw on the banks of the Orinoco and the Atabapo, and subsequently in the Andes, at the pass of Quindiu, as high as 3200 feet above the level of the sea, is almost as peculiar among palms as is the leaf-form of the Gingko among trees. The habitus and physiognomy of palms are expressive of a grandeur of character which it is difficult to describe in words. The stem (*caudex*) is simple. and very rarely divided into branches after the manner of the Dracæna, as in Cucifera thebaica (the Doom Palm), and in Hyphæne coriacea. It is sometimes disproportionately thick, as in Corozo del Sinu, our Alfonsia oleifera; of a reed-like feebleness, as in Piritu, (*Kunthia montana*), and the Mexican Corypha nana; of a somewhat fork-like and protuberant form towards the lower part, as in Cocos; sometimes smooth and sometimes scaly, as in the Palma de Covijaó de Sombrero, in the Llanos; or, lastly, prickly, as in Corozo de Cumana and Macanilla de Caripe, having the thorns very regularly arranged in concentric rings.

Characteristic differences also manifest themselves in the roots, which, in some cases, project about a foot or a foot and a half from the ground, raising the stem on a scaffolding, as it were, or coiled round it in a padded-like roll. I have seen viverras and even very small monkeys pass under the scaffolding formed by the roots of the Caryota. Occasionally the stem is swollen only in the middle, being smaller above and below, as in the Palma Real of the island of Cuba. The green of the leaves is either dark and shining, as in Mauritia Cocos, or of a silvery white on the under side, as in the slender fan-palm, *Corypha Miraguama*, which we saw in the harbour of Trinidad de Cuba. Sometimes the middle of the fan-like leaf is adorned with concentric yellow and blue stripes, in the manner of a peacock's tail, as in the prickly Mauritia, which Bonpland discovered on the Rio Atabapo.

"The direction of the leaves is a no less important characteristic than their form and colour. The leaflets (foliola) are either ranged in a comb-like manner close to one another, with a stiff parenchyma (as in *Cocos Phœnix*), to which they owe the beautiful reflections of solar light that play over the surface of the leaves, which shine with a brilliant verdure in *Cocos*, and with a fainter and ashy-coloured hue in the date-palm; or sometimes the foliage assumes a reed-like appearance, having a thinner and more flexible texture, and being curled near the extremity (as in *Jagua*, *Palma Real del Sinu*, *Palma Real de Cuba*, and *Piritu del Orinoco*). This direction of the leaves, together with the lofty stem, gives to the palms their character of high majesty. It is a characteristic of the physiognomical beauty of the palm that its leaves are directed aspiringly upwards throughout the whole period of its duration, (and not only in the youth of the tree, as is the case with the Date-Palm, which is the only one introduced into Europe.) The more acute the angle made by the leaves with the upper part of the stem (that is, the nearer they approach the perpendicular,) the grander and nobler is the form of the tree. How different is the aspect of the pendent leaves of the *Palma de Covija del Orinoco y de los Llanos de Calabozo* (Corypha tectorum), from the more horizontal leaves of the Date and Cocoa-nut palms, and the lofty heavenward-pointing branches of the *Jagua*, the *Cucurito*, and *Pirijao*.

"Nature seems to have accumulated all the beauties of form in the Jagua palm, which, intermingled with the Cucurito or Vadgihai, whose stem rises to a height of 80 or even more than 100 feet, crowns the granite rocks at the cataracts of Atures and Maypures, and which we also occasionally saw on the lonely banks of the Cassiquiare. Their smooth and slender stems rise to a height of from 64 to 75 feet, projecting like a colonnade above the dense mass of the surrounding foliage. These aërial summits present a marked and beautiful contrast with the thickly-leaved species of *Ceiba*, and with the forest of *Laurineæ*, *Calophyllum*, and the different species of *Amyris* which surround them. Their leaves, which seldom exceed seven or eight in number, incline vertically upwards to a height of 16 or 17 feet, and are curled at the extremities in a kind of feathery tuft. The parenchyma of the leaf is of a thin grass-like texture, causing the leaflets to wave with graceful lightness on the gently oscillating leafstalk. The floral buds burst forth, in all species of palms, from the stem immediately beneath the leaves; and the mode in which this takes place modifies their physiognomical character. Thus in some, as in *Corozo del Sinu*, the sheath is perfectly erect, and the fruit rises like a thyrsus, resembling the fruits of the Bromelia. In the greater number, the sheaths, which in some species are smooth, and in others very prickly and rough, incline downwards. In some, again, the male blossoms are of a dazzling white, and it may then be seen shining from a great distance; but in most species of palms they are yellow, closely compressed, and of an almost faded appearance, even when they first burst from the spathe.

In palms with feathery leaves the leaf-stalks either burst from the dry, rough, ligneous portion of the stem (as in *Cocos, Phœnix, Palma Real del Sinu*), or there rises in the rough part of the stem a grass-green, smooth, and thinner shaft, like one column above another, from which the leaf-stalk springs, as in *Palma Real de la Havana, Oreodoxa regia*, which excited the admiration of Columbus. In the fan-palms (*foliis palmatis*), the leafy crown often rests on a layer of dry leaves, which imparts to the tree a character of melancholy solemnity and grandeur (as in *Moriche, Palma de sombrero de la Havana*). In some umbrella-palms, the crown consists of a very few scattered leaves, raised on slender stalks (as in *Miraguama*).

"The form and colour of the fruit also present more variety than is generally supposed to be the case in Europe. *Mauritia flexuosa* has egg-shaped fruits, whose smooth, brown, and scaly surface gives them the appearance of young pine cones. How great is the difference between the large triangular cocoa-nut, the berry of the date, and the small stone-fruit of the Corozo! But of all the fruits of the palm, none can be compared for beauty with those of the Pirijao (*Pihiguao*) of San Fernando de Atabapo and of San Balthasar. They are oval, and of a golden colour (one-half being of a purplish red); are mealy, without seed, two or three inches in thickness, and hang in clusters like grapes from the summits of their majestic palm-trunks." I have already spoken in the earlier part of this work of these beautiful fruits, of which there are seventy or eighty clustered together in one bunch, and which can be prepared in a variety of ways like bananas and potatoes.

The spathe enclosing the blossom bursts suddenly open in some species of palms, with an audible report. Richard Schomburgh has like myself observed this phenomenon* in the flowering of the Oreodoxa oleracea. This first opening of the blossoms of the palm accompanied with noise, reminds us of Pindar's Dithyrambus on Spring, and of the moment when in the Argive Nemæa, "the first opening shoot of the date-palm announces the coming of balmy spring."†

Palms, bananas, and arborescent ferns constitute three forms of especial beauty peculiar to every portion of the

* Schomburgk, *Reisen in Britisch Guiana*, Th. i. S. 50.

† *Cosmos*, vol. ii. p. 376. (Bohn's Edition.)

tropical zone; wherever heat and moisture co-operate, vegetation is most exuberant and vegetable forms present the greatest diversity. Hence South America is the most beautiful portion of the palm world. In Asia the palm form is rare, in consequence perhaps of a considerable part of the Indian continent beneath the equator having been destroyed and covered by the ocean in some earlier revolution of our planet. We know scarcely anything of the African palms between the Bay of Benin and the coast of Ajan; and we are, generally speaking, as already observed, acquainted with only a very small number of African palm-forms.

Palms, next to Coniferæ, and some species of Eucalyptus belonging to the family of the Myrtaceæ, afford examples of the loftiest growth. Stems of the Cabbage-palm (*Areca oleracea*) have been seen from 160 to 170 feet in height.* The Wax-palm, our Ceroxylon andicola, which we discovered in the Montaña de Quindiu on the side of the Andes, between Ibague and Carthago, attains the enormous height of 180 to 190 feet. I was able to make an accurate measurement of the trunks of some of these trees, which had been felled in the woods. Next to the Wax-palm, the Oreodoxa Sancona, which we found in flower in the valley of Cauca, and which affords a very hard and admirable wood for building, appeared to me to be the highest of all American palms. The fact, that notwithstanding the enormous mass of fruit yielded by some single palms, the number of individuals of each species growing wild is not very considerable, can only be explained by the frequent abortive development of the fruit, and by the voracity of the enemies by whom they are assailed from all classes of animals. In the basin of the Orinoco, however, whole tribes find the means of subsistence for many months together in the fruit of the palm. "In palmetis, Pihiguao consitis, singuli trunci quotannis fere 400 fructus ferunt pomiformes, tritumque est verbum inter Fratres S. Francisci, ad ripas Orinoci et Guainiæ degentes, mire pinguescere Indorum corpora, quoties uberem Palmæ fructum fundant."†

* Aug. de Saint-Hilaire, *Morphologie végétale*, 1840, p. 176.

† "In the Palm groves at Pihiguao, single trees annually bear as 400 fruit of an apple shape; and it is well known among the Brothers of San Francisco, who live on the banks of the Orinoco and Guania, that the Indians become very fat at the time that the Palms put forth their unctuous fruit."—Humboldt, *de distrib. geogr. Plant.*, p. 240.

(16) p. 224—"*From the earliest infancy of human civilization.*"

We find, as far as history and tradition extend, that the Banana has constantly been cultivated in all continents within the tropical zone. The fact of African slaves having, in the course of centuries, brought some varieties of the Banana fruit to America is as certain as that of the cultivation of this vegetable product by the natives of America prior to its discovery by Columbus. The Guaikeri Indians in Cumana assured us that on the coast of Paria, near the Golfo Triste, the Banana will occasionally produce germinating seeds, if the fruit be suffered to ripen on the stem. It is from this cause, that wild Bananas are occasionally found in the recesses of the forests, in consequence of the ripe seeds being scattered abroad by birds. At Bordones also, near Cumana, perfectly formed and matured seeds have been occasionally found in the fruit of the Banana.*

I have already remarked, in another work,† that Onesicritus and other companions of the great Macedonian, make no mention of high arborescent ferns, although they speak of the fan-leaved umbrella palms and of the tender evergreen verdure of the banana-plantations. Among the Sanscrit names given by Amarasinha for the Banana (the *Musa* of botanists) we find *bhanu-phala* (sun-fruit), *varana-buscha*, and *moko*. Phala signifies fruit generally. Lassen explains Pliny's words (xii. 6), "Arbori nomen palæ, pomo arienæ," to this effect, that "The Roman mistook the word *pala*, fruit, for the name of the tree, whilst *varana*, changed in the mouth of a Greek to *ouarana*, was transformed into *ariena*. The Arabic *mauza*, our Musa, may have been formed from *moko*. The Bhanu fruit seems to approach to Banana fruit."‡

(17) p. 224—"*Form of the Malvaceæ.*"

Larger forms of the Mallow appear, as soon as we have crossed the Alps; *Lavatera arborea*, near Nice and in Dalmatia; and *L. olbia*, in Liguria. The dimensions of the

* Compare my *Essai sur la Géographie des Plantes*, p. 29, and my *Rélat. hist.* t. i. pp. 104, 587, t. ii. pp. 355, 367.

† *Cosmos*, vol. ii. p. 524 (Bohn's Edition).

‡ Compare Lassen, *Indische Alterthumskunde*, bd. i. s. 262, with my *Essai politique sur la Nouvelle Espagne*, t. ii. p. 382, and *Rélat. hist.*, t. i. p. 491.

Baobab (monkey bread-tree) have already been given. (See pp. 270—272.) With the form of the Malvaceæ are associated the botanically allied families of the Byttneriaceæ, (*Sterculia*, *Hermannia*, and the blossoms of the large-leaved *Theobroma Cacao*, whose flowers break forth from the bark of the trunk as well as from the roots); the Bombaceæ (*Adansonia*, *Helicteres*, and *Cheirostemon*); and, lastly, the Tiliaceæ (*Sparmannia Africana*). Our *Cavanillesia plantanifolia* of Turbaco, near Carthagena in South America, and the celebrated Ochroma-like Hand-tree, the *Macpalxochiquahuitl* of the Mexicans, (from Macpalli, the flat of the hand,) *Arbol de las manitas* of the Spaniards, our *Cheirostemon platanoides*, are splendid representatives of the mallow form. In the last named, the anthers are connected together in such a manner as to resemble a hand or claw rising from the beautiful purplish-red blossoms. There is in all the Mexican free states only one individual remaining, one single primæval stem of this wonderful genus. It is supposed not to be indigenous, but to have been planted by a king of Toluca, about five hundred years ago. I found that the spot where the Arbol de las Manitas stands is 8825 feet above the level of the sea. Why is there only one tree of the kind? Whence did the kings of Toluca obtain the young tree or the seed? It is equally enigmatical, that Montezuma should not have possessed one of these trees in his botanical gardens of Huaxtepec, Chapoltepec, and Iztapalapan, which were used as late as by Philip the Second's physician, Hernandez, and of which gardens traces still remain; and it appears no less striking that the Hand-tree should not have found a place among the drawings of subjects connected with natural history, which Nezahual Coyotl, king of Tezcuco, caused to be made, half a century before the arrival of the Spaniards. It is asserted that the Hand-tree grows wild in the forests of Guatimala.* We found two Malvaceæ, *Sida Phyllanthos* (Cavan.), and *Sida Pichinchensis*, rising in the equatorial region to the great height of 13,430, and 15,066 feet on the mountain of Antisana and at the volcano of Rucu Pichincha.‡ The *Saxifraga Boussingaultii* rises from

* Humboldt et Bonpland, *Plantes équinoxiales*, t. i. p. 82, pl. 24; *Essai polit. sur la Nouv. Esp.* t. i. p. 98.

† See our *Plantes équin.* t. ii. p. 113, pl. 116.

600 to upwards of 700 feet higher, on the declivity of Chimborazo.

(18) p. 225—"*Form of the Mimosæ.*"

The delicate and feathery foliage of the Mimosæ, Acaciæ, Schrankiæ, and Desmanthus, may be regarded as peculiarly characteristic of tropical vegetation; although some representatives of this form may also be found without the tropics. In the Old Continent of the northern hemisphere, and indeed in Asia, I can instance only one low shrub, described by Marshal von Biberstein as *Acacia Stephaniana*, but which, according to Kunth's more recent investigations, is a species of the genus *Prosopis*. This social plant covers the arid plains of the province of Schirvan on the Kur (Cyrus), near New Schamach, as far as the ancient Araxes. Olivier found it also in the neighbourhood of Bagdad. It is the *Acacia foliis bipinnatis* mentioned by Buxbaum, and which extends towards the north as far as 42° lat.* In Africa the *Acacia gummifera* (Willd.), extends to Mogador, and therefore as far as 32° north lat.

In the New Continent, *Acacia glandulosa* (Michaux), and *A. brachyloba* (Willd.), adorn the banks of the Mississippi and Tenessee, and the Savannahs of the Illinois. The *Schrankia uncinata* was found by Michaux to penetrate from Florida northwards to Virginia (therefore as far as 37° north lat.). *Gleditschia triacanthos* is met with, according to Barton, to the east of the Alleghany mountains, as far as 38° north lat., and west of the same range even to 41° north lat. The extreme northern limit of *Gleditschia monosperma* is two degrees further southward. Such are the boundaries of the Mimosa form in the northern hemisphere, while in the southern hemisphere, beyond the tropic of Capricorn, simple-leaved Acaciæ are found as far as Van Dieman's Land; the Acacia cavenia described by Claude Gay being even found in Chili between 30° and 37° south lat.† Chili has no true Mimosa, but three species of Acacia; and even in the north of Chili the Acacia cavenia grows only to a height of 12 or 13 feet, whilst in the south, as it approaches the sea-coast, it

* See his *Tableau des Provinces situées sur la côte occidentale de la Mer Caspienne, entre les fleuves Terek et Kour*, 1798, pp. 58, 120.

† See Molina's *Storia naturale del Chili*, 1782, p. 174.

scarcely rises a foot above the ground. The most sensitive of the Mimosas which we saw in the northern portion of South America, are (next to the Mimosa pudica,) *M. dormiens*, *M. somnians*, and *M. somniculosa*. The irritability of the African sensitive plant was already noticed by Theophrastus (iv. 3), and by Pliny (xiii. 10); but I find the first description of the South American sensitive plants (Dormideras) in Herrera (Decad. ii. lib. iii. cap. 4). The plant first attracted the attention of the Spaniards, in 1518, in the Savannahs on the isthmus round Nombre de Dios ("parece como cosa sensible"), and it was pretended that the leaves ("de echura de una pluma de pajaros,") only contracted together when they were touched with the finger, and not when brought in contact with a piece of wood. In the small swamps which surround the town of Mompox on the Magdalena River, we discovered a very beautiful aquatic Mimosa (*Desmanthus lacustris*), a representation of which is given in our "Plantes équinoxiales" (t. i. p. 55, pl. 16). In the chain of the Andes of Caxamarca we found two Alpine Mimosas (Mimosa montana and Acacia revoluta) growing at elevations of from 9000 to nearly 9600 feet above the level of the sea.

As yet no true Mimosa, (in the meaning of the word as established by Willdenow,) nor even any Inga, has been found in the temperate zone. Amongst all the Acacias the Oriental *Acacia Julibrissin*, which Forskäl has confounded with *Mimosa arborea*, endures the greatest degree of cold. In the Botanical Garden of Padua there is a high stem of considerable thickness growing in the open air, although the mean temperature of Padua is below 56° Fahrenheit.

(19) p. 225.—"*Heaths.*"

We do not, in these physiognomical considerations, by any means comprehend, under the name of Heaths, the whole natural family of the Ericaceæ, which, on account of the similarity and analogy in the flowering parts of the plant, include Rhododendrum, Befaria, Gaultheria, and Escallonia; we limit ourselves to the very accordant and characteristic form of the species of Erica, including Calluna (*Erica vulgaris*, L.).

"Whilst in Europe Erica carnea, E. tetralix, E. cinerea,

and Calluna vulgaris, cover large tracts of country, extending from the plains of Germany, and from France and England, to the extremity of Norway; Southern Africa presents the most varied assortment of species. One single species, Erica umbellata, which is indigenous in the southern hemisphere, at the Cape of Good Hope, is again found in Northern Africa, Spain, and Portugal. Erica vagans and E. arborea also belong to the opposite coasts of the Mediterranean. The former is met with in Northern Africa, in the neighbourhood of Marseilles, in Sicily and Dalmatia, and even in England; the second in Spain, Istria, Italy, and the Canaries."* The common heath, *Calluna vulgaris* (Salisbury), which is a social plant, covers large tracts from the mouth of the Scheldt to the western declivity of the Ural. Beyond the Ural both Oaks and Heaths disappear. Both are wanting in the whole of Northern Asia, and in all Siberia, as far as the Pacific. Gmelin† and Pallas‡ have expressed their astonishment at this disappearance of Calluna vulgaris; which, on the eastern declivity of the Ural chain is even more decided and more sudden than one might be led to conclude, from the words of the last-named great naturalist. Pallas merely says, "ultra Uralense jugum sensim deficit, vix in Isetensibus campis rarissime apparet, et ulteriori Sibiriæ plane deest." Chamisso, Adolph Erman, and Heinrich Kittlitz collected Andromedas but no Calluna in Kamtschatka and on the north-west coast of America. The accurate knowledge which we at present possess of the mean temperature of different portions of Northern Asia, as well as of the distribution of annual heat throughout the different seasons, in no way explains the non-advance of the Heath to the east of the Ural. Dr. Joseph Hooker has treated with much ingenuity, in a note to his "Flora Antarctica," of the two contrasting phenomena of the distribution of plants, "uniformity of surface accompanied by a similarity of vegetation", and again, "instances of a sudden change in the vegetation, unaccompanied with any diversity of geological

* Klotsch, *Ueber die geographische Verbreitung der Erica-Arten mit bleibender Blumenkrone. Manuscr.*

† *Flora Sibirica*, t. iv., p. 129.

‡ *Flora Rossica*, t. i., pars 2, p. 53.

and other feature."* Is there an Erica in Central Asia? That which Saunders, in Turner's "Travels to Thibet,"* has described in the highlands of Nepaul, besides other European plants (Vaccinium Myrtillus, and V. oxycoccus), as Erica vulgaris, is, according to the opinion communicated to me by Robert Brown, probably the Andromeda fastigiata of Wallich. The absence of Calluna vulgaris and of all species of Erica, throughout the whole of the continental part of America is an equally striking fact, since Calluna is met with in the Azores and in Iceland. It has not hitherto been found in Greenland, but it was discovered some years ago in Newfoundland. The natural family of the Ericaceæ is also almost entirely wanting in Australia, where its place is supplied by the Epacrideæ. Linnæus described only 102 species of the genus Erica, but, according to Klotzsch's observations, this genus comprises 440 true species, after the varieties have been carefully excluded.

(20) p. 226—"*The Cactus form.*"

When the natural family of the Opuntiaceæ is separated from the Grossulariaceæ (species *Ribes*), and is confined within the limits indicated by Kunth,‡ we may regard the whole as exclusively American. I am not ignorant, that Roxburgh, in the *Flora indica* (inedita), mentions two species of Cactus which he regards as peculiar to the south-east of Asia, viz., Cactus indicus, and C. chinensis. Both are widely diffused, originally wild or having become so, and different from Cactus opuntia and C. Coccinellifer; but it is remarkable that this Indian plant should have no ancient Sanscrit name. The so-called Chinese Cactus has been introduced by cultivation into the island of St. Helena. Modern investigations, prosecuted at a period when a more general interest has been awakened in relation to the original distribution of plants, will unquestionably remove the doubts that have frequently been advanced against the existence of Asiatic Opuntiaceæ. We see, in a similar manner, certain vital forms appear separately in the animal world.

* *Botany of the Antarctic Voyage of the Erebus and Terror*, 1844, p. 210.

† *Philos. Transact.*, vol. lxxix. p. 86.

‡ *Handbuch der Botanik*, s. 609.

How long did the Tapir continue to be regarded as a characteristic form of the New Continent! And yet the American Tapir is, as it were, repeated in that of Malacca (*Tapirus indicus*, Cuv.).

Although the Cactus form belongs, properly speaking, to the tropical regions, there are some species in the New Continent, that are indigenous to the temperate zone on the Missouri and in Louisiana; as, for instance, Cactus missuriensis and C. vivipara. Back, in his northern expedition, saw with astonishment, the banks of the Rainy Lake in lat. 48° 40′ (long. 92° 53′) entirely covered with C. Opuntia. South of the equator the Cactus does not advance further than Rio Itata (lat. 36°) and Rio Biobio (lat. 37¼°) In the part of the chain of the Andes lying within the tropics, I have found species of Cactus (*C. sepium*, *C. chlorocarpus*, *C. bonplandii*) on elevated plains from 9000 to upwards of 10,600 feet above the level of the sea; but in Chili, in the temperate zone, a far more strongly marked Alpine character is exhibited by Opuntia Ovallei, whose upper and lower limits have been accurately determined through barometric measurements by the learned botanist, Claude Gay. The yellow-flowering Opuntia Ovallei, which has a creeping stem, does not descend below 6746 feet, advancing as high as the line of perpetual snow; and even above it, wherever a few masses of rock remain uncovered. These little plants have been gathered at spots lying at an elevation of 13,663 feet above the level of the sea.* Some species of Echinocactus are also true alpine plants in Chili. A counterpart to the much admired fine-haired Cactus senilis is presented by the thick-wooled Cereus lanatus, called by the natives *Piscol*, which has a fine red fruit. We found it near Guancabamba, in Peru, on our journey to the Amazon river. The dimensions of the Cactaceæ (a group on which the Prince of Salm-Dyck was the first to throw considerable light) present the most striking contrasts. Echinocactus Wislizeni, which has a circumference of seven feet and a half, with a height of four feet and a quarter, is only third in size, being surpassed by E. ingens, (Zucc.) and E. platyceras, (Lem.)† The Echinocactus Stainesii attains a diameter of from two feet to two and a-half; E

* Claudio Gay, *Flora Chilensis*, 1848, p. 30.

† Wislizenus, *Tour to Northern Mexico*, 1848, p. 97.

visnago, belonging to Mexico, has a diameter of upwards of three feet, with a height of more than four feet, and weighs as much as from 700 to 2000 lbs.; while the Cactus nanus, which we collected near Sondorillo, in the province of Jaen, is so small and so loosely rooted in the sand, that it gets between the toes of dogs. The Melocactuses, which are full of juice even in the driest season, as the Ravenala of Madagascar (wood-leaf in the language of the country from *rave*, *raven*, a leaf, and *ala*, the Javanese *halas*, a wood), are vegetable springs, which the wild horses and mules open by stamping with their hoofs—a process in which they frequently injure themselves.* Cactus Opuntia has spread during the last quarter of a century in a remarkable manner through Northern Africa, Syria, Greece, and the whole of Southern Europe; penetrating from the coasts of Africa far into the interior, where it associates with the native plants.

After being accustomed to see Cactuses only in our hot-houses, we were astonished at the density of the woody fibres in old cactus stems. The Indians are aware that cactus wood is indestructible, and admirably adapted for oars and the thresholds of doors. There is hardly any physiognomical character of exotic vegetation that produces a more singular and ineffaceable impression on the mind of the traveller, than an arid plain densely covered with columnar or candelabra-like stems of cactuses, similar to those near Cumana, New Barcelona, Coro, and in the province of Jaen de Bracamoros.

### (21) p. 226—"*Orchideæ.*"

The almost animal-like form occasionally observed in blossoms of the Orchideæ is most strongly marked in Anguloa grandiflora, celebrated in South America as the Torito; in the Mosquito (our Restrepia antennifera); in the Flor del Espiritu Santo (likewise an Anguloa, according to *Floræ Peruvianæ Prodrom.* p. 118, tab. 26); in the ant-like flower of Chiloglottis cornuta;† in the Mexican Bletia speciosa; and in the whole host of our remarkable European species of Ophrys: *O. muscifera*, *O. apifera*, *O. aranifera*, *O. arachnites*, *&c.* The taste for these splendidly flowering plants has so much increased, that the number of species cultivated by Messrs. Loddige,

* See p. 15.
† Hooker, *Flora antarctica*, p. 69.

which, in 1813, was only 115, was upwards of 1650 in 1843, and in 1848, the number was estimated at no fewer than 2360. What a treasure of sumptuously flowering and unknown Orchideæ may be inclosed in the interior of Africa wherever there is an abundant supply of water! Lindley, in his beautiful work, *On the Genera and Species of Orchideous Plants*, 1840, counted exactly 1980 species; whilst Klotzsch at the close of the year 1848 counted 3545.

Whilst the temperate and cold zone possess only terrestrial Orchideæ, growing close to the ground, both forms, the terrestrial, as well as the parasitical, growing on the trunks of trees, are indigenous in the beautiful regions of the tropics. To the former class belong the tropical genera Neottia, Cranichis, and most Habenarias. But we have found both these forms as alpine plants on the declivity of the Andes of New Granada and Quito, viz., the parasitical (*Epidendreæ*) Masdevallia uniflora (at an elevation of 10,231 feet), Cyrtochilum flexuosum (at 10,103 feet), and Dendrobium aggregatum (at 9485 feet); and the terrestrial forms of Altensteinia paleacca, near Lloa Chiquito, at the foot of the volcano of Pichincha. Claude Gay is of opinion that the Orchideæ supposed to have been found growing on trees in the Island of Juan Fernandez and even at Chiloe, were probably only parasitical Pourretiæ, which advance as far south at least as 40°. In New Zealand, the tropical form of Orchideæ, hanging from trees, is still to be seen as far south as 45°. But the Orchideæ of Auckland and Campbell Islands (Chiloglottis, Thelymitra, and Acianthus), grow on level ground in moss. In the animal world there is at least one tropical form that penetrates further south. The Island of Macquarie (lat. 54° 39′) has an indigenous parrot, which lives therefore in a region nearer to the south pole than Dantzig is to the north pole.*

(22) p. 226—"*Form of the Casuarinæ.*"

Acacias, in which the place of the leaves is supplied by phyllodia, Myrtaceæ (Eucalyptus, Metrosideros, Melaleuca, Leptospermum), and Casuarinæ, constitute the sole characteristics of the vegetable world of Australia (New Holland) and Tasmania (Van Diemen's Land). Casuarinæ with their

* Compare the section *Orchideæ* in my work, *De distrib. geogr. Plant.*, pp. 241—247.

leafless, thin, thread-like, articulated branches, and their joints furnished with membranous, toothed spathes, have been compared by travellers,* according to differences of species, either with arborescent Equisetaceæ (Horsetails) or with our Scotch firs. I have been much struck with the singular appearance of leaflessness presented by the small thickets of Colletia and Ephedra in South America, near the coast of Peru. Casuarina quadrivalvis penetrates, according to Labillardière, as far south as 43° in Tasmania. The mournful form of the Casuarina is not unknown in the East Indies and even on the eastern coast of Africa.

(23) p. 227—"*Acicular-leaved trees.*"

The family of the Coniferæ (including the genera of Dammara, Ephedra, and Gnetum of Java and New Guinea, which are essentially allied to it, though distinctly separated by the form of the leaf and the whole conformation), plays so important a part in consequence of the number of individuals in each species, and by its geographical diffusion, while it covers in the northern temperate zone, as a social plant, such extensive districts, that we are almost compelled to wonder at the inconsiderable number of the species. We are not acquainted with so many Coniferæ by three-fourths as there are Palms already described, nay, the Coniferæ are numerically less than the Aroideæ. Zuccarini, in his "Contributions to the Morphology of the Coniferæ,"† enumerates 216 species, of which 165 belong to the Northern and 51 to the Southern hemisphere. These proportional numbers must now, in consequence of my researches, be differently expressed, since, with the species of Pinus, Cupressus, Ephedra, and Podocarpus, which Bonpland and I discovered in the tropical part of Peru, Quito, New Granada, and Mexico, the number of the cone-bearing trees flourishing between the tropics amounts to 42. The excellent and latest work of Endlicher‡ contains 312 species of Coniferæ now living, and 178 of a primeval mundane period which are now buried in the coal formation, in variegated sandstone, in keuper, and in

* See Darwin, *Journal of Researches*, p. 449.

† See his *Abhandl. der Wiss. zu München*, bd. iii. 1837–1843, s. 752.

‡ *Synopsis Coniferarum*, 1847.

Jura limestone. The vegetation of the eocene world presents especially to us forms which, by their coëval relationship with several families of the present world, remind us that with it many intervening members have disappeared. The Coniferæ, so frequent in the primeval world, accompany, in particular, the ligneous remains of Palms and Cycadeæ; but in the most recent beds of lignite or brown coal we again find Coniferæ, our Pines and Firs, associated with Cupuliferæ (or Mastworts), Maples and Poplars.*

If the surface of the earth did not rise to great altitudes within the tropics, the strikingly characteristic form of acicular-leaved trees would have remained wholly unknown to the inhabitants of that zone. I took great pains, in common with Bonpland, to trace out, in the Mexican Highlands, the *lower* and *upper* boundary line of the Coniferæ and Oaks. The heights, at which both begin to grow (los Pinales y Encinales, Pineta et Querceta), are hailed with joy by those who come from the sea coast, because they announce a climate not yet invaded, as far as experience has hitherto shown, by that mortal disease called the black vomit (vomito prieto, a form of the yellow fever). For the oaks, especially the Quercus Xalapensis (one of the twenty-two Mexican species of oak which we first described), the lower line of vegetation, on the way from Vera Cruz to the capital of Mexico, somewhat below the Venta del Encero, is 3048 feet above the sea. At the western slope of the plateau, between the South Sea and Mexico, the inferior line for oaks is something lower; it begins near a hut named Venta de la Moxonera, between Acapulco and Chilpanzingo, at the absolute height of 2481 feet. I found a similar difference in the lower boundary line of the pine-forest. This boundary, towards the South Sea, in the Alto de los Caxones, north of Quaxinquilapa, is for the Pinus Montezumæ (Lamb.), which we at first had considered to be the Pinus occidentalis (Swartz), at the height of 4092 feet; but towards Vera Cruz, at the Cuesta del Soldado, it rises to 5979 feet. Both these kinds of tree, therefore, the oaks and firs as specified above, descended lower towards the Pacific than towards the Caribbean Gulf. During my ascent of the Cofre di Perote, I found the superior boundary line of the oaks to be 10,353 feet; that of the Pinus Montezumæ 12,936 feet (about 2000 feet higher than the summit of Mount Ætna)

* See *Cosmos*, vol. i. pp. 282–287 (Bohn's edition).

and here, in February, considerable masses of snow had already fallen.

The greater the heights at which the Mexican cone-bearing trees begin to show themselves, the more singular is it, in the island of Cuba (where, at the border of the tropical zone the air, it is true, is cooled down during northerly winds to 46°.6 Fahr.), to see another kind of fir (*P. Occidentalis*, Swartz), in the plain itself, or on the gentle hills of the Isle of Pines, growing among palms and mahogany trees (*Swietenia*). Columbus even makes mention of a fir-wood (*Pinal*) in the journal of his first voyage (Diario del 25 de Nov., 1492), at Caya de Moya, north-east of Cuba. At Haiti, too (St. Domingo), the Pinus occidentalis near Cape Samana descends from the mountains down to the very beach. The stems of these firs, wafted by the gulf-stream to the two Azores, Graciosa and Fayal, were among the principal signs that proclaimed to the great discoverer the existence of unknown lands in the West.* Is it positively ascertained that the Pinus occidentalis is entirely absent from Jamaica, notwithstanding its lofty mountains? We may be permitted to inquire also, what kind of Pinus grows on the eastern coast of Guatimala, since the P. tenuifolia (Benth.) is assuredly found only on the mountains near Chinanta.

On taking a general view of the species of plants which form the upper tree-boundary in the northern hemisphere from the frigid zone to the equator; I find, for Lapland, according to Wahlenberg, in the Sulitelma Mountains (lat. 68°), not acicular-leaved trees but birches (Betula alba), far above the upper limit of the Pinus sylvestris; and for the temperate zone I find in the Alps (lat. 45° 45′) Pinus picea (Du Roi), advanced beyond the birches. In the Pyrenees (lat. 42° 30′), we find Pinus uncinata (Ram.) and P. sylvestris, var. rubra; within the tropics in Mexico (lat. 19°—20°), Pinus Montezumæ extends far beyond Alnus toluccensis, Quercus spicata, and Q. crassipes; and in the snow-crowned mountains of Quito, beneath the equator, Escallonia myrtilloides, Aralia avicennifolia, and Drymis Winteri attain the highest limits. This last species of tree, identical with the Drymis granatensis (Mut.), and the Wintera aromatica of Murray, presents, as Dr. Joseph Hooker has shown,† the most singular instance of the unin-

* See my *Examen crit.*, t. ii. pp. 246–259.

† *Flora Antarctica*, p. 229.

terrupted dissemination of the same species of tree from the southernmost part of Tierra del Fuego and Hermit Island, where it was discovered as early as 1577 by Drake's expedition, up to the northern Highlands of Mexico, over a meridian extent of 86° of latitude or 5160 miles. Where the acicular or needle-leaved trees, as in the Swiss Alps and the Pyrenees, and not the birch as in the extreme north, form the boundary of arborescent vegetation on the loftiest mountains, which they picturesquely encircle, they are immediately followed in their ascent towards the snow-crowned summits, in Europe and Western Asia by the Alpine roses, Rhododendra, and at the Silla de Caracas, and the Peruvian Paramo de Saraguru, by the purplish-red blossoms of the graceful Befariæ. In Lapland the Rhododendron laponicum immediately follows the Coniferous trees; in the Swiss Alps, the Rhododendron ferrugineum and R. hirsutum, and in the Pyrenees the R. ferrugineum alone; and in the Caucasus the R. caucasicum. But R. caucasicum has also been found isolated by De Candolle in the Jura mountains (in the Creux de Vent), 5968 feet lower down, at the inconsiderable height of from 3303 to 3730 feet. If we would trace out the last zone of vegetation near the snow-line we must name, according to our personal observation, in tropical Mexico, Cnicus nivalis and Chelone gentianoides; in the cold mountainous tracts of New Granada, the woolly Espeletia grandiflora, E. corymbosa, and E. argentea; in the Andes chain of Quito, Culcitium rufescens, C. ledifolium, and C. nivale;—yellow-blossomed Compositæ, which replace the somewhat more northerly lanose herbs of New Granada, and the Epeletiæ, with which they have so much physiognomical resemblance. This substitution or repetition of similar and almost identical forms in regions that are separated from each other by seas or wide intervening tracts, is a wonderful law of nature. It prevails even in the rarest forms of the floras. In Robert Brown's family of the Rafflesiæ, separated from the Cytineæ, the two Hydnoræ in Southern Africa (H. Africana and H. Triceps), described by Thunberg and Drege, have, in South America, their counterpart in the H. Americana of Hooker.

Far above the regions of Alpine herbs, of the grasses and the lichens, nay, beyond the boundary of perpetual snow, there occasionally appears a phanerogamic plant, growing sporadically, and as it were isolated, to the astonishment of botanists; and this occurs both within the tropics and in the temperate zone, on fragments of rock which remain free from snow and are probably warmed by open fissures. I have already mentioned the Saxifraga Boussingaulti, which is found at a height of 15,773 feet on the Chimborazo; in the Swiss Alps the Silene acaulis, a clovewort or caryophyllea, has been seen at a height of 11,382 feet. The former vegetates at 640, the latter at 2621 feet above the respective local limits of snow, heights which were determined when both the plants were discovered.

In our European Coniferous woods the Red Pine (or Norway Spruce), and the White (or Silver) Pine show great and remarkable variations as regards their geographical dispersion on the slopes of mountains. Whilst in the Swiss Alps the Red Pine (*Pinus picea*, Du Roi, *foliis compresso-tetragonis;* unfortunately named by Linnæus and by most botanists of our time the *Pinus abies!*), forms the limit of tree vegetation at the mean height of 5883 feet, and only here and there does the lowly alder (*Alnus viridis*, Dec., *Betula viridis*, Vill.), advance higher towards the snow-limit; the White Pine (*Pinus abies*, Du Roi, *Pinus picea*, Linn., foliis planis, pectinato-distichis, emarginatis), has its limit, according to Wahlenberg, about 1000 feet lower. The Red Pine does not grow at all in Southern Europe, in Spain, the Apennines, and Greece; and, as Ramond remarks, it is only seen on the slope of the northern Pyrenees at great heights, and is entirely wanting in the Caucasus. The Red Pine extends further to the north in Scandinavia than the White, which latter tree appears in Greece (on the Parnassus, the Taygetus, and the Œta), as a variety with long acicular leaves, *foliis apice integris, breviter mucronatis*, the Abies Apollinis of the acute observer Link.*

On the Himalaya the acicular-leaved form of trees is distinguished by the mighty thickness and height of the stem as well as by the length of the leaf. The chief ornament of the mountain range is the Cedar Deodwara (*Pinus deodara*, Roxb.), which word is, in Sanscrit, dêwa-dâru, *i.e.* timber for the gods, its stem being nearly from 13 to 14 feet in diameter. It ascends in Nepaul to more than 11,700 feet above the level of the sea. More than 2000 years ago the Deodwara

* See *Linnæa*, bd. xv. 1841, s. 529, and Endlicher's *Synopsis Coniferarum*, p. 96.

cedar near the River Behut, that is, the Hydaspes, furnished the timber for the fleet of Nearchus. In the valley of Dudegaon, north of the copper mines of Dhunpoor in Nepaul, Dr. Hoffmeister, so early lost to science, found in a forest the Pinus longifolia (Royle), or the Tschelu Fir, mixed with the lofty stems of a palm—Chamærops martiana (Wallich).* Such an interspersion of the *pineta* and *palmeta* had already, in the new continent, excited the astonishment of the companions of Columbus, as a friend and contemporary of the admiral's, Petrus Martyr Anghiera, relates.† I myself saw, for the first time, this blending of pines with palms on the road from Acapulco to Chilpanzingo. The Himalaya, like the Mexican highlands, besides its genera of pine and cedar, possesses also forms of the Cypress (*Cupressus torulosa*, Don.); of the Yew (*Taxus Wallichiana*, Zuccar.); of the Podocarpus (*Podocarpus nereifolia*, Brown); and the Juniper (*Juniperus squamata*, Don., and *J. excelsa*, Bieberst.; the latter species occurring also at Schipke in Thibet, in Asia Minor, Syria, and the Grecian Islands; on the other hand, Thuja, Taxodium, Larix, and Araucaria, are forms of the New Continent, which are wanting in the Himalaya.

Besides the twenty species of pine with which we are acquainted in Mexico, the United States of North America, in their present extension to the Pacific, present forty-five described species, whilst all Europe can only enumerate fifteen. The same difference between abundance and paucity of forms is shown in the oaks, in favour of the New Continent (a quarter of the world the most connected and most elongated in a meridional direction). It has, however, been very recently demonstrated by the extremely accurate researches of Siebold and Zuccarini to be an erroneous assertion, that many European species of pine, in consequence of their wide distribution throughout Northern Asia, passed over to the Japanese islands, and there mingled with a genuine Mexican species, the Weymouth pine (*Pinus strobus*, L.), as Thunberg asserts. What Thunberg considered to be European species of pine, are species entirely different. Thunberg's Red Pine (*Pinus abies*, Linn.) is *P. polita*, Sieb., and often planted near Buddhist temples; his northern common fir (*Pinus sylvestris*) is P.

* See Hoffmeister's *Briefe aus Indien während der Expedition des Prinzen Waldemar von Preussen*, 1847, s. 351.

† *Dec.* iii. lib. x. p. 68.

Massoniana, Lamb.; his P. cembra, the German and Siberian stone pine-tree, is P. parviflora, Sieb.; his common larch (*P. larix*) is the P. leptolepis, Sieb.; his Taxus baccata, the fruit of which the Japanese courtiers eat as a precautionary measure when attending long ceremonies,* forms a special genus and is Cephalotaxus drupacea, Sieb. The Japanese islands, despite the proximity of the Asiatic Continent, have a very different character of vegetation. Thunberg's Japanese Weymouth pine, which would present an important phenomenon, is moreover a naturalized tree, that differs entirely from the indigenous pines of the New World. It is Pinus korajensis, Sieb., which has migrated from the peninsula of Corea and Kamtschatka to Nipon.

Of the 114 species now known of the genus Pinus, there is not one in the whole southern hemisphere, for the Pinus Merkusii, described by Junghuhn and De Vriese, still belongs to that part of the island of Sumatra which is north of the equator, that is, to the district of the Battas. The P. insularis, Endl., belongs to the Philippines, although at first it was introduced into Loudon's *Arboretum* as P. timoriensis. From our present increasing knowledge of the geography of plants, we know that there are excluded also from the southern hemisphere, in addition to the genus Pinus, all the races of Cupressus, Salisburia (*Ginkgo*), Cunninghamia (*Pinus lanceolata*, Lamb.), Thuja, one species of which (*Th. gigantea*, Nutt.) at the Columbia river rises as high as 180 feet, Juniperus, and Taxodium (Mirbel's *Schubertia*). I can introduce this last genus here with the greater certainty, inasmuch as a Cape plant, Sprengel's Schubertia capensis, is no Taxodium, but forms a special genus, Widringtonia, Endl., in quite another division of the Coniferæ.

This absence from the southern hemisphere of the true Abietineæ, of the Juniperineæ, Cupressineæ, and all the Taxodineæ, as likewise of the Torreya, of the Salisburia

* Thunberg, *Flora Japonica*, p. 275. The allusion is somewhat amusing; we annex a translation of Thunberg's note:—"This fruit resembles acorns, and is of an astringent nature. For this reason the Japanese interpreters, when constrained to remain in the royal presence longer than usual, chew it, as an antidiuretic. It is brought to table at the second course with Acrodrya, and is said to be very wholesome, and to relax the bowels although it constricts the mouth. The expressed oil is in request for the kitchen, especially among the Chinese monks who live at Nagasacca."—Ed.

adiantifolia, and of the Cephalotaxus among the Taxineæ, vividly reminds us of the enigmatical and still obscure conditions which determined the original distribution of vegetable forms. This distribution can by no means be satisfactorily explained either by the similarity or diversity of the soil, by thermal relations, or by meteorological conditions. I have long since directed attention to the fact, that the southern hemisphere possesses, for instance, many plants of the natural family of the Rosaceæ, but not a single species of the genus Rosa itself. Claude Gay informs us, that the Rosa Chilensis, described by Meyen, is a variety that has become wild of the Rosa centifolia, Linn., which has been naturalized in Europe for thousands of years. Such wild-growing varieties occupy large tracts in Chili near Valdivia and Osorno.*

In the whole tropical region of the northern hemisphere we only found one single indigenous rose, our Rosa Montezumæ, and this was on the Mexican highland, near Moran, at a height of 9336 feet. We may count among the strange phenomena observed in the distribution of plants, the total absence of the Agave from Chili, though it possesses Palms, Pourretias, and many species of Cactus; and although A. americana flourishes luxuriantly in Roussillon, at Nice, at Botzen, and in Istria, where it was probably introduced from the New Continent since the sixteenth century, and where it forms one connected line of vegetation from the north of Mexico, across the isthmus of Panama, as far as Southern Peru. With respect to the Calceolarias, I long believed that, like the roses, they were only to be found exclusively on the northern side of the equator. In fact, among the twenty-two species that we brought with us, not one was gathered to the north of Quito and the volcano of Pichincha; but my friend Professor Kunth remarks that Calceolaria perfoliata, which Boussingault and Capt. Hall found near Quito, advances also as far as New Granada, and that this species, as well as C. integrifolia, was sent by Mutis from Santa Fé de Bogotá to the great Linnæus.

The species of Pinus, which are so abundant in the wholly inter-tropical Antilles, as well as in the tropical mountain regions of Mexico, do not cross the isthmus of Panama, and are wholly wanting in the equally mountainous parts of tro-

* Gay, *Flora Chilensis*, p. 340.

pical South America, that lie north of the equator; they are equally unknown on the elevated plains of New Granada, Pasto, and Quito. I have advanced in the plains and on the mountains from the Rio Sinu, near the isthmus of Panama, as far as 12° south lat.; and in this territorial extent, of nearly 1600 miles in length, the only forms of needle-leaved trees that I saw, were the taxoid Podocarpus (P. taxifolia), 64 feet high, in the Andes pass of Quindiu and in the Paramo de Saraguru, in 4° 26′ north and 3° 40′ south latitude, and an Ephedra (E. americana) near Guallabamba, north of Quito.

Among the group of the Coniferæ, the following are common to the northern and southern hemispheres: Taxus, Gnetum, Ephedra, and Podocarpus. Long before l'Heritier, the last genus had been very properly distinguished from Pinus by Columbus on the 25th of November, 1492. He says, "Pinales en la Serrania de Haiti que no llevan piñas, pero frutos que parecen azeytunos del Axarafe de Sevilla."* Species of yew extend from the Cape of Good Hope to 61° north lat. in Scandinavia, consequently through more than 95 degrees of latitude. Podocarpus and Ephedra are almost as widely distributed; and even from among the Cupuliferæ, the species of the oak genus, usually termed by us a northern form, though they do not cross the equator in South America, reappear in the southern hemisphere, at Java, in the Indian archipelago. To this latter hemisphere ten genera of the cone-bearing trees exclusively appertain, of which we will here cite only the most important: Araucaria, Dammara (*Agathis*, Sal.), Frenela (comprising about 18 Australian species), Dacrydium and Lybocedrus, whose habitat is both in New Zealand and the Straits of Magellan. New Zealand possesses one species of the genus Dammara (*D. australis*), but no Araucaria. The contrary, by a singular contrast, is the case in New Holland.

In the form of acicular-leaved trees, Nature presents us with the greatest length of stem existing in arborescent productions. I use the term arborescent, for, as we have already remarked, among the Laminariæ (the oceanic algæ) Macrocystis pyrifera, between the coast of California and 68° south lat., often attains a length of more than 400 feet. If we exclude the six Araucarias of Brazil, Chili, New Holland,

* See my *Examen crit.* t. iii. p. 24.

the Norfolk Islands and New Caledonia, then those Coniferæ are the highest, whose habitat is the temperate zone of the North. As we have found among the family of the palms the most gigantic of all, the Ceroxylon andicola, about 192 feet high, in the temperate Alpine climate of the Andes, so in like manner do the loftiest cone-bearing trees belong, in the *northern* hemisphere, to the temperate north-western coast of America and to the Rocky Mountains (lat. from 40° to 52°), in the *southern* hemisphere to New Zealand, Tasmania or Van Dieman's Land, to Southern Chili and Patagonia, (where the lat. is again from 43° to 50°). The most gigantic forms among the genus Pinus are Sequoia (Endl.), Araucaria, and Dacrydium. I only name those species whose height not merely reaches but often exceeds 200 feet. That the reader may have a standard of comparison, he is reminded that in Europe the loftiest Red and White Pines, especially the latter, reach a height of from 160 to 170 feet; for instance, in Silesia, the pine in the Lampersdorf forest, near Frankenstein, long famous for its altitude, is only 158 feet high, although 17 feet in girth.*

We give the following examples:—

Pinus Grandis (Dougl.), in New California, attains a height of 202—224 feet.

Pinus Frémontiana (Endl.), also there, and probably of the same height.†

Dacrydium Cupressinum (Solander), in New Zealand, above 213 feet.

Pinus Lambertiana (Dougl.), in North-western America, 223—234 feet.

Araucaria Excelsa (R. Brown), the Cupressus columnaris of Forster, in Norfolk Island and the surrounding rocks, 182—223 feet. The six Araucariæ hitherto known fall into two groups, according to Endlicher:

*a.* The American (Brazil and Chili), A. brasiliensis [Rich.], between 15° and 25° south lat., and A. imbricata [Pavon], between 35° and 50° south lat.; the latter 234—260 feet;

*β.* The Australian (A. Bidwilli [Hook.] and A. Cunninghami [Ait.] on the eastern side of New Holland, A. excelsa

* See Ratzeburg, *Forstreisen,* 1844, s. 287.

† Torrey and Frémont, *Report of the Exploring Expedition to the Rocky Mountains in* 1844, p. 319.

of Norfolk Island, and A. Cookii [R. Brown] of New Caledonia). Corda, Presl, Göppert, and Endlicher have already found five fossil Araucariæ in lias, in chalk, and in lignite.*

Pinus Douglasii (Sab.) in the valleys of the Rocky Mountains and at the Columbia River (north lat. 43°—52°). That meritorious Scotch botanist, whose name this tree bears, suffered a dreadful death in 1833, when he came from New California to collect plants on the Sandwich Islands. He inadvertently fell into a pit, into which one of the wild bulls of that country, always viciously disposed, had previously fallen. This traveller has described from accurate measurements a stem of P. Douglasii, which at three feet from the ground was 57½ feet round, and 245 feet high.†

Pinus Trigona (Rafinesque), on the western slope of the Rocky Mountains.‡ This "gigantic fir" was measured with great care; the girth of the stem at 6¼ feet above the ground was often from 38 to 45 feet. One stem was 300 feet high, and without branches for the first 192 feet.

Pinus Strobus (in the eastern part of the United States of North America, especially on this side of the Mississippi, but also again in the Rocky Mountains, from the source of the Columbia to Mount Hood, from 43° to 54° north lat.), in Europe called the Weymouth Pine, and in North America the White Pine, commonly no more than 160 to 190 feet high, but several have been seen in New Hampshire of 250 and 266 feet.§

Sequoia Gigantea (Endl.; the Condylocarpus, Sal.), of New California, like the Pinus trigona, about 300 feet high.

The nature of the soil and the conditions of heat and moisture, on which the nourishment of plants simultaneously depends, promote, it must be admitted, the development and the increase of the number of the individuals in a species; but the gigantic height attained by the stems of a few among the many nearly allied species of the same

* Endlicher, *Coniferæ fossiles*, p. 301.

† See *Journal of the Royal Institution*, 1826, p. 325.

‡ See description in Lewis and Clarke's *Travels to the Source of the Missouri River and across the American Continent to the Pacific Ocean* (1804–6), 1814, p. 456.

§ Dwight, *Travels*, vol. i. p. 36, and Emerson, *Report on the Trees and Shrubs growing naturally in the Forests of Massachusetts*, 1846, p. 60–66.

genus is not dependent on soil and climate but on a specific organization, on internal natural disposition, common alike to the vegetable and to the animal world. With the Araucaria imbricata of Chili, the Pinus Douglasii of the Columbia River, and the Sequoia gigantea of New California (245—300 feet) contrasts most strongly—not the Willow (*Salix arctica*) stunted by cold or mountain height, and only two inches high,—but a little phanerogamic plant in the beautiful climate of the southern tropical region, in the Brazilian province of Goyaz. The moss-like Tristicha hypnoides, of the Monocotyledonous family of the Podostemeæ, hardly attains the height of three lines. "While crossing the Rio Clairo in the province of Goyaz," says an excellent observer, "I perceived on a stone a plant, the stalk of which was not more than three lines high, and which I considered at first to be a moss. It was, however, a phanerogamic plant, supplied with sexual organs like our oaks, and those gigantic trees which raised their majestic heads around."*

Besides the height of the stem, the length, breadth, and position also of the leaves and fruit, the aspiring or horizontal, almost umbellate ramification, the gradation of the colour from fresh or silver-greyish green to dark brown, give a peculiar physiognomical character to the Coniferæ. The acicular leaves of Pinus Lambertiana (Douglas) in North-Western America are five, those of the P. excelsa (Wallich) on the southern slope of the Himalaya near Katmandu, seven, and those of P. longifolia (Roxb.) on the mountain range of Cashmere, more than twelve inches long. Moreover, in one and the very same species, these acicular leaves vary in the most remarkable manner, from the combined influence of the nourishment derived from soil and air, and of the height above the level of the sea. I found these variations in the length of the leaves of our common wild pine (*Pinus sylvestris*) so great, while travelling in a west and east direction over an extent of 80° of longitude (more than 3040 miles) from the Scheldt, through Europe and Northern Asia, to Bogoslowsk, in the Northern Ural, and Barnaul beyond the Obi, that occasionally, deceived by the shortness and rigidity of the leaves, I have mistaken it for another species of pine, allied to the mountain fir, *P. rotundata*, Link, (*Pinus uncinata*, Ram.)

* Auguste de St. Hilaire, *Morphologie végétale*, 1840, p. 98.

These are, as Link correctly observes,* transitions to Ledebour's P. sibirica of the Altai.

The delicate and pleasing green though deciduous foliage of the Ahuahuete (*Taxodium distichum*, Rich., *Cupressus disticha*, Linn.) on the Mexican plateau especially delighted me. In this tropical region the tree, swelling out to a portly bulk, and the Aztec name of which signifies "water-drum" (from atl, water, and huehuetl, drum), flourishes from 5750 to 7670 above the level of the sea, whilst it descends towards the plain in the marshy district (Cypress swamps) of Louisiana as far as 43° lat. In the southern States of North America the Taxodium distichum (*Cyprès chauve*), as well as in the lofty plains of Mexico, attains a height of 128 feet, with an enormous girth, the diameter being from 30 to nearly 40 feet, when measured near the ground.† The roots, too, present a very remarkable phenomenon, for they have woody excrescences, which are sometimes of a conical and rounded, sometimes of a tabular shape, and project three and even nearly five feet above the ground. Travellers have compared these woody excrescences, in spots where they are numerous and frequent, to the grave-tablets of a Jewish churchyard. Auguste de St. Hilaire remarks, with much acuteness: "These excrescences of the bald cypress, which resemble boundary-posts, may be regarded as exostoses, and like these live in the air; adventitious buds would doubtless escape from them, if the nature of the tissue of the coniferous plants did not oppose itself to the development of those concealed germs that give birth to these kinds of buds."‡ In addition to the above, a remarkably enduring vitality is manifested in the roots of cone-bearing trees by the phenomenon which, under the name of "Effervescence," (aftergrowth?) has attracted, in many ways, the attention of botanical physiologists, and which phenomenon, it appears, rarely displays itself in other dicotyledonous plants. The stumps of the felled white Pine, left in the ground, form, during a succession of several years, new layers of wood, and continue to increase in thickness, without throwing out shoots, branches, or leaves. The excellent observer Göppert believes, that this takes place solely

* *Linnæa,* bd. xv. 1841, s. 489.
† Emerson, *Report on the Forests,* pp. 49, 101.
‡ *Morphologie végétale,* p. 91.

through nourishment derived from the roots, which the extremity of the stem receives from a neighbouring living tree of the same species. The roots of the living tree he conceives are organically incorporated with those of the stump.* Kunth, in his excellent new *Lehrbuch der Botanik*, is opposed to this explanation of a phenomenon, which was even known, though imperfectly, to Theophrastus.† According to him, this process is perfectly analogous to that by which metallic plates, nails, carved letters, nay, even stags' horns become imbedded within the body of wood. "The cambium, that is, the thin, walled cellular tissue, conducting muco-granular sap, from which new formations alone proceed, continues without any relation to the buds (being perfectly independent of them) to deposit new layers of wood on the outermost layer."‡

The relation above alluded to, between the absolute height of the ground and the geographical as well as isothermal latitude, shows itself often, no doubt, when one compares the arborescent vegetation of the tropical part of the Andes chain with the vegetation of the north-west coast of America, or the banks of the Canadian lakes. The same remark was made by Darwin and Claude Gay in the southern hemisphere, when they, in their descent from the plateau of Chili, advanced towards Eastern Patagonia, and the Archipelago of Tierra del Fuego; here woods of Drymis Winteri, together with Fagus antarctica and Fagus Forsteri, cover every thing with long uniform rows in a northern and southern direction down to the low lands. Trifling deviations from the law of constant *station-ratios* between *mountain height* and *geographical latitude*, depending or local causes, not sufficiently investigated, occur even in Europe. I would call to mind the limits of altitude for the birch and common fir in a part of the Swiss Alps, on the Grimsel. The fir (*Pinus sylvestris*) flourishes there up to 6330; and the birch (*Betula alba*) up to 6906 feet; beyond them again there is a belt of stone pines (*Pinus cembra*), whose upper boundary is 7343 feet. The birch, in consequence, lies there between two belts of Coniferæ.

* Göppert, *Beobachtungen über das sogenannte Umwallen der Tannenstöcke*, 1842, s. 12.

† *Hist. Plant.*, lib. iii. cap. 7, pp. 59, 60. Schneidei.

‡ Th. i. s. 143, 166.

According to the excellent observations of Leopold von Buch, and the more recent ones of Martius, who also visited Spitzbergen, the limits of the geographical distribution in the high Scandinavian north (in Lapland) are as follows: "The Fir extends to 70°; the White Birch (*Betula alba*) to 70° 40′; the Dwarf-Birch (*B. nana*) to 71° at least: Pinus cembra is entirely wanting in Lapland."*

As the length and the position of the acicular leaves define the physiognomic character of the coniferæ, this is still more designated by the specific difference of the leaf-breadth, and the parenchymatous development of the appendicular organs. Several species of Ephedra may be said to be almost leafless; but in Taxus, Araucaria, Dammara, (Agathis), and the Salisburia adiantifolia of Smith (*Gingko biloba*, Linn.), the breadth of the leaf gradually increases. I have here arranged the genera morphologically. Even the names of the species, as first chosen by botanists, indicate such an arrangement. Dammara orientalis of Borneo and Java, often 11 feet in diameter, was at first named loranthifolia: Dammara australis (Lamb.), in New Zealand, rising to 150 feet high, was originally named zamæfolia. Neither of these has acicular leaves, but "folia alterna oblongo lanceolata, opposita, in arbore adultiori sæpe alterna, enervia, striata." The lower surface of the leaf is densely covered with stomata. These transitions of the appendicular system, from the greatest contraction to a broad leaf surface, possess, like every advance from simple to compound, both a morphological and a physiognomical interest.† The short-stalked, broad, split leaf of the Salisburia (Kämpfer's Ginkgo), has also the breathing pores (stomata) only on the inferior side. The original habitat of the tree is not known. It became distributed from the Chinese temples to the gardens of Japan, in consequence of the intercourse that existed in olden times between the congregations of Buddha.

I was a witness of the singularly painful impression, which the first sight of a pine-forest at Chilpanzingo made on one

* Compare Unger, *Ueber den Einfluss des Bodens auf die Vertheilung der Gewächse*, s. 200; Lindblom, *Adnot. in geographicam plantarum intra Sueciam distributionem*, p. 89; Martius, in the *Annales des Sciences naturelles*, t. xviii. 1842, p. 195.

† Link, *Urwelt*, Th. i. 1834, s. 201–211.

of our companions in travelling from a port in the South Sea through Mexico to Furope. Born in Quito, under the equator, he had never seen needle-leaved trees and folia acerosa. The trees appeared to him to be leafless, and because we were journeying towards the cold north, he thought he recognised already, in the extreme contraction of the organs, the impoverishing influence of the Pole. The traveller, whose impressions I am here describing, and whose name neither Bonpland nor myself can mention without regret, was an excellent young man, the son of the Marquis de Selvalegre, Don Carlos Montufar, whose noble and ardent love of freedom courageously led him, a few years later, to a violent, though not dishonourable, death, in the war of independence, waged by the Spanish colonies.

(24) p. 227—"*Pothos plants, Aroideæ.*"

Caladium and Pothos are forms appertaining exclusively to the tropical world, whilst the different species of Arum belong more to the temperate zone. Arum italicum, A. dracunculus, and A. tenuifolium advance as far as Istria and Friuli. No Pothos has hitherto been discovered in Africa. The East Indies possess several species of this genus (P. scandens and P. pinnata), which have a less beautiful physiognomy and are of less luxuriant growth than the American Pothos plants. We discovered a beautiful true arborescent Aroidea (Caladium arboreum), having a stem from 16 to more than 21 feet in height, near the convent of Caripe, east of Cumana. Beauvois found a singular Caladium (Culcasia scandens) in the kingdom of Benin.* In the Pothos form the parenchyma occasionally expands to so great a degree that the leaf-surface becomes perforated with holes, as in Calla pertusa (Kunth), and Dracontium pertusum (Jacquin), which we collected in the forests of Cumana. It was the Aroideas which first drew attention to the remarkable phenomenon of the *fever-heat* evolved by certain plants during the period of their inflorescence, and which even sensibly affects the thermometer, and is connected with a great and temporary increase in the absorption of oxygen from the atmosphere. Lamarck, in 1789, observed this increase of temperature in the Arum italicum. According to Hubert and Bory de St. Vincent,

* Palisot de Beauvois. *Flore d'Oware et de Benin,* t. i. 1804, p. 4, pl. III.

the vital heat of the Arum cordifolium rises in the Isle of France to 110° or 120°, whilst the temperature of the surrounding air is only 66°.2 Fahr. Even in Europe, Becquerel and Breschet found a difference of 39°.4. Dutrochet observed a paroxysm,—a rhythmical decrease and increase of vital heat,—which appeared by day to attain a double maximum. Théodore de Saussure remarked analogous augmentations of heat, although only of 1°.1 and 1°.8 Fahr., in other families of plants; as, for instance, in Bignonia radicans and Cucurbita pepo. In the latter, the male plant exhibited a greater increase of temperature than the female, when measured by a very sensitive thermoscopic apparatus. Dutrochet—whose early death is greatly to be regretted, on account of the important services he rendered to physics and vegetable physiology—likewise observed,* by means of thermo-magnetic multiplicators, a vital heat of 0°.25 to 0°.67 Fahr. in many young plants (Euphorbia lathyris, Lilium candidum, Papaver somniferum), and even among funguses, in many species of Agaricus and Lycoperdon. This vital heat disappeared at night, but not by day, even when the plants were placed in the dark.

The contrast presented by the physiognomy of the Casuarineas, acicular-leaved trees, and the almost leafless Peruvian Colletias and Pothos plants (Aroideas), is still more striking when we compare these types of extreme contraction in the leaf form with Nymphæaceæ and Nelumboneæ. Here we again meet, as in the Aroideæ, with leaves in which the cellular tissue is excessively expanded upon long, fleshy, succulent petioles,—as Nymphæa alba, N. lutea, N. thermalis (formerly called N. lotus, from the hot spring of Pecze, near Groswardein in Hungary), the species of Nelumbo, Euryale amazonica (Pöppig), and Victoria Regina, allied to the prickly Euryale, although of a very different genus, according to Lindley, and discovered in 1837 by Sir Robert Schomburgk in the river Berbice, in British Guiana. The round leaves of this splendid aquatic plant are from 5 to 6 feet in diameter, and surrounded by upright margins from 3 to 5 inches in height, which are light green on the inner side, but of a bright crimson on the outside. These agreeably perfumed flowers, of which 20 or 30 may be seen together in a small space, are about 15 inches in diameter, of a white or rose colour, and

* *Comptes rendus de l'Institut,* t. viii. 1839, p. 454, t. ix. pp. 614—781.

have many hundred petals.* Pöppig also gives to the leaves of his Euryale amazonica, which he found at Tefé, a diameter of about 6 feet.† Whilst Euryale and Victoria present a greater parenchymatous expansion of the leaf-form in all its dimensions than other genera, the most gigantic development of the blossoms occurs in a parasitical Cytinea, which Dr. Arnold discovered in Sumatra in 1818. This flower, Rafflesia Arnoldi (R. Brown), has a stemless blossom measuring three feet in diameter, surrounded by large leaf-like scales. Like funguses, it has an animal odour, and smells something like beef.

(25) p. 227—"*Lianes, Creeping Plants, (Span. Vejuccos.)*"

According to Kunth's division of Bauhinias, the true genus Bauhinia belongs to the New Continent. The African Bauhinia, B. rufescens (Lam.), is a Pauletia (Cav.), a genus of which we also discovered some new species in South America. In the same manner the Banisterias of the Malpighiaceæ are actually an American form. Two species are indigenous to the East Indies, and one—described by Cavanilles as B. leona —to Western Africa. In the tropical zone, and in the Southern hemisphere, species of the most different families belong to the climbing plants which in those regions render the forests so impenetrable to man and so accessible and habitable to the whole monkey family (Quadrumana), the Cercoleptes, and the small tiger cats. The Lianes thus afford whole flocks of gregarious animals an easy means of rapidly ascending high trees, passing from one tree to another, and even of crossing brooks and rivulets.

In the south of Europe and in the north of America, Hops from the Urticeæ, and the species of Vitis from the Ampelideæ, belong to Climbing Plants; while this form is represented in the tropics by climbing and trailing grasses. We found on the elevated plains of Bogota, in the pass of Quindiu in the Andes, and in the Cinchona forests of Loxa, a Bambusa allied to Nastus, our Chusquea scandens, twined round powerful trunks of trees, adorned at the same time with flowering Orchideæ. Bambusa scandens (Tjankorreh), which Blume found in Java, belongs probably to Nastus, or to the

* Robert Schomburgk, *Reisen in Guiana und am Orinoko,* 1841, s. 233.

† Pöppig, *Reise in Chile, Peru, und auf dem Amazonenstrome.* Bd. ii. 1836, s. 432.

grass-genus Chusquea, the Carrizo of the Spanish settlers. In the pine forests of Mexico, Climbing Plants seem to be entirely wanting; but in New Zealand a fragrant Pandanus, Freycinetia Banksii, together with one of the Smilaceæ, Ripogonum parviflorum (R. Brown), which renders the forests almost impenetrable, winds round a gigantic fir-tree more than 200 feet high, Podocarpus dacryoides (Rich.), called Kakikatea in the language of the country.*

A striking contrast to these Climbing Grasses and Creeping Pandaneas is afforded by the splendid many-coloured blossoms of the Passion flowers (among which, however, we ourselves found one arborescent, upright, species (Passiflora glauca) in the Andes of Popayan, at an elevation of nearly 10,500 feet, and by the Bignoniaceæ, Mutisiæ, Alströmeriæ, Urvilleæ, and Aristolochiæ. Among the latter, our Aristolochia cordata has a coloured (purplish red) calyx, about seventeen inches in diameter; "flores gigantei, pueris mitræ instar inservientes." Owing to the quadrangular form of their stalks, their flattening, which is not occasioned by any external pressure, and a band-like undulatory motion, many of these climbing plants have a peculiar physiognomy. The diagonal intersections of the stems of Bignonias and Banisterias form, by means of furrows in the ligneous substance, and through its clefts, where the bark penetrates to some depth, cruciform or mosaic-like figures.†

(26) p. 228—"*The form of Aloes.*"

To this group of plants, which is characterised by a great similarity, belong Yucca aloifolia, which penetrates as far north as Florida and South Carolina; Y. angustifolia (Nutt.), which advances to the banks of the Missouri; Aletris arborea; the Dragon-tree of the Canaries, and two other Dracænas belonging to New Zealand; arborescent Euphorbias; and Aloe dichotoma, Linn., (formerly the genus Rhipidodendrum of Willdenow), the celebrated Koker-boom, whose stem is four feet in thickness, about twenty feet high, and has a crown measuring 426 feet round.‡ The forms which I have

Ernest Dieffenbach, *Travels in New Zealand,* 1843, vol. i. p. 426.

† See the very correct delineations in Adrien de Jussieu, *Cours de Botanique,* pp. 77—79, figs. 105—108.

‡ Patterson, *Reisen in das Land der Hottentotten und der Kaffern,* 1790, s. 55.

here associated together belong to very different families: as, for instance, to the Liliaceæ, Asphodeleæ, Pandaneæ, Amaryllideæ, and Euphorbiaceæ; and are therefore, with the exception of the last named, all included under the great division of Monocotyledons. One of the Pandaneæ, Phytelephas macrocarpa (Ruiz), which we found on the banks of the Magdalena river in New Granada, exactly resembles with its feathery leaves a small palm-tree. The Tagua (as it is called by the Indians) is moreover, as Kunth has observed, the only Pandanea of the New Continent. The singular Agave-like and high-stemmed Doryanthes excelsa of New South Wales, which the intelligent Correa de Serra was the first to describe, belongs to the Amaryllideæ, like our low-growing Narcissuses and Jonquils.

In the candelabra-like form of Aloes, the branches of the main-trunk must not be confounded with the flower-stalks. In the American aloe, Agave Americana (Maguey de Cocuyza), which is entirely wanting in Chili, and in the Yucca acaulis (Maguey de Cocuy), the leaf-stalks present a candelabra-like arrangement of the blossoms during the excessively rapid and gigantic development of the inflorescence, which, as is well known, is but too transient a phenomenon. In some arborescent Euphorbias the physiognomical character depends, however, on the branches and their arrangement. Lichtenstein describes,* with much animation, the impression made upon him by the appearance of an Euphorbia officinarum which he saw in the "Chamtoos Rivier," near Cape Town. The form of the tree was so symmetrical, that it repeated itself on a small scale, like a candelabrum, to a height of more than 30 feet. All the branches were furnished with sharp thorns.

Palms, Yucca and Aloe plants, arborescent Ferns, some Aralias, and the Theophrasta, where I have seen it in a state of luxuriant growth, present to the eye a certain physiognomical resemblance of character by the nakedness of the stems (there being no branches) and the beauty of their summits or crowns, however they may otherwise differ in the structure of the inflorescence.

Melanoselinum decipiens, (Hofm.), which has been introduced into our gardens from Madeira, and is sometimes from 10 to 12 feet high, belongs to a peculiar group of

* See his *Reisen im südlichen Afrika*, th. i. s. 370.

arborescent umbelliferæ allied to the Araliaceæ, to which other species, as yet undiscovered, will undoubtedly at some future time be added. Ferula, Heracleum, and Thapsia likewise attain a considerable height, but they are still herbaceous shrubs. Melanoselinum stands almost entirely alone as an arborescent umbelliferous plant; Bupleurum (*Tenoria*) fruticosum, Linn., of the shores of the Mediterranean, Bubon galbanum of the Cape, and Crithmum maritimum of our sea-coasts, are only shrubs. Tropical countries, where, as Adanson long since very correctly remarked, Umbelliferæ and Cruciferæ are almost wholly wanting in the plains, exhibit, as we ourselves observed, the most dwarfish of all the umbelliferous family on the lofty mountain ridges of the South American and Mexican Andes. Among the thirty-eight species which we collected on elevations whose mean temperature was below 54°.5 Fahr., we found Myrrhis andicola, Fragosa arctioïdes, and Pectophytum pedunculare, interspersed with an equally dwarfish Alpine Draba, growing moss-like close to the rock and the frequently frozen earth, at a height of 13,428 feet above the level of the sea. The only tropical umbelliferous plants which we found on the plain in the New Continent were two species of Hydrocotyle (*H. umbellata* and *H. leptostachya*) between the Havannah and Batabano, and therefore at the extreme limit of the torrid zone.

(27) p. 228—"*The form of Grasses.*"

The group of the arborescent grasses which Kunth has collected under the head of Bambusaceæ, in his great work on the plants collected by Bonpland and myself, constitutes one of the most beautiful adornments of tropical vegetation. Bambu, called also Mambu, occurs in the Malay language, although according to Buschmann merely as an isolated expression, the ordinary term in use being buluh, whilst the only name for this species of cane in Java and Madagascar is wuluh, voulou. The numbers of the genera and species included in this group have been extraordinarily increased by the industry of botanical travellers. It has been found that the genus Bambusa is entirely wanting in the New Continent, to which region, however, the gigantic Guaduas, discovered by us, and which attain a height of from 50 to 64 feet, together with the Chusquea, exclusively belong; that Arundinaria (Rich.)

occurs in both continents, although differing specifically in each; that Bambusa and Beesha (Rheed.), occur in India and the Indian Archipelago; and that Nastus grows in the islands of Madagascar and Bourbon. With the exception of the high-climbing Chusquea, these forms morphologically replace each other in different parts of the earth. In the northern hemisphere far beyond the limits of the torrid region, in the valley of the Mississippi, the traveller is gladdened by the sight of a species of Bamboo, the Arundinaria macrosperma, formerly called also Miegia and Ludolfia. In the southern hemisphere, in the south of Chili, between the parallels of 37° and 42°, Gay found one of the Bambusacea more than 20 feet high (not a climbing, but a still undescribed arborescent self-supporting Chusquea), growing, mingled with Drymis Chilensis, in a region clothed with an uniform forest-covering of Fagus obliqua.

Whilst in India the Bambusa flowers so frequently that in Mysore and Orissa the seeds are mixed with honey, and eaten like rice,* in South America the Guadua blossoms so very seldom that in the course of four years we were only twice able to procure the flowers; once on the solitary banks of the Cassiquiare, the arm connecting the Orinoco with the Rio Negro and the Amazon, and again in the province of Popayan, between Buga and Quilichao. It is a very striking fact that some plants grow with the greatest vigour in certain localities without flowering; as is the case with the European olive-trees introduced into America centuries ago, and growing between the tropics, near Quito, at elevations of about 9600 feet above the level of the sea; and in like manner the walnuts, hazel-nut bushes, and the fine olive-trees (*Olea Europea*) of the Isle of France.†

As some of the Bambusaceæ (arborescent grasses) advance into the temperate zone, so also they do not suffer in the torrid zone from the temperate climate of mountain districts. They are certainly more luxuriant as social plants between the sea-shore and elevations of about 2558 feet in the Province de las Esmeraldas, west of the volcano of Pichincha, where Guadua angustifolia (Bambusa Guadua of our *Plantes équinoxiales,* t i. tab. xx) generates in its interior

* Buchanan, *Journey through Mysore,* vol. ii. p. 341; and Stirling, in the *Asiat. Res.* vol. xv. p. 205.

† See Bojer, *Hortus Mauritianus,* 1837, p. 201.

large quantities of the siliceous Tabaschir (Sanscrit *tvakkschira*, cow-milk). We saw the Guadua advance in the pass of Quindiu, in the chain of the Andes, to a height of 5755 feet above the level of the sea, as determined by barometric measurements. Nastus borbonicus has been called a true Alpine plant by Bory de St. Vincent, and according to him it does not descend lower than 3840 feet on the declivity of the volcano in the island of Bourbon. This appearance or the repetition at great elevations of certain forms belonging to torrid plains calls to mind the group of Alpine palms (Kunthia montana, Ceroxylon andicola, and Oreodoxa frigida) of which I have already spoken, and a grove of Musaceæ (Heliconia, perhaps Maranta), 16 feet high, which I found growing isolated on the Silla de Caracas, at a height of more than 7000 feet above the level of the sea.* While the form of gramineæ, with the exception of some few herbaceous dicotyledons, constitutes the highest phanerogamic zone on the snow-crowned summits of mountains, so the grasses mark the boundary of phanerogamic vegetation in a horizontal direction, towards the northern and southern polar regions.

Many admirable general results, no less than a great mass of important materials, have been yielded to the geography of plants by my young friend, Joseph Hooker, who, after having but recently returned with Sir James Ross from the frozen antarctic regions, is now engaged in exploring the Thibetian Himalayà. He draws attention to the fact that phanerogamic flowering plants (grasses) advance 17½° nearer to the north than to the south pole. In the Falkland Islands, near the thick knots of Tussac grass, Dactylis cæspitosa, Forster, (a Festuca, according to Kunth), and in Tierra del Fuego, under the shade of the birch-leaved Fagus antarctica, there grows the same Trisetum subspicatum, which spreads over the whole range of the Peruvian Andes, and across the Rocky Mountains, to Melville Island, Greenland, and Iceland, and is also found in the Swiss and Tyrolese Alps as well as in the Altai, in Kamtschatka, and in Campbell's Island, south of New Zealand, extending therefore over 127 degrees of latitude, or from 54° south to 72° 50′ north lat. "Few grasses," says Joseph Hooker,† "have so wide a range as Trisetum subspicatum (Beauv.), nor am I

* *Relat. hist.* t. i. pp. 605—606.

† *Flora antarctica*, p. 97.

acquainted with any other arctic species which is equally an inhabitant of the opposite polar regions." The South Shetland Islands, which are separated by Bransfield Straits from d'Urville's "Terre de Louis-Philippe" and from Peak Haddington, a volcano, 7046 feet high, and situated in 64° 12′ south lat., have recently been visited by Dr. Eights, a botanist from the United States. He found there (probably in 62° or 62¼° south lat.) a small grass, Aira antarctica,* which is "the most antarctic flowering plant hitherto discovered."

Even in Deception Island, belonging to the same group, 62° 50′, only lichens are met with, and no longer any species of grass; and in like manner further south-east, in Cockburn's Island (64° 12′) near Palmer's Land, only Lecanoras, Lecideas, and five foliaceous Mosses, among which is our German Bryum argenteum, were gathered. "This appears to be the Ultima Thule of antarctic vegetation," for further south even terrestrial cryptogamia are wanting. In the great bay formed by Victoria Land, on a small island lying opposite to Mount Herschel (in 71° 49′ lat.), and on Franklin Island, 92 miles north of the volcano, Erebus, (12,366 feet in height), and in 76° 7′ south lat , Hooker found no trace of vegetation. In extreme northern latitudes, the distribution of even the higher organisms is very different; for here phanerogamic plants advance 18½° nearer to the pole than in the southern hemisphere. Walden Island (80½° north lat.) possesses still ten species of phanerogamia. Antarctic phanerogamic vegetation is also poorer in species at equal distances from the pole; thus Iceland has five times more phanerogamia than the southern group of Auckland and Campbell Islands, but the uniform vegetation of the antarctic regions is, from climatic causes, both more succulent and more luxuriant.†

(28) p. 229—"*Ferns.*"

If we estimate the whole number of the cryptogamia hitherto described at 19,000 species, as has been done by Dr. Klotzsch, a naturalist possessing a profound acquaintance with the Agamic plants, we shall have for Fungi 8000 (of which Agarici constitute the eighth part); for Lichens, according to J. von Flotow of Hirschberg, and Hampe of

* Hooker, *Icon. plant.* vol. ii. tab. 150.

† Compare Hooker, *Flora antarctica,* pp. vii. 74, 215, with Sir James Ross, *Voyage in the Southern and Antarctic Regions,* 1839—1843, vol. ii. pp. 335—342.

Blankenburg, at least 1400; for the Algæ 2580; for Mosses and Liverworts, according to Carl Müller of Halle, and Dr. Gottsche of Hamburgh, 3800; and for Ferns 3250. For this last important result we are indebted to the profound investigations made by Professor Kunze of Leipzig, on this group of plants. It is a striking fact that the family of the Polypodiaceæ alone includes 2165 of the whole number of described Filices, whilst other forms, as the Lycopodiacæ and Hymenophyllaceæ, number only 350 and 200. There are therefore nearly as many described species among Ferns as among Grasses.

It is singular that no mention of the beautiful arborescent ferns is to be found in the classic authors of antiquity, Theophrastus, Dioscorides, and Pliny; while, from the information given by the companions of Alexander, Aristobulus, Megasthenes, and Nearchus, reference is made* to Bamboos, "quæ fissis internodiis lembi vice vectitabant navigantes;" to the Indian trees "quarum folia non minora clypeo sunt;" to the Fig-tree which takes root from its branches, and to Palms, "tantæ proceritatis, ut sagittis superjici nequeant." I find the first mention of arborescent ferns in Oviedo.† "Among ferns," says this experienced traveller, who had been appointed by Ferdinand the Catholic, Director of the Gold-washings in Haiti, "there are some which I class with trees, because they are as thick and high as Pine-trees. (Helechos que yo cuento por arboles, tan gruesos como grandes pinos y muy altos). They mostly grow among the mountains and where there is much water." This estimate of their height is exaggerated, for in the dense forests near Caripe even our Cyathea speciosa only attains a height of 32 to 37 feet; and an admirable observer, Ernst Dieffenbach, did not see in the most northern of the three islands of New Zealand any trunks of Cyathea dealbata exceeding 42½ feet. In the Cyathea speciosa and the Meniscium of the Chaymas missions, we observed in the midst of the most shady part of the primeval forest, that the scaly stems of some of the most luxuriantly developed of these trees were covered with a shining carbonaceous powder, which appeared to be owing to a singular decomposition of the fibrous parts of the old leaf stalks.‡

* Humboldt, *de distrib. geogr. Plant.*, pp. 178, 213.
† *Historia de las Indias*, 1535, fol. xc.
‡ Humboldt, *Relat. hist.*, t. i. p. 437.

Between the tropics, where, on the declivities of the Cordilleras, climates are superimposed in strata, the true region of arborescent ferns lies between about 3200 and 5350 feet above the level of the sea. In South America and in the Mexican highlands they seldom descend lower towards the plains than 1280 feet. The mean temperature of this happy region is between 64°.6 and 70°.8 Fahr. It reaches the lowest stratum of clouds (which floats the nearest to the surface of the sea and the plain), and it therefore enjoys uninterruptedly a high degree of humidity, together with a great equality in its thermal relations.* The inhabitants, who are of Spanish descent, call this region "Tierra templada de los helechos."

The Arabic designation for ferns is *feledschun*, filix, (from which the *f* has been changed, according to Spanish usage, into *h*,) and perhaps the term may be connected with the verb *faladscha*, "it divides," from the finely cut margin of the frond.†

The conditions of genial mildness in an atmosphere charged with aqueous vapour and of great uniformity in respect to moisture and warmth, are fulfilled on the declivities of the mountains in the valleys of the Andes, and more especially in the southern milder and more humid hemisphere, where arborescent ferns advance not only to New Zealand and Van Diemen's Land (Tasmania), but even as far as the Straits of Magellan and Campbell Island, and therefore to a southern latitude almost identical in degrees with the parallel in which Berlin is situated north of the equator. From among the family of arborescent ferns there flourishes the vigorous Dicksonia squarrosa, in 46° south lat. in Dusky Bay, New Zealand; D. antarctica of Labillardière in Tasmania; a Thyrsopteris in the Island of Juan Fernandez; an undescribed Dicksonia, whose stem is from 12 to 16 feet high, near Valdivia in Southern Chili; and a Lomaria, somewhat less in height, in the Straits of Magellan. Campbell Island is still nearer to the south pole, in 52½° lat., but even there the leafless stem of the Aspidium venustum rises to a height of more than four feet.

The climatic relations under which Ferns (*Filices*) in general flourish, are manifested in the numerical laws of their

* Robert Brown, *In Expedition to Congo*, Append. p. 423.

† Abu Zacaria Ebn el Awam, *Libro de Agricultura*, traducido por J. A. Banqueri, t. ii. Madr. 1802, p. 736.

quotients of distribution. In the plains within the tropical regions of large continents this quotient is, according to Robert Brown, and from more recent investigations on the subject, $\frac{1}{20}$ of all the phanerogamia, and in mountainous districts of large continents $\frac{1}{6}$ to $\frac{1}{8}$. This ratio is quite different on the small islands scattered over the ocean; for here the proportion borne by the number of ferns to the sum total of all the phanerogamic plants increases so considerably, that in the South-Sea Islands the quotient rises to $\frac{1}{4}$, while in the sporadic islands, St. Helena and Ascension, the number of ferns is almost equal to half of the whole phanerogamic vegetation.* In receding from the tropics (where on the large continents d'Urville estimates the proportional number at $\frac{1}{20}$), the *relative* frequency of ferns decreases rapidly as we advance into the temperate zone. The quotients are for North America and the British Islands $\frac{1}{35}$, for France $\frac{1}{58}$, for Germany $\frac{1}{52}$, for the dry parts of Southern Italy $\frac{1}{74}$, for Greece $\frac{1}{84}$. The *relative* frequency again increases considerably towards the frigid north. Here the family of ferns decreases much slower in the number of its species than does that of phanerogamic plants. The luxuriantly aspiring character of the species, and the number of individuals contained in each, augment the deceptive impression of *absolute* frequency. According to Wahlemberg's and Hornemann's catalogues, the relative numbers of the Filices are for Lapland $\frac{1}{25}$, for Iceland $\frac{1}{18}$, for Greenland $\frac{1}{12}$.

Such are, according to our present knowledge, the natural laws that manifest themselves in the distribution of the graceful form of Ferns. But it would seem as if in the family of Ferns, which have so long been regarded as cryptogamic, we had lately acquired evidence of the existence of another natural law,—the morphological law of propagation. Count Leszczyc-Suminski, who happily combines the power of microscopic investigation with a very remarkable artistic talent, has discovered an organisation capable of effecting fructification in the prothallium of ferns. He distinguishes two sexual apparatuses, of which the female portion is situated in hollow ovate cells in the middle of the sporangium, and the male in the ciliated antheridia, or the organs producing spiral threads, which have already been examined by Nägeli. Fructification

* See a valuable Treatise by d'Urville, *Distribution géographique des fougères sur la surface du Globe*, in the *Annales des Sciences nat.*, t. vi. 1825, pp. 51, 66, 73.

is supposed to be effected by means of moveable ciliated spiral threads and not by pollen tubes.* According to this view, Ferns would be, as Ehrenberg remarks,† products of a microscopic fructification taking place on the prothallium, which here serves as a fertilizing receptacle, while throughout the whole course of their often arborescent development they would be flowerless and fruitless plants, having a bud-formation. The spores lying as sori on the under side of the frond are not seeds but flower-buds.

(29) p. 229—"*The Liliaceæ.*"

Africa is the principal seat of this form; there the greatest diversity obtains; there they form masses and determine the natural character of the region. The New Continent exhibits also, it is true, magnificent Alströmeriæ and species of Pancratium, Hæmanthus, and Crinum. We have enriched the first of these genera with nine, and the second with three species; but these American liliaceous plants are more diffused and of less social habits than the European Irideæ.

(30) p. 229—"*The Willow Form.*"

Nearly 150 different species of the main representatives of this form, or rather of the Willow itself, are already known. They cover the northern parts of the earth from the equator to Lapland. Their number and their varieties of form increase between the 46th and 70th degrees of latitude, more especially in that part of northern Europe which has been so remarkably indented by the early revolutions of our planet. I am acquainted with ten or twelve species of inter-tropical Willows, and these, like the Willows of the southern hemisphere, are deserving of special attention. As nature appears to delight in all zones in a wondrous multiplication of certain animal forms, as for instance, Anatidæ (*Lamellirostres*), and Pigeons; so likewise are Willows, Pines, and Oaks, widely diffused; the latter always exhibiting a similarity in their fruit, although various differences exist in the form of the leaves. In Willows belonging to the most widely different climates the similarity of the foliage, of the ramification, and of the whole physiognomical conformation, is almost greater

* Count Suminski, *Zur Entwickelungs-Geschichte der Farrnkräuter* 1848, S. 10—14.

† *Monatl. Berichte der Akad. zu Berlin,* Januar, 1848, S. 20.

than in Coniferæ. In the more southern part of the temperate zone, north of the equator, the number of the species of Willows decreases considerably; although (according to the "Flora atlantica" of Desfontaines) Tunis has still its own species, resembling Salix caprea; whilst Egypt, according to Forskäl, numbers five species, from the catkins of whose male blossoms is distilled the remedial agent Moie chalaf (*aqua salicis*), so much used in the East. The Willow which I saw in the Canarics is also, according to Leopold von Buch and Christian Smith, a peculiar species (*S. canariensis*), although common to those islands and to Madeira. Wallich's catalogue of the plants of Nepaul and the Himalaya already gives 13 species belonging to the subtropical zone of the East Indies, and which have in part been described by Don, Roxburgh, and Lindley. Japan has its own species, of which one, S. japonica, (Thunb.), is also met with in Nepaul as an Alpine plant.

There was not, as far as I am aware, any species of Willow known as belonging to the tropical zone before my expedition, with the exception of S. tetrasperma. We collected seven new species, three of them on the plateaux of Mexico, at an elevation of 8500 feet above the level of the sea. Still higher, as for instance on the Alpine plains, between 12,000 and 15,000 feet, which we frequently visited, we saw nothing in the Andes of Mexico, Quito, and Peru, to remind us of the many small creeping Alpine Willows of the Pyrenees, the Alps, or of Lapland (*S. herbacea*, *S. lanata*, and *S. reticulata*). In Spitzbergen, whose meteorological relations have so much analogy with those of the snow-crowned summits of Switzerland and Scandinavia, Martius described two Dwarf-Willows, whose small woody stems and branches trail along the ground, and are so concealed in the turf-bogs that it is with difficulty their diminutive leaves can be discovered under the moss. The Willow species which I found in 4° 12′ south lat., at the entrance of the Cinchona or Peruvian Bark forests, near Loxa in Peru, and which has been described by Willdenow as Salix Humboldtiana, is most widely diffused over the western part of South America. A Beach-Willow (*S. falcata*), which we discovered on the sandy shores of the Pacific, near Truxillo, is, according to Kunth, probably a mere variety of the former. In like manner the beautiful and frequently pyramidal Willow, which we constantly saw on the banks of the Magdalena river, from Mahates to Bojorque, and which, according to the report of the natives, had only spread thus far within a few years, may also be identical with S. Humboldtiana. At the confluence of the Magdalena with the Rio Opon, we found all the islands covered with Willows, many of which had stems 64 feet high, with a diameter of from only 8 to 10 inches.* Lindley has made us acquainted with a species of Salix belonging to Senegal, and therefore to the equinoctial region of Africa.† Blume also found two species of Willow near the equator in Java, one wild and indigenous in the island (*S. tetrasperma*), and another cultivated (*S. Sieboldiana*). I am only acquainted with the two Willows belonging to the south temperate zone, which have been described by Thunberg (*S. hirsuta* and *S. mucronata*). They grow interpersed with Protea argentea, which has the same physiognomy as the Willow, and their leaves and young branches constitute the food of the hippopotamus of the Orange River. The family of Willows is entirely wanting in Australia and the neighbouring islands.

### (31) p. 229—"*The Myrtle Form.*"

The Myrtle is a graceful plant, with stiff, shining, crowded, and generally entire and small leaves marked with dots. Myrtles impart a peculiar character to three regions of the earth, viz., to southern Europe, more especially to the islands composed of calcareous rocks and trachytic stone, which project from the basin of the Mediterranean; to the continent of New Holland, which is adorned with Eucalyptus, Metrosideros, and Leptospermum; and to an intertropical region in the Andes of South America, part of which is a low plain, while the remainder lies at an elevation of from 9000 to more than 10,000 feet above the level of the sea. This Alpine region, called in Quito the Paramos, is entirely covered with trees having a Myrtle-like aspect, even though they may not all belong to the Myrtaceæ. At this elevation grow Escalonia myrtilloides, E. tubar, Simplocos Alstonia, species of Myrica, and the lovely Myrtus microphylla, of which we have given a drawing in our *Plantes équinoxiales*, t. i. p. 21, pl. iv.; it grows on micaceous schist, at an elevation of 10,000 feet on the Paramo de Saraguru, (near Vinayacu and Alto de

* Humboldt et Kunth, *Nova Gen. Plant.*, t. ii. p. 22, Tab. 99.
† Lindley, *Introd. to the Natural System of Botany*, p. 99.

Pulla,) which is adorned with so many beautiful flowering Alpine plants. M. myrsinoides ascends in the Paramo de Guamani as high as 11,200 feet. By far the greater number of the 40 species of the genus Myrtus which we collected in the equinoctial zone, and of which 37 were undescribed, belong to the plains and the less elevated mountain spurs. We brought only a single species (*M. xalapensis*) from the mild tropical climate of the mountains of Mexico; but the Tierra templada, in the direction of the Volcano of Orizaba, no doubt possesses many yet undescribed varieties. We found M. maritima near Acapulco, on the very shore of the Pacific.

The *Escalloniæ*,—among which *E. myrtilloides*, *E. tubar*, *E. floribunda* are the ornaments of the Paramos, and remind us strongly (by their physiognomical aspect) of the myrtle-form, —formerly constituted, together with the European and South American Alpine roses (Rhododendrum and Befaria), with Clethra, Andromeda, and Gaylussacia buxifolia, the family of the *Ericeæ*. Robert Brown* has arranged them in a special family, which Kunth has placed between the Philadelphiæ and Hamamelideæ. Escallonia floribunda affords by its geographical distribution one of the most striking examples of the relation existing between distance from the equator and vertical elevation above the level of the sea. I would here again borrow support from the testimony of the accurate observer, my friend Auguste de St. Hilaire.† "MM. Humboldt and Bonpland in their expedition discovered Escallonia floribunda in 4° south lat. at an elevation of 8952 feet. I found the same plant in 21° south lat. in Brazil, which although elevated is very much less so than the Andes of Peru. This plant is of common occurrence between 24° 50′ and 25° 55′ in the Campos Geraes, and I also met with it again on the Rio de la Plata in about 35° lat., on a level with the sea."

The group of the Myrtaceæ,—to which belong Melaleuca, Metrosideros, and Eucalyptus, commonly classed under the general denomination of Leptospermeæ,—produce partially, wherever the true leaves are supplied by phyllodia (petiole-leaves), or where the direction of the leaves is inclined towards the unexpanded petiole, a distribution of streaks of light

* See the additions to *Franklin's Narrative of a Journey to the shores of the Polar Sea*, 1823, p. 765.

† *Morphologie végétale*, 1840, p. 52.

and shade wholly unknown in our deciduous-leaved forest. We find that the earliest botanical travellers who visited New Holland were astonished at the singular effect thus produced. Robert Brown was the first to show that this phenomenon depends on the vertical direction of the expanded petioles (the phyllodia of Acacia longifolia and Acacia suaveolens), and on the circumstance, that the light, instead of falling on horizontal surfaces, passes between vertical ones.* Morphological laws in the development of the leaves determine the peculiar character of the varying light and shade. "Phyllodia," says Kunth, "can in my opinion merely occur in families which have compound pinnate leaves; and in fact they have as yet only been met with in Leguminosæ (in the Acacias). In Eucalyptus, Metrosideros, and Melaleuca, the leaves are simple (simplicia), and their edgewise position depends on a half-turn of the leaf-stalk (petiolus); moreover, it must be remarked, that both surfaces of the leaves are of a similar character." In the scantily shaded forests of New Holland the optical effects here alluded to are the more frequent, since two groups of Myrtaceæ and Leguminosæ, species of Eucalyptus and Acacia, there constitute nearly one-half of all the greyish-green tree vegetation. Moreover, between the bast-layers of Melaleuca, there are formed easily soluble membranes, which force their way outwards, and by their whiteness reminds us of our birch bark.

The sphere of distribution of the Myrtaceæ is very different in the two continents. In the New Continent, and especially in its western parts, this family, according to Joseph Hooker,† scarcely extends beyond the parallel of 26° north lat., while in the Southern Hemisphere, there are in Chili, according to Claude Gay, ten species of Myrtle and twenty-two of Eugenia, which mixed with Proteaceæ (Embothrium and Lomatia) and with Fagus obliqua, there constitute forests. The Myrtaceæ become more frequent from the 38th degree of south lat.; in the island of Chiloe, where a metrosideros-like species (Myrtus stipularis) forms almost impenetrable underwood, which is there named Tepuales; and in Patagonia to the extremity of Tierra del Fuego in 56° lat.

* Adrien de Jussieu, *Cours de Botanique*, pp. 106, 120, and 700; Darwin, *Journal of Researches*, 1845, p. 433.

† *Flora antartica*, p. 12.

While in Europe the Myrtaceæ do not extend northward further than 46° lat., they penetrate in Australia, Tasmania, New Zealand and the Auckland Islands to 50½° south latitude.

(32) p. 229—" *Melastomaceæ.*"

This group comprises the genera Melastoma (Fothergilla and Tococa Aub. and Rhexia (Meriana and Osbeckia), of which we have collected no less than sixty new species in tropical America alone, on both sides of the equator. Bonpland has published a splendid work on the Melastomaceæ, in two volumes, with coloured plates. There are species of Rhexia and Melastoma which ascend in the chain of the Andes, as Alpine or Paramos shrubs, to 9600 and even more than 11,000 feet above the level of the sea; as for instance Rhexia cernua, R. stricta, Melastoma obscurum, M. aspergillare, and M. lutescens.

(33) p. 229—" *The Laurel-form.*"

To this form belong Laurus, Persea, the Ocoteæ, so numerous in South America, and,—on account of their physiognomic similarity,—Calophyllum, also the splendidly aspiring Mammea from the Guttiferæ.

(34) p. 229—" *How instructive to the landscape-painter would be a work which should illustrate the leading forms of vegetation.*"

In order to define with more distinctness what I have here only briefly referred to, I may be permitted to incorporate the following considerations from my sketch of a history of landscape painting, and of a graphical representation of the physiognomy of plants.*

" All that relates to the expression of the passions and the beauty of the human form has perhaps attained its fullest development in the temperate northern zone under the skies of Greece and Italy. The artist, drawing from the depths of imagination, no less than from the contemplation of beings of his own species, derives the types of historical painting alike from unfettered creation and from truthful imitation. Landscape painting, though scarcely a more imitative art, has a more material basis, and a more earthly tendency. It requires for its development a greater amount of various and distinct

* *Cosmos,* vol. ii. p. 453 (Bohn's edition.)

impressions, which, when imbibed from external contemplation, must be fertilized by the powers of the mind in order to be presented to the senses of others as a creative work of art. The grander style of heroic landscape-painting is the combined result of a profound appreciation of nature, and of this inward process of the mind.

"Everywhere, in every separate portion of the earth, nature is indeed only a reflex of the whole. The forms of organization recur again and again in different combinations. Even the icy north is cheered for months together by the presence of herbs and large Alpine blossoms covering the earth, and by a mild azure sky. Hitherto landscape painting among us has pursued her graceful labours familiar only with the simpler forms of our native floras, but not therefore without depth of feeling and richness of creative fancy. Dwelling only on the native and indigenous form of our vegetation, this branch of art, notwithstanding that it has been circumscribed by such narrow limits, has yet afforded sufficient scope for highly-gifted painters, such as the Caracci, Gaspar Poussin, Claude Lorraine, and Ruysdael, to produce the happiest and most varied creations of art, by their magical power of managing the grouping of trees, and the effects of light and shade. That progress which may still be expected in art, from a more animated intercourse with the tropical world, and from ideas engendered in the mind of the artist by the contemplation of Nature in her grandest forms, will never diminish the fame of the old masters. I have alluded to this, to recal the ancient bond which unites a knowledge of Nature with poetry and a taste for art. For in landscape painting, as in every other branch of art, a distinction must be drawn between the elements generated by a limited field of contemplation and direct observation, and those which spring from the boundless depth of feeling, and from the force of idealising mental power. The grand conceptions which landscape painting, as a more or less inspired branch of the poetry of nature, owes to the creative power of the mind, are, like man himself, and the imaginative faculties with which he is endowed, independent of place. These remarks especially refer to the gradations in the form of trees from Ruysdael and Everdingen, through the works of Claude Lorraine, to Poussin and Annibal Caracci. In the great masters of art there is no indication of local limitation. But an extension of the visible horizon, and an acquaintance with the nobler and grander forms of nature, and with the luxuriant fulness of life in tropical regions, afford the advantage of not simply enriching the material groundwork of landscape-painting, but also of inducing more vivid impressions in the minds of less highly gifted painters, and thus heightening their powers of artistic creation."

(35) p. 230—"*From the thick and rough bark of the Crescentiæ and Gustaviæ.*"

In *Crescentia Cujete* (the Tutuma tree, whose large fruit-shells are so indispensable to the natives as household utensils), in *Cynometra*, the Cacao-tree (*Theobroma*), and the *Perigara Gustavia* (Linn.), the tender blossoms burst forth from the half-carbonized bark. When children eat the fruit of the *Pirigara speciosa* (the *Chupo*), their whole bodies become tinged with yellow; and this jaundice, after a continuance of from twenty-four to thirty-six hours, disappears without the use of medicine.

An indelible impression was produced on my mind by the luxuriant power of vegetation in the tropical world, when, on entering a Cacao plantation (*Caca hual*), in the Valles de Aragua, after a damp night, I saw for the first time large blossoms springing from the root of a *Theobroma*, deeply imbedded in the black soil. This is one of the most instantaneous manifestations of the activity of the vegetative force of organisation. Northern nations speak of "the awakening of Nature at the first genial breath of Spring;"—expressions that strongly contrast with the imaginative complaint of the Stagirite, who regarded vegetable forms as buried in a "still sleep, from which there is no awakening, and free from the desires that excite to spontaneous motion."*

(36) p. 230—"*Draw on their heads as caps.*"

These are the flowers of our *Aristolochia cordata*, to which reference has been made in Illustration 25. The largest flowers in the world, besides those belonging to the Compositæ (the Mexican *Helianthus annuus*), are produced by *Rafflesia Arnoldi*, *Aristolochia*, *Datura*, *Barringtonia*, *Gustavia*, *Carolinea*, *Lecythis*, *Nymphæa*, *Nelumbium*, *Victoria Regina*, *Magnolia*, *Cactus*, the Orchideæ, and the Liliaceous forms.

* Aristot. *De Generat. Animal.* v. i. p. 778, and *De Somno et Vigil.* cap. i. p. 455, Bekker.

(37) p. 231—" *The luminous worlds which spangle the firmament from pole to pole.*"

The more magnificent portion of the southern sky, in which shine the constellations of the Centaur, Argo, and the Southern Cross, where the Magellanic clouds shed their pale light, is for ever concealed from the eyes of the inhabitants of Europe. It is only under the equator that man enjoys the glorious spectacle of *all* the stars of the southern and northern heavens revealed at one glance. Some of our northern constellations,—as, for instance, Ursus Major and Ursus Minor, —owing to their low position when seen from the region of the equator, appear to be of a remarkable, almost fearful magnitude. As the inhabitant of the tropics beholds *all* stars, so too, in regions where plains, deep valleys, and lofty mountains are alternated, does Nature surround him with representatives of every form of vegetation.

---

In the foregoing sketch of a "Physiognomy of Plants," I have endeavoured to keep in view three nearly allied subjects, —*the absolute diversity of forms*; their *numerical* relations, *i.e.* their local preponderance in the whole number of phanerogamic floras; and their *geographical and climatic distribution.* If we would rise to a general view regarding vital forms;—the physiognomy, the study of the numerical relations (the arithmetic of botany), and the geography of plants (the study of the local zones of distribution), cannot, as it seems to me, be separated from one another. The study of the physiognomy of plants must not be exclusively directed to the consideration of the striking contrasts of form which the larger organisms present, when considered separately; but it must rise to the recognition of the laws which determine *physiognomy of nature generally*, the picturesque character of vegetation over the whole surface of the earth, and the vivid impression produced by the grouping of contrasted forms in different zones of latitude and elevation. It is when concentrated into this focus that we first clearly perceive the close and intimate connection existing between the subjects treated of in the preceding pages. We have here entered upon a field of inquiry hitherto but little cultivated. I have ventured to follow the method first propounded with such brilliant results in Aristotle's zoological works, and which is so especially adapted to establish scientific confidence,—a method in which the incessant effort to arrive at a generalisation of ideas supported by individual illustrations, is associated with an endeavour to penetrate to the specialities of phenomena.

The enumeration of forms is, from the physiognomical difference of their nature, incapable of any strict classification. Here, as everywhere in the consideration of external forms, there are certain main types which present the strongest contrasts,—as the groups of the Arborescent Grasses, the Aloe form and the species of Cactus, Palms, Acicular-leaved trees, Mimosaceæ, and Bananas. Even scantily dispersed individuals belonging to these groups determine the character of a district, and produce a lasting impression on the mind of the unscientific but susceptible beholder. Other forms, perhaps more numerous and preponderating, may not appear equally marked either by the shape or position of the leaves; the relation of the stem to the branches, luxuriant vigour, animation, and grace; or even by the melancholy contraction of the leaf-organs.

As, therefore, a physiognomical classification, or a distribution into groups according to external appearance, does not admit of being applied to the whole vegetable kingdom collectively, the basis on which such a classification should be grounded must necessarily be wholly different from that which has been so happily chosen for the establishment of our comprehensive systems of the natural families of plants. Vegetable physiognomy grounds its divisions and the choice of its types on all that possesses mass,—as the stem, branches, and appendicular organs (the form, position, and size of the leaf, the character and brilliancy of the parenchyma), and consequently on all that is now included under the special term, *the organs of vegetation*, and on which depend the preservation (nourishment and development) of the individual; while systematic botany, on the other hand, bases the arrangement of the natural families of plants on a consideration of the organs of propagation, on which depends the preservation of the species.* It was already taught in the school of Aristotle,† that the generation of seed is the ultimate aim of the being and life of a plant. The process of development in the organs of fructification has become, since Caspar Fried. Wolf,‡ and

* Kunth, *Lehrbuch der Botanik,* 1847. Th. i. s. 511; Schleiden, *Die Pflanze und ihr Leben,* 1848, s. 100.

† *Probl.* 20, 7.

‡ *Theoria Generationis,* § 5—9.

our great poet Goëthe, the morphological basis of all systematic botany.

This science and that also of vegetable physiognomy proceed, I would here again observe, from two different points of view; the former depending upon an accordance in the inflorescence and in the reproduction of the delicate sexual organs; the latter on the conformation of the parts constituting the axes (the stem and branches) and on the outline of the leaves, which are mainly determined by the distribution of the vascular bundles. As, moreover, the stem and branches, together with their appendicular organs, predominate by mass and volume, they determine and strengthen the impression we receive, while they individualize the physiognomical character of the vegetation, as well as that of the landscape or the zone in which some distinguished types occur. The law is here expressed by the accordance and affinity in the marks appertaining to the vegetative, *i.e.* the nutritient organs. In all European colonies the inhabitants have been led by resemblances of physiognomy (*habitus, facies*) to apply the names of European forms to certain tropical plants, which bear wholly different flowers and fruits from the genera to which these designations originally referred. Everywhere in both hemispheres, the northern settler has believed he could recognise Alders, Poplars, Apple and Olive trees; being misled for the most part by the form of the leaves and the direction of the branches. The charm associated with the remembrance of native forms has strengthened the illusion, and European names of plants have thus been perpetuated from generation to generation in the slave colonies, where they have been further enriched by denominations borrowed from the negro languages.

A remarkable phenomenon is presented by the contrast frequently observed to arise from a striking accordance in physiognomy, coupled with the greatest difference in the organs of inflorescence and fructification—between the external form as determined by the appendicular or leaf-system, and the sexual organs on which are based the various groups of the natural systems of botany. One would be disposed *à priori* to believe that the aspect of vegetative organs (leaves) exclusively so called, must depend upon the structure of the organs of reproduction, but this dependence has only been observed in a very small number of families, as Ferns, Grasses, Cyperaceæ, Palms, Coniferæ, Umbelliferæ, and

Aroideæ. In the Leguminosæ this accordance between the physiognomical character and the inflorescence can scarcely be recognized, excepting where they are separated into groups (as Papilionaceæ, Cæsalpinineæ, and Mimosaceæ.) The types which exhibit, when compared together, a very different structure of inflorescence and fructification, notwithstanding external accordance in physiognomy, are Palms and Cycadeæ, the latter being most nearly allied to the Coniferæ; *Cucusta*, belonging to the Convolvulaceæ, and the leafless *Cassytha*, a parasitical Laurinea; *Equisetum* (from the division of the Cryptogamia) and *Ephedra* (a coniferous tree). The Grossulareæ (*Ribes*) are so nearly allied by their efflorescence to Cactuses, *i. e.* the family of the Opuntiaceæ, that it is only very lately that they have been separated from them! One common family (that of the Asphodeleæ) comprises the gigantic tree, *Dracæna Draco*, the Common Asparagus, and the coloured flowering *Aletris*. Simple and compound leaves frequently belong not only to the same family, but even to the same genus. We found in the elevated plateaux of Peru and New Granada among twelve new species of *Weinmannia*, five with simple, and the remainder with pinnate leaves. The genus *Aralia* exhibits yet greater independence in the leaf-form, which is either simple, entire, lobed, digitate, or pinnate.*

Pinnate leaves appear to me to belong especially to those families which occupy the highest grade of organic development, as for instance, the *Polypetalæ*; among *perigynic* plants, the Leguminosæ, Rosaceæ, Terebinthaceæ, and Juglandeæ; among *hypogynic* plants the Aurantiaceæ, Cedrelaceæ, and Sapindaceæ. The elegant form of the doubly pinnate leaf, which constitutes so great an adornment of the torrid zone, is most frequently met with among the Leguminosæ; among the Mimosaceæ, and also among some Cæsalpinias, Coulterias and Gleditschias; but never, as Kunth has observed, among the Papilionaceæ.

The form of pinnate, and more especially of compound leaves, is unknown in Gentianeæ, Rubiaceæ, and Myrtaceæ. In the morphological development presented by the richness and varied aspect of the appendicular organs of dicotyledons, we are only able to recognize a very small number of general laws.

* See Kunth, *Synopsis Plantarum quas in itinere collegerunt* Al. de Humboldt et Am. Bonpland, t. iii. pp. 87, 360.

ON THE

# STRUCTURE AND MODE OF ACTION

OF

# VOLCANOS

## IN DIFFERENT PARTS OF THE EARTH.

(This Memoir was read at a Public Meeting of the Academy, at Berlin, on the 24th January, 1823.)

---

WHEN we consider the influence exerted on the study of nature during the last few centuries, by the extension of geographical knowledge and by means of scientific expeditions to remote regions of the earth, we are at once made sensible of the various character of this influence, according as the investigations have been directed to the forms of the organic world, the study of the inorganic crust of the earth, or to the knowledge of rocks, their relative ages, and their origin. Different vegetable and animal developments exist in every division of the earth, whether it be on the plains, where, on a level with the sea, the temperature varies with the latitude and with the various inflections of the isothermal lines, or on the steep declivity of mountain ranges, warmed by the direct rays of the sun. Organic nature imparts to every region of the globe its own characteristic physiognomy. But this does not apply to the inorganic crust of the earth divested of its vegetable covering, for everywhere, in both hemispheres, from the equator to the poles, the same rocks are found grouped with some relation to each other, either of attraction or repulsion. In distant lands, surrounded by strange forms of vegetation, and beneath a sky beaming with other stars than those to which his eye had been accustomed, the mariner often recognises, with joyful surprise, argillaceous schists and rocks familiar to him in his native land.

This independence of geological relations on the actual condition of climates does not diminish the beneficial influence exercised on the progress of mineralogy and physical geognosy by the numerous observations instituted in distant regions of the earth, but simply gives a particular direction to them. Every expedition enriches natural history with new genera of plants and animals. At one time we acquire a knowledge of new organic forms which are allied to types long familiar to us, and which not unfrequently, by furnishing links till then deficient, enable us to establish, in all its original perfection, an uninterrupted chain of natural structures. At another time we become acquainted with isolated structures, which appear either as the remains of extinct genera, or members of unknown groups, the discovery of which stimulates further research. It is not, however, from the investigation of the earth's crust that we acquire these manifold additions to our knowledge, for here we meet rather with an uniformity in the constituent parts, in the super-position of dissimilar masses, and in their regular recurrence, which cannot fail to excite the surprise and admiration of the geologist. In the chain of the Andes, as in the mountains of Central Europe, one formation appears, as it were, to call forth another. Masses identical in character assume the same forms; basalt and dolerite compose twin mountains; dolomite, sandstone, and porphyry form abrupt rocky walls; while vitreous trachyte, containing a large proportion of feldspar, rises in bell-shaped and high-vaulted domes. In the most remote regions large crystals are separated in a similar manner from the compact texture of the fundamental mass, and, blending and grouping together into subordinate strata, frequently announce the commencement of new and independent formations. It is thus that the inorganic world may be said to reflect itself, more or less distinctly, in every mountain of any great extent. It is necessary, however, in order perfectly to understand the most important phenomena of the composition, relative age, and origin of formations, to compare together the observations made in regions of the earth most widely remote from each other. Problems which have long baffled the geologist in his own northern region, find their solution in the vicinity of the equator. If, as we have already observed, remote regions do not present us with new formations, that is to say, with unknown groupings of simple substances, they at least help us to unravel the great and universal laws of nature, by showing how different strata of the crust of the earth are mutually superimposed on, and intersect, each other in the form of veins, or rise to different elevations in obedience to elastic forces.

Although our geological knowledge may be thus extensively augmented by researches over vast regions, it can hardly be a matter of surprise that the class of phenomena constituting the principal subject of this address should have been so long examined in an imperfect manner, since the means of comparison were of difficult, and almost, it may be said, of laborious access.

Until towards the close of the eighteenth century all that was known of the form of volcanos and of the action of their subterranean forces was derived from observations made on two volcanic mountains of Southern Italy, Vesuvius and Etna. As the former of these was the more accessible, and (like all volcanos of slight elevation) had frequent eruptions, a hill became to a certain degree the type according to which a whole world—the mighty volcanos of Mexico, South America, and the Asiatic Islands—was supposed to be formed. Such a mode of reasoning involuntarily calls to mind Virgil's shepherd, who believed that in his own humble cot he saw the image of the eternal city, Imperial Rome.

This imperfect mode of studying nature might indeed have been obviated by a more attentive examination of the whole Mediterranean, and especially of its eastern islands and littoral districts, where mankind first awoke to intellectual culture and to a higher standard of feeling. Among the Sporades, trachytic rocks have risen from the bottom of the sea, and have formed islands similar to those of the Azores, which in the course of three centuries have appeared periodically at three almost equal intervals of time. Between Epidaurus and Trœzene, near Methone, in the Peloponnesus, there is a Monte Nuovo, described by Strabo and since by Dodwell. Its elevation is greater than that of the Monte Nuovo of the Phlegræan fields near Baiæ, and perhaps even than that of the new volcano of Xorullo, in the plains of Mexico, which I found to be surrounded by many thousand small basaltic cones, upheaved from the earth, and still emitting smoke. It is not only in the basin of the Mediterranean, that volcanic fires escape from the permanent craters of isolated mountains having a constant communication with the interior of the earth, as Stromboli, Vesuvius, and Etna; for at Ischia, and on Mount Epomeus, and also, according to the accounts of the ancients, in the Lelantine plain, near Chalcis, lavas have flowed from fissures which have suddenly opened on the surface of the earth. Besides these phenomena, which fall within historical periods, that is, within the narrow bounds of authentic tradition, and which Ritter purposes collecting and explaining in his masterly work on geography, the shores of the Mediterranean present numerous remains of the earlier action of fire. The south of France exhibits in Auvergne a distinct and peculiar system of volcanos, linearly arranged, trachytic domes alternating with cones of eruption, emitting lava streams in the form of bands. The plains of Lombardy, which are on a level with the sea, and constitute the innermost bay of the Adriatic, inclose the trachyte of the Euganean Hills, where rise domes of granular trachyte, obsidian, and pearl-stone. These masses are developed from each other, and break through the lower chalk formations and nummulitic limestone, but have never been emitted in narrow streams. Similar evidence of former revolutions of our earth, is afforded in many parts of the Greek Continent and in Western Asia, countries which will undoubtedly some day yield the geologist ample materials for investigation, when the light of knowledge shall again shine on those lands whence it first dawned on our western world, and when oppressed humanity shall cease to groan beneath the weight of Turkish barbarism.

I allude to the geographical proximity of such numerous and various phenomena in order to show that the basin of the Mediterranean, with its series of islands, might have enabled the attentive observer to note all those phenomena which have recently been discovered under various forms and structures in South America, Teneriffe, and in the Aleutian islands, near the Polar region. The materials for observation were, no doubt, accumulated within a narrow compass; but it was yet necessary that travels in distant countries and comparisons between extensive tracts of land, both in and out of Europe, should be undertaken, in order to obtain a correct idea of the resemblance between volcanic phenomena and of their dependence on each other.

Language, which so frequently imparts permanence and authority to first, and often also erroneous views, but which points, as it were, instinctively to the truth, has applied the term *volcanic* to all eruptions of subterranean fire and molten matter; to columns of smoke and vapour which ascend sporadically from rocks, as at Colares, after the great earthquake of Lisbon; to Salses, or argillaceous cones emitting moist mud, asphalt, and hydrogen, as at Girgenti in Sicily, and at Turbaco in South America; to hot Geyser springs, which rise under the pressure of elastic vapours; and, in general, to all operations of impetuous natural forces which have their seat deep in the interior of our planet. In Central America (Guatimala) and in the Philippine Islands, the natives even formally distinguish between *Volcanes de agua y de fuego*, volcanos emitting water, and those emitting fire; designating by the former appellation, mountains from which subterranean waters burst forth from time to time, accompanied by a dull hollow sound and violent earthquakes.

Without denying the connection, which undoubtedly exists among the phenomena just referred to, it would seem advisable to apply more definite terms to the physical as well as to the mineralogical portion of the science of geology, and not at one time to designate by the word *volcano* a mountain terminating in a permanent fire-emitting mouth, and at another to apply it to any subterranean cause, be it what it may, of volcanic action. In the present condition of our earth, the form of isolated conical mountains (as those of Vesuvius, Etna, the Peak of Teneriffe, Tunguragua and Cotopaxi) is certainly the shape most commonly observed in volcanos. I have myself seen such volcanos varying in height from the most inconsiderable hill to an elevation of more than 19,000 feet above the level of the sea. Besides such conical forms, however, we continually meet with permanent fire-emitting mouths, in which the communication with the interior of the earth is maintained on far-extended jagged ridges, and not even always from the centre of their mural summits, but at their extremity towards their slope. Such, for instance, is Pichincha, situated between the Pacific and the city of Quito, which has acquired celebrity from Bouguer's earliest barometric formulæ, and such are the volcanos on the Steppe de los Pastos, situate at more than 10,000 feet above the level of the sea. All these variously shaped summits consist of trachyte, formerly known as trap-porphyry; a granular stone full of narrow fissures, composed of different kinds of feldspar (labradorite, oligoklase, and albite), augite, hornblende, and sometimes interspersed mica, and even quartz. Wherever the evidences of the first eruption, the ancient structures—if I may use the expression —remain complete, the isolated cone is surrounded, circus-like, with a high wall of rock consisting of different superimposed strata, encompassing it like an outer sheath. Such walls or circular inclosures are termed *craters of elevation*, and constitute a great and important phenomenon, upon which that eminent geologist, Leopold von Buch, from whose writings I have borrowed many facts advanced in this treatise, presented so remarkable a paper to our Academy five years ago.

Volcanos which communicate with the atmosphere by means of fire-emitting mouths, such as conical basaltic hills, and dome-like craterless trachytic mountains, (the latter being sometimes low, like the Sarcouy, and sometimes high, like the Chimborazo,) form various groups. Comparative geography draws our attention, at one time, to small Archipelagos or independent mountain-systems, with craters and lava streams, like those in the Canary Isles and the Azores, and without craters or true lava streams, as in the Euganean hills, and the Siebengebirge near Bonn; at another time, it makes us acquainted with volcanos arranged in single or double chains, and extending for many hundred miles in length, either running parallel with the main direction of the range, as in Guatimala, Peru, and Java, or intersecting its axis at right angles, as in tropical Mexico. In this land of the Aztecs fire-emitting trachytic mountains alone attain the high snow limit: they are ranged in the direction of a parallel of latitude, and have probably been upheaved from a chasm extending over upwards of 420 miles, intersecting the whole continent from the Pacific to the Atlantic.

This crowding together of volcanos, either in rounded groups or double lines, affords the most convincing proof that their action does not depend on slight causes located near the surface, but that they are great and deep-seated phenomena. The whole of the eastern portion of the American continent, which is poor in metals, has in its present condition no fire-emitting openings, no trachytic masses, and perhaps no basalt containing olivine. All the volcanos of America are united in the portion of the continent opposite to Asia, along the chain of the Andes, which runs nearly due north and south over a distance of more than 7200 miles.

The whole elevated table-land of Quito, which is surmounted by the high mountains of Pichincha, Cotopaxi, and Tunguragua, constitutes one sole volcanic hearth. The subterranean fire bursts sometimes from one and sometimes from another of these openings, which have generally been regarded as independent volcanos. The progressive movement of the fire has, for three centuries, inclined from north to south. Even the earthquakes, which so fearfully devastate this portion of the globe, afford striking evidence of the existence of subterranean communications, not only between countries where there are no volcanos—as has long been known—but likewise between volcanic apertures situated at a distance from each other. Thus the volcano of Pasto, east of the river Guaytara, continued during three months of the year 1797, to emit, uninterruptedly, a lofty column of smoke, until it suddenly ceased at the moment of the great earthquake of Riobamba, (at a distance of 240 miles,) and the mud eruption of the "Moya," in which from thirty to forty thousand Indians perished.

The sudden appearance, on the 30th of January, 1811, of the island of Sabrina, in the group of the Azores, was the precursor of the dreadful earthquakes which, further westward, shook, from May, 1811, to June, 1813, almost uninterruptedly, first the Antilles, then the plains of the Ohio and Mississippi, and lastly, the opposite coasts of Venezuela or Caracas. Thirty days after the total destruction of the beautiful capital of the province, there was an eruption of the long inactive volcano of St. Vincent, in the neighbouring islands of the Antilles. A remarkable phenomenon accompanied this eruption: at the moment of this explosion, which occurred on the 30th of April, 1811, a terrible subterranean noise was heard in South America, over a district of more than 35,000 square miles. The inhabitants of the banks of the Apure, at the confluence of the Rio Nula, and those living on the remote sea-coast of Venezuela, agreed in comparing this sound to the noise of heavy artillery. The distance from the confluence of the Rio Nula with the Apure (by which I entered the Orinoco) to the volcano of St. Vincent, measured in a straight line, is no less than 628 miles. This noise was certainly not propagated through the air, and must have arisen from some deep-seated subterranean cause; its intensity was, moreover, hardly greater on the shores of the Caribbean sea, near the seat of the raging volcano, than in the interior of the country in the basin of the Apure and the Orinoco.

It would be useless to multiply examples of this nature, by adducing others which I have collected: I will therefore only refer to one further instance, namely, the memorable earthquake of Lisbon, an important phenomenon in the annals of Europe. Simultaneously with this event, which took place on the 1st of November, 1755, not only were the Lakes of Switzerland and the sea off the Swedish coasts violently agitated, but in the eastern portion of the Antilles, near the islands of Martinique, Antigua, and Barbadoes, the tide, which never exceeds thirty inches, suddenly rose upwards of twenty feet. All these phenomena prove, that subterranean forces are manifested either dynamically, expansively, and attended by commotion, in earthquakes; or possess the property of producing, or of chemically modifying substances in volcanos; and they further show, that these forces are not seated near the surface in the thin crust of the earth, but deep in the interior of our planet, whence through fissures and unfilled veins they act simultaneously at widely distant points of the earth's surface.

The more varied the structure of volcanos, that is to say, of elevations inclosing a channel through which the molten masses of the interior of the earth reach the surface, the more important it is to form a correct idea of these structures by careful measurement. The interest derived from measurements of this kind, which I made a special subject of inquiry in the western hemisphere, is increased by the consideration, that the objects to be measured vary in magnitude at different points. A philosophical study of nature seeks, in considering the changes of phenomena, to connect the present with the past.

In order to ascertain the periodic recurrence, or the laws of the progressive changes in nature, we require certain fixed points, and carefully conducted observations, which, by their connection with definite epochs, may serve as a basis for numerical comparisons. If the mean temperature of the atmosphere and of the earth in different latitudes, or the mean height of the barometer at the sea level, had been determined only once in every thousand years, we should know to what extent the heat of climates has increased or diminished, and whether any changes have taken place in the height of the atmosphere. Such points of comparison are especially required to determine the inclination and declination of the magnetic needle, and the intensity of those electro-magnetic forces on which Seebeck and Erman, two admirable physicists belonging to this Academy, have thrown so much light. If it be a meritorious undertaking on the part of learned societies to investigate with perseverance the cosmical changes in the heat and pressure of the atmosphere, and particularly the magnetic direction and intensity, it is no less the duty of the travelling geologist to direct attention to the varying height of volcanos in determining the inequalities of the earth's surface. The observations which I formerly made in the Mexican mountains, at the volcano of Toluca, at Popocatepetl, at the Cofre de Perote, or Nauhcampatepetl, and Xorullo, and in the Andes of Quito at Pichincha, I have had opportunities since my return to Europe of repeating, at different periods, on Mount Vesuvius. Where complete trigonometric or barometric measurements are wanting, their place may be supplied by angles of altitude laid down with precision, and taken at points accurately determined. The comparison of such determinations, made at different periods of time, may sometimes be even preferable to the complication of more complete operations.

Saussure measured Vesuvius in 1773, and at that time both the north-western and south-eastern margins of the crater appeared to him to be equal in height. He found their elevation above the level of the sea to be 3894 feet. The eruption of 1794 occasioned a falling in towards the south, and an inequality in the margins of the crater, which may be distinguished from a considerable distance even by the most unpractised eye. Leopold von Buch, Gay Lussac, and myself, measured Mount Vesuvius three times in the year 1805, and found that the elevation of the northern margin, la Rocca del Palo, opposite the Somma, was exactly as it had been given by Saussure, while the southern margin was 479 feet lower than it had been in 1773. The elevation of the volcano itself towards Torre del Greco (the side towards which, for thirty years, the volcanic action has been principally directed) had, at that time, decreased one-eighth. The cone of cinders bears to the total height of Vesuvius the relation of 1 : 3; in Pichincha, the ratio is as 1 : 10, and at the Peak of Teneriffe, as 1 : 22. Of these three volcanic mountains, Vesuvius has, therefore, comparatively, the highest cone of cinders; probably because, being a volcano of inconsiderable height, it has chiefly acted through its summit.

A few months ago, in the year 1822, I succeeded not only in repeating my earlier barometric measurements of Mount Vesuvius, but also in determining more completely all the margins of the crater (1) during three ascents of the mountain.

These determinations are, perhaps, deserving of some degree of attention, since they embrace the long period of the great eruptions between 1805 and 1822, and are probably the only measurements hitherto published of any volcano which admit of comparison in all their parts. They prove, that the margins of the crater should be regarded as a much more permanent phenomenon than has hitherto been supposed, from the hasty observations made on the subject; and that this character appertains to them everywhere, and not merely in those instances where, as at the Peak of Teneriffe, and in all the volcanos of the Andes, they evidently consist of trachyte. According to my latest determinations it would seem, that since the time of Saussure, a period of forty-nine years, the north-western margin of Vesuvius has probably not changed at all, and that the south-eastern one, in the direction of Bosche Tre Case, which in 1794 had become 426 feet lower, has since then only altered about 64 feet.

If, in the newspaper reports of great eruptions, we often find assertions made of an entire change of form in Mount Vesuvius, and if these assertions appear to be confirmed by the picturesque views of the volcano made at Naples, the cause of the error arises from the outlines of the margins of the crater having been confounded with those of the cones of eruption accidentally formed in its centre, the bottom of which has been raised by the force of vapours. A cone of eruption of this kind, formed by the accumulation of masses of rapilli and scoriæ, gradually came to view, above the south-eastern margin of the crater, between the years 1816 and 1818. The eruption in the month of February, 1822, increased this cone to such an elevation, that it projected from 107 to 117 feet above the north-western margin of the crater (the Rocca del Palo). This remarkable cone, which was at length regarded at Naples as the actual summit of Vesuvius, fell in with a fearful crash at the last eruption, on the night of the 22nd of October; in consequence of which, the bottom of the crater, which had continued uninterruptedly accessible from the year 1811, is now nearly 800 feet below the northern and 213 feet below the southern margin of the volcano. The varying form and relative position of the cones of eruption, the apertures of which must not, as they sometimes are, be confounded with the crater of the volcano, give to Vesuvius at different epochs a peculiar physiognomy; so much so, that the historiographer of this volcano, by a mere inspection of Hackert's landscapes in the Palace of Portici, might guess the exact year in which the artist had made his sketch, by the outline of the summit of the mountain, according as the northern or southern side is represented in respect to height.

Twenty-four hours after the fall of the cone of scoriæ, which was 426 feet high, and when the small but numerous streams of lava had flowed off, on the night between the 23rd and 24th of October, there began a fiery eruption of ashes and rapilli, which continued uninterruptedly for twelve days, but was most violent during the first four days. During this period the explosions in the interior of the volcano were so loud that the mere vibrations of the air caused the ceilings to crack in the Palace of Portici, although no shocks of an earthquake were then or had previously been experienced. A remarkable phenomenon was observed in the neighbouring villages of Resina, Torre del Greco, Torre del' Annunziata, and Bosche Tre Case. Here the atmosphere was so completely saturated with ashes that the whole region was enveloped in complete darkness during many hours in the middle of the day. The inhabitants were obliged to carry lanterns with them through the streets, as is often done in Quito during the eruptions of Pichincha. Never had the flight of the inhabitants been more general, for lava streams are less dreaded even than an eruption of ashes, a phenomenon unknown here in any degree of intensity, and one which fills the imaginations of men with images of terror from the vague tradition of the manner in which Herculaneum, Pompeii, and Stabiæ were destroyed.

The hot aqueous vapour which issued from the crater during the eruption, and diffused itself through the atmosphere, formed, on cooling, a dense cloud, which enveloped the column of ashes and fire, that rose to an elevation of between 9000 and 10,000 feet above the level of the sea. So sudden a condensation of vapour, and, as Gay Lussac has shown, the formation of the cloud itself, tended to increase electric tension. Flashes of forked lightning darted in all directions from the column of ashes, while the rolling thunder might be clearly distinguished from the deep rumbling sounds within the volcano. In no other eruption had the play of the electric forces been so powerfully manifested as on this occasion.

On the morning of the 26th of October the strange report was circulated that a stream of boiling water was gushing from the crater, and pouring down the cone of cinders. Monticelli, the zealous and learned observer of the volcano, soon perceived that this erroneous report originated in an optical illusion, and that the supposed stream of water was a great quantity of dry ashes which issued like drift sand from a crevice in the highest margin of the crater. The long drought, which had parched and desolated the fields before this eruption of Vesuvius, was succeeded, towards the termination of the phenomenon, by a continued and violent rain, occasioned by the *volcanic storm* which we have just described. A similar phenomenon characterizes the termination of an eruption in all zones of the earth. As the cone of cinders is usually wrapped in clouds at this period, and as the rain is poured forth with most violence near this portion of the volcano, streams of mud are generally observed to descend from the sides in all directions. The terrified peasant looks upon them as streams of water that rise from the interior of the volcano and overflow the crater, while the deceived geologist believes that he can recognise in them either sea-water or muddy products of the volcano, the so-called *eruptions boueuses*, or, in the language of the old French systematisers, products of an igneo-aqueous liquefaction.

Where, as is generally the case in the chain of the Andes, the summit of the volcano penetrates beyond the snow-line, attaining sometimes an elevation twice as great as that of Mount Etna, the inundations we have described are rendered very frequent and destructive, owing to the melting and permeating snow.

These are phenomena which have a meteorological connection with the eruptions of volcanos, and are variously modified by the heights of the mountains, the circumference of the summits which are perpetually covered with snow, and the degree to which the walls of cinder cones become heated; but they cannot be regarded in the light of true volcanic phenomena. Subterranean lakes, communicating by various channels with the mountain streams, are frequently formed in deep and vast cavities, either on the declivity or at the base of volcanos. When the whole mass of the volcano is powerfully shaken by those earthquakes which precede all eruptions of fire in the Andes, the subterranean vaults open, and pour forth streams of water, fishes, and tuffaceous mud. This singular phenomenon brings to mind the *Pimelodes Cyclopum*, or the Silures of the Cyclops, which the inhabitants of the plateau of Quito call Preñadilla, and of which I gave a circumstantial account soon after my return to Europe. When, on the night between the 19th and 20th of June, 1698, the summit of Mount Carguairazo, situated to the north of Chimborazo, and having an elevation of more than 19,000 feet, fell in, all the country for nearly 32 square miles was covered with mud and fishes. A similar eruption of fish from the volcano of Imbaburu was supposed to have caused the putrid fever, which, seven years before this period, raged in the town of Ibarra.

I refer to these facts because they throw some light on the

difference between the eruption of dry ashes and mud-like inundations of tuff and trass, investing fragments of wood, charcoal, and shells. The quantity of ashes recently erupted from Mount Vesuvius, like every phenomenon connected with volcanos and other great and fearful natural phenomena, has been greatly exaggerated in the public papers; and two Neapolitan chemists, Vicenzo Pepe and Guiseppe di Nobili, even asserted that the cinders were mixed with given proportions of gold and silver, notwithstanding the counter-statements of Monticelli and Covelli. According to my researches the stratum of ashes which fell during the twelve days was only three feet in thickness in the direction of Bosche Tre Case, on the declivity of the cone, where they were mixed with rapilli, while in the plains its greatest thickness did not exceed from 16 to 19 inches. Measurements of this kind must not be made at spots where the ashes have been drifted by the wind, like snow or sand, or where they have been accumulated in pulp-like heaps by means of water. The times are passed in which, after the manner of the ancients, nothing was regarded in volcanic phenomena save the marvellous, and when men would believe, like Ctesias, that the ashes from Etna were borne as far as the Indian peninsula. A portion of the Mexican gold and silver veins is certainly found in trachytic porphyry, but in the ashes of Vesuvius which I myself collected, and which were, at my request, examined by that distinguished chemist Heinrich Rose, no trace of either gold or silver was to be discovered.

However much these results, which perfectly correspond with the more exact observations of Monticelli, may differ from those recently announced, it cannot be denied that the eruption of ashes, which continued from the 24th to the 28th of October, is the most memorable that has been recorded, on unquestionable evidence, in reference to Mount Vesuvius, since the death of the elder Pliny. The quantity of ashes erupted on this occasion was probably three times as great as the whole quantity which has fallen since volcanic phenomena have been observed with attention in Italy. A stratum from 16 to 19 inches in thickness does certainly, at first sight, seem very inconsiderable, when compared with the mass with which we find Pompeii covered. But, without taking into account the heavy rains and the inundations which must have increased the bulk of this stratum in the course of ages, and without reviving the animated contention maintained with much scepticism on the other side of the Alps, regarding the causes of the destruction of the Campanian cities, it may, at any rate, be here observed that the eruptions of a volcano, at widely remote epochs, cannot be compared with respect to their intensity. All conclusions must be insufficient that are based on mere analogies of quantitative relations of the lava and ashes, the height of the column of smoke, and the intensity of the explosions.

We learn from the geographical description of Strabo, and from the opinion expressed by Vitruvius on the volcanic origin of pumice, that, until the year of Vespasian's death, that is to say, until the eruption which buried Pompeii, Vesuvius appeared more like an extinct volcano than a Solfatara. When, after a long-continued repose, subterranean forces suddenly opened for themselves new channels, penetrating through strata of primitive rock and trachyte, effects must have been produced to which no analogy is afforded by those of subsequent occurrence. We clearly learn from the well-known letter in which Pliny the younger informs Tacitus of the death of his uncle, that the renewal of the eruptions, or, one might almost say, the revival of the slumbering volcano, began with an outbreak of ashes. The same phenomenon was observed at Xorullo, when the new volcano, in the month of September, 1759, breaking through strata of syenite and trachyte, was suddenly upheaved in the plain. The country people fled in terror on finding their cottages covered with ashes thrown up from the earth, which was bursting in every direction.

In the ordinary periodical manifestations of volcanic activity a shower of ashes usually terminates each partial eruption. The letter of the younger Pliny contains, moreover, a passage which clearly shows that the dry ashes falling from the air immediately attained a height of four or five feet, independent of accumulation by drifts. "The court," the narrative continues, "which led to the apartment in which Pliny took his siesta, was so filled wlth ashes and pumice that, had the sleeper tarried longer, he would have found the passage wholly blocked up." Within the inclosed limits of a court the wind cannot have exercised any very considerable influence on the drifting of the ashes.

I have interrupted my comparative view of volcanos by different observations in relation to Vesuvius, partly on account of the great interest excited by its recent eruption, and partly because every great outpouring of ashes almost involuntarily recalls to mind the classic soil of Pompeii and Herculaneum. In a note, not adapted to be read to the audience to whom this lecture is addressed, I have collected all the elements of the barometric measurements which I made during the close of last year at Mount Vesuvius, and in the Campi Phlegræi.

We have hitherto considered the form and effects of those volcanos which are permanently connected, by means of a crater, with the interior of the earth. The summits of such volcanos are upheaved masses of trachyte and lava intersected by numerous veins. The permanency of their effects indicates a highly complex structure. They have, so to say, a certain individuality of character, which remains unaltered for long periods of time. Contiguous mountains generally yield wholly different products; for instance: leucitic and feldspathic lavas, obsidian with pumice, and basaltic masses containing olivine. They belong to the more recent phenomena of the earth, usually breaking through all the strata of the floetz formation, and their lava currents and products are of subsequent origin to our valleys. Their life, if I may be permitted to use a figurative expression, depends upon the mode and the duration of their connection with the interior of the earth. After continuing for centuries in a state of repose, their activity is often suddenly revived, and they then become converted into Solfataras, emitting aqueous vapours, gases, and acids. Occasionally, as at the Peak of Teneriffe, their summits have already become a laboratory of regenerated sulphur, while considerable lava currents, being basaltic near the base, and mixed with obsidian and pumice at greater elevations, where the pressure is less, continue to flow from the sides of the mountain (2).

Besides volcanos which have permanent craters, there is another kind of volcanic phenomena less frequently observed than the former, but especially instructive to the geologist, as they remind us of the primitive world, that is, of the earliest revolutions of our planet. Trachytic mountains suddenly open, and after throwing up ashes and lava, close again never perhaps to re-open. Such has been the case with the mighty volcano of Antisana in the chain of the Andes, and with Mount Epomæus in Ischia, in the year 1302. Occasionally such an eruption has occurred even in the plains, as on the table-land of Quito, in Iceland at a distance from Hecla, and in the Lelantine plains of Eubœa. Many upheaved islands belong to this class of transitory phenomena. In these cases, the connection with the interior of the earth is not permanent, the action ceasing as soon as the fissure, or channel of communication, is again closed. Veins of basalt, dolerite, and porphyry, which traverse almost all formations in different parts of the earth; and the masses of syenite, augitic porphyry, and amygdaloid, which characterise the most recent strata of transition rock, and the oldest stratum of the floetz formation; have all probably been formed in a similar manner. In the youthful period of our planet, the substances that had continued in a fluid condition within the earth, broke through its crust, everywhere intersected with fissures, and became solidified as granular veins, or were spread out in broad superimposed strata. The products that may be termed exclusively volcanic, which have come down to us from the primitive ages of the world, have not flowed in streams or bands like the lava of our isolated conical mountains. The mixtures of augite, titanic iron, feldspar, and hornblende, may have been the same at different periods, sometimes allied to basalt, sometimes to trachyte; while chemical substances, (as we learn from Mitscherlich's important labours and the analogies presented by artificial igneous products,) may have ranged themselves in layers according to some definite laws of crystallization. In all cases we perceive that substances similarly composed have come to the surface of the earth by very different means, either by being simply upheaved, or escaping through temporary fissures; and that breaking through the older rocks, that is to say, through the earlier oxidized earth's crust, they have flowed in the form of lava streams from conical mountains having a permanent crater. If we do not sufficiently distinguish between these various phenomena, our knowledge of the geology of volcanos will again be shrouded in that obscurity, from which numerous comparative experiments are now beginning gradually to release it.

The questions have often been asked, what is it that burns in volcanos, what generates the degree of heat capable of mixing earths and metals together in a state of fusion? Modern chemistry has attempted to reply that it is the earths, metals, and alkalies themselves, that is to say, the metalloids of these substances, which burn. The solid and already oxidized crust of the earth separates the surrounding atmosphere, with the oxygen it contains, from the combustible unoxidized substances in the interior of our planet. By the contact of these metalloids with the atmospheric oxygen the disengagement of caloric ensues. The celebrated and talented chemist, who advanced this explanation of volcanic phenomena, soon himself relinquished it. The experiments which have been made in mines and caverns in all parts of the earth, and which M. Arago and myself have collected in a separate treatise, prove that even at an inconsiderable depth, the temperature of the earth is much higher than the mean temperature of the atmosphere at the same place. This remarkable, and almost universally confirmed fact, is connected with what we learn from volcanic phenomena. The depth at which we might regard the earth as a fused mass, has been calculated. The primitive cause of this subterranean heat is, as in all planets, the formative process itself, the separation of the spherically conglomerating mass from a cosmical aëriform fluid, and the cooling of the terrestrial strata at different depths by the radiation of heat. All volcanic phenomena are probably the result of a permanent or transient connection between the interior and the exterior of our planet. Elastic vapours press the fused oxidizing substances upwards through deep fissures. Volcanos therefore are intermittent earth-springs, from which the fluid mixtures of metals, alkalies, and earths, which become consolidated into lava currents, flow gently and calmly, when being upheaved they find a vent. In a similar manner, according to Plato's Phædon, the ancients regarded all volcanic streams of fire as effusions of the Pyriphlegethon.

I would fain be permitted to add one yet bolder observation to those I have already ventured to advance. May not the cause of one of the most wonderful phenomena presented by the study of petrifactions, be dependent on the condition of the inner heat of our planet, which is indicated by thermometric experiments on springs (3) rising from different depths, and by observations on volcanos? We find tropical animals, arborescent ferns, palms, and bamboos, buried in the cold north, and everywhere the primitive world presents a distribution of organic structures wholly at variance with existing climatic relations. Many hypotheses have been advanced in elucidation of so important a problem, such as the approximation of a comet, the altered obliquity of the ecliptic, and the increased intensity of the sun's light; but none of these have satisfied at once the astronomer, the physicist, and the geologist. I, for my part, would willingly leave undisturbed the axis of the earth or the light of the sun's disk, (from whose spots a celebrated astronomer explained fruitfulness and failure of crops,) yet it appears to me that in every planet there exist, independently of its relations to a central body and its astronomical position, numerous causes for the development of heat, in processes of oxidation, in precipitation, in the chemically altered capacity of bodies, the increase of electro-magnetic tension, and in the channels of communication opened between its internal and external parts.

Wherever, in the primitive world, heat was radiated from the deeply fissured crust of the earth, palms, arborescent ferns, and all the animals of the torrid zone, could perhaps have flourished for centuries over extensive tracts of land. According to this view, which I have already published in my work entitled *Geognostischer Versuch über die Lagerung der Gebirgsarten in beiden Hemisphären*,* the temperature of volcanos would be that of the interior of our earth itself, and the same causes which now occasion such fearfully devastating results, may have been able to produce, in every zone, the most luxuriant vegetation on the newly oxidized crust of the earth and on the deeply fissured strata of rocks.

Should it be assumed, for the purpose of explaining the wonderful distribution of tropical forms in their ancient mausolea, that the long-haired elephantine animals, which are now found embedded in ice, were once indigenous to northern lati-

* *Geognostical Essay on the superposition of Rocks in both Hemispheres.* 8vo. Lond. 1803.

tudes, and that animals of similar forms, belonging to the same type, as, for instance, lions and lynxes, were capable of living in wholly different climates, such a mode of explanation would at all events not admit of being extended to vegetable products. From causes developed by the physiology of vegetation, palms, bananas, and arborescent monocotyledons, are unable to endure the deprivation of their appendicular organs, by the northern cold; and in the geological problem which we are here considering, it seems to me a matter of difficulty to admit any distinction between vegetable and animal structures One and the same mode of explanation must be applied to both forms.

In concluding this treatise, I have added some uncertain and hypothetical conjectures to the facts which have been collected in widely remote regions of the earth. The philosophical study of nature rises above the requirements of mere delineation, and does not consist in the sterile accumulation of isolated facts. The active and inquiring spirit of man may therefore be occasionally permitted to escape from the present into the domain of the past, to conjecture that which cannot yet be clearly determined, and thus to revel amid the ancient and ever-recurring myths of geology.

---

## EXPLANATORY ADDITIONS.

(1) p. 363.—" *A more complete determination of the margins of the Crater of Mount Vesuvius.*"

My astronomical fellow-labourer, Oltmanns, who was unhappily too early lost to science, has re-calculated the barometric measurements I made on Mount Vesuvius (from the 22nd to the 25th of November, and on the 1st of December, 1822), and compared the results with those yielded by the measurements given to me in manuscript by Lord Minto, Visconti, Monticelli, Brioschi, and Poulett Scrope.

A. *Rocca del Palo, the highest northern margin of the Crater of Vesuvius, was estimated by*—

| | Feet. |
|---|---|
| Saussure, in 1773, barometrically, probably according to Deluc's formula . . . .. . | 3894 |
| Poli (1794), barometrically . . . . . . | 3875 |
| Breislak (1794), barometrically, although, as in the case of Poli, it is uncertain what formula was used . | 3920 |
| Gay-Lussac, Leopold von Buch, and Humboldt (1805), barometrically, according to the formula of Laplace, as in all the following barometric results . . | 3856 |
| Brioschi (1810), trigonometrically . . . . . | 4079 |
| Visconti (1816), trigonometrically . . . . . | 3977 |
| Lord Minto (1822), barometrically, and frequently repeated . . . . . . . . . | 3971 |
| Poulett Scrope (1822). This calculation is somewhat uncertain, owing to the unknown relation of the diameters of the tubes to those of the cistern . . | 3862 |
| Monticelli and Covelli (1822) . . . . . . | 3990 |
| Humboldt (1822) . . . . . . . . . | 4022 |

The most probable final result is 2026 feet above the hermitage, or 3996 feet above the level of the sea.

B. *The lowest south-eastern margin of the Crater, opposite Bosche Tre Case.*

| | |
|---|---|
| After the eruption of 1794, this margin was 426 feet lower than the Rocca del Palo, consequently, if the latter be estimated at 3996 feet, it would be . . | 3570 |
| Gay-Lussac, Leopold von Buch, and Humboldt (1805), barometrically . . . . . . . . | 3414 |
| Humboldt (1822), barometrically . . . . . | 3491 |

C. *The elevation of the cone of scoriæ that fell into the Crater on the 22nd October*, 1822.

| | Feet. |
|---|---|
| Lord Minto, barometrically . . . . . . | 4156 |
| Brioschi, trigonometrically, according to different combinations— | |
| Either . . . . . . . . . . | 4067 |
| Or . . . . . . . . . . | 4099 |

The most probable final result for the height of the cone of scoriæ that fell in during the year 1822, is 4131 feet.

D. *Punta Nasone, the highest summit of the Somma.*

| | |
|---|---|
| Schuckburgh (1794), barometrically, probably according to his own formula . . . . . . | 3734 |
| Humboldt (1822), barometrically, according to the formula of Laplace . . . . . . | 3747 |

E. *Plain of the Atrio del Cavallo.*

| | |
|---|---|
| Humboldt (1822), barometrically . . . . . | 2577 |

F. *Base of the cone of ashes.*

| | |
|---|---|
| Gay-Lussac, Leopold von Buch, and Humboldt (1805), barometrically . . . . . . . . | 2366 |
| Humboldt (1822), barometrically . . . . . | 2482 |

G. *Hermitage of Salvatore.*

| | |
|---|---|
| Gay-Lussac, Leopold von Buch, and Humboldt (1805), barometrically . . . . . . . . | 1918 |
| Lord Minto (1822), barometrically . . . . . | 1969 |
| Humboldt (1822), again barometrically . . . . | 1974 |

Some of my measurements have appeared in Monticelli's *Storia de' fenomeni del Vesuvio, avvenuti negli anni* 1821—1823, p. 115, but owing to the correction of the height of the mercury in the cistern having been omitted, the numbers are not given with perfect exactness. When it is remembered that the results contained in the above table were obtained with barometers of very different construction, at different hours of the day, during the prevalence of various winds, and on the unequally heated declivity of a volcano, in a locality where the decrease of the atmospheric temperature differs very considerably from that assumed in our barometrical formulæ, the amount of correspondence between the various results will appear sufficiently satisfactory.

My measurements of 1822, at the time of the Congress of Verona, when I accompanied the late King to Naples, were conducted with more care and under more favourable circumstances than those of 1805. Differences of elevations are moreover always preferable to absolute elevations. These differences show, that since 1794, the relative condition of the margins of the Rocca del Palo and of that towards Bosche Tre Case had remained almost the same. I found, in 1805, for the height, 441, and in 1822, nearly 524 feet. A distinguished geologist, Mr. Poulett Scrope, obtained 473 feet, although his absolute heights for these two margins of the crater appear somewhat too low. So inconsiderable a variation in a period of twenty-eight years, and during violent disturbances in the interior of the mountain, is undoubtedly a remarkable phenomenon.

The height to which the cones of scoriæ rise from the bottom of the crater at Vesuvius also deserves special attention. Shuckburgh found in 1776 a cone of this nature to be 3932 feet above the level of the Mediterranean; and, according to Lord Minto—a remarkably exact observer—the cone of scoriæ which fell in on the 22nd of October, 1822, was even 4156 feet high. On both occasions therefore the cone of scoriæ in the crater exceeded the highest point of the margin of the crater. On comparing the measurements of Rocca del Palo from 1773 to 1822, one is almost involuntarily led to hazard the bold conjecture that the northern margin of the crater has been gradually upheaved by subterranean forces. The correspondence of the three measurements made between 1773 and 1805 is almost as striking as in those between 1816 and 1822. No doubt can be entertained as to the height being from 3970 to 4021 feet during the latter period. Ought less confidence to be attached to the measurements made thirty or forty years previously, and which only gave from 3875 to 3894 feet? After a longer lapse of time the question may be decided, as to how much is attributable to errors of measurement, and how much to the upheaval of the margin of the crater. There is here no accumulation of loose masses from above; if therefore the solid trachytic lava strata of the Rocca del Palo actually rise, we must assume that they are upheaved from below by volcanic forces.

My learned and indefatigable friend, Oltmanns, has published the details of all these measurements with critical remarks.* Would that this work might incite geognosists to enter upon a series of hypsometric observations, by which, in the course of time, Vesuvius, which is, excepting Stromboli, the most accessible of all European volcanos, may be thoroughly understood in all periods of its development.

(2) p. 371—" *At elevations where the pressure is less.*"

Compare Leopold von Buch on the Peak of Teneriffe, in his *Physikalische Beschreibung der canarischen Inseln*, 1825, s. 213, and in the *Abhandlungen der königl. Akademie zu Berlin, aus den J.* 1820—21, s. 99.

(3) p. 373—" *Springs which rise from different depths.*"

Compare Arago in the *Annuaire du Bureau des Longitudes pour* 1835, p. 234. The increase of the temperature is in our latitudes 1° Fahr. for nearly every 54 feet. In the Artesian boring at the New Salt-works (Oeynhausen's Bath) near Minden, which is the greatest known depth that has been reached below the surface of the sea, the temperature of the water at 2231 feet, is fully 91° Fahrenheit, whilst the mean upper temperature of the air may be assumed at 49°·3 Fahr. It is very remarkable that, even in the third century, Saint Patricius, bishop of Pertusa, should have been led, from the thermal springs near Carthage, to form a very correct view of such an increase of heat.†

* See *Abhandl. der Königl. Akademie der Wissenschaften zu Berlin.* Jahr 1822 und 1823, s. 3—20.

† *Acta S. Patricii,* p. 555, ed. Ruinart; *Cosmos,* vol. i. p. 220, (Bohn's edition).

# VITAL FORCE, OR THE RHODIAN GENIUS.

THE Syracusans, like the Athenians, had their Poecile, where representations of gods and heroes, the works of Grecian and Italian art, adorned the richly decorated halls of the Portico. Incessantly the people streamed thither; the young warrior to feast his eyes upon the deeds of his forefathers, the artist to contemplate the works of the great masters. Among the numerous paintings which the active enterprise of the Syracusans had collected from the mother country, there was but one which for full a century had continued to attract the attention of every visitor. Even when the Olympian Jupiter, Cecrops, the founder of cities, and the heroic courage of Harmodius and Aristogiton, failed to attract admirers, a dense crowd still pressed round this one picture. Whence this preference? Was the painting a rescued work of Apelles, or did it bear the impress of the school of Callimachus? No! although it possessed both grace and beauty, yet neither in the blending of the colours, nor in the character and style of its composition, could it be compared with many other paintings in the Poecile.

The crowd—and how numerous are the classes included in this denomination—ever admires and wonders at what it does not understand! For more than a century had that painting been publicly exhibited, and yet, although Syracuse contained within its narrow limits more artistic genius than all the

* A Portico in Athens containing a picture gallery painted chiefly by Polygnotus, with the assistance of Micon and Panænus. Zeno taught his doctrines there, and was in consequence called the Stoic, from stoa, a portico, and his school the Stoic-school.—ED.

rest of sea-girt Sicily, the riddle of its meaning still remained unsolved. It was not even known to what temple it had formerly belonged, for it had been saved from a stranded vessel, which was only conjectured, from the freight it carried, to have come from Rhodes.

The foreground of the picture was occupied by a numerous group of youths and maidens, whose uncovered limbs, although well formed, were not cast in that slender mould which we so much admire in the statues of Praxiteles and Alcamenes. The fuller development of their limbs, which bore indications of laborious exercise,—the human expression of passion and of care stamped on their features,—all seemed to divest them of a heavenly or God-like type, and to fix them as creatures of the earth. Their hair was simply adorned with leaves and wild flowers. Their arms were extended towards each other with impassioned longing, but their earnest and mournful gaze was rivetted on a Genius, who, surrounded by a brilliant halo, hovered in the midst of the group. On his shoulder was a butterfly, and in his right hand he held aloft a flaming torch. His limbs were moulded with child-like grace; his eye radiant with celestial light. He looked imperiously upon the youths and maidens at his feet. No other characteristic traits could be distinguished in the picture. Some, however, thought they could perceive at his foot the letters ζ and ς, and as antiquarians were then no less bold than they are now, they inferred, though far from happily, that the artist was called Zenodorus, the name borne at a later date by the modeller of the Colossus of Rhodes.

"The Rhodian Genius," for so this mysterious painting was called, did not however want for interpreters in Syracuse. Virtuosi, especially the younger of them, on their return from a flying visit to Corinth or Athens, would have deemed themselves deficient in all pretensions to connoisseurship, had they not immediately advanced some new explanation. Some regarded the Genius as the personification of spiritual

Love, forbidding the enjoyment of sensual pleasures; others were of opinion that the dominion of Reason over the Passions was here signified. The wiser preserved silence, and while they conjectured that the painting was intended to represent something of a sublimer character, delighted to linger in the Poecile to admire the simple composition of the group.

The question continued to remain undecided. Copies of the painting, with various additions, were sent to Greece, but without eliciting any explanation respecting its origin. At length, however, when at the early rising of the Pleiades the Ægean Sea was again opened to navigation, ships from Rhodes entered the port of Syracuse. They contained a treasure of statues, altars, candelabras, and pictures, which a love of art had caused the Dionysii to collect in Greece. Among the paintings there was one which was instantly recognised as the companion to the "Rhodian Genius." It was of the same size, and exhibited a similar tone of colouring, although in a better state of preservation.

The Genius stood as before in the centre, but without the butterfly; his head was drooping, his torch extinguished and reversed. The group of youths and maidens thronged simultaneously around him in mutual embrace; their looks were no longer sad and submissive, but announced a wild emancipation from restraint, and the gratification of long-nourished passion.

The Syracusan antiquaries had already begun to accommodate their former explanations of the "Rhodian Genius" to the newly arrived painting, when the Tyrant ordered it to be conveyed to the house of Epicharmus. This philosopher of the school of Pythagoras dwelt in the remote part of Syracuse called Tyche. He seldom visited the court of the Dionysii, not but that learned men from all the Greek colonies assembled there, but because proximity to princes is apt to rob the most intellectual of their spirit and freedom. He occupied himself unceasingly in studying the nature of things and their forces, the origin of plants and animals, and those harmonious laws by which the celestial bodies on a large, and the snow-flake and the hail-stone on a small scale, assume a globular form. Decrepid with age, he caused himself to be carried daily to the Poecile, and thence to the harbour of Nasos, where, as he said, the wide ocean presented to his eye an image of the Boundless and the Infinite, which his mind strove in vain to comprehend. He was honoured alike by the lower classes and by the tyrant, but he avoided the latter, while he joyfully cultivated and often assisted the former.

Epicharmus lay weak and exhausted on his couch, when the newly arrived work of art was brought to him by the command of Dionysius. He was furnished at the same time with a faithful copy of the "Rhodian Genius," and the philosopher now caused both paintings to be placed before him. He gazed on them long and earnestly, then called together his scholars, and in accents of emotion thus addressed them:

"Remove the curtain from the window, that I may once more feed my eyes with the sight of the richly animated and living earth. Sixty years long have I pondered on the internal springs of nature and on the differences inherent in matter, but it is only this day that the 'Rhodian Genius' has taught me to see clearly that which before I had only conjectured. While the difference of sexes in all living beings beneficently binds them together in prolific union, the crude matters of inorganic nature are impelled by like instincts. Even in the darkness of chaos, matter was accumulated or separated according as affinity or antagonism attracted or repelled its various parts. The celestial fire follows the metals, the magnet, the iron; amber when rubbed attaches light bodies; earth blends with earth; salt separates from the waters of the sea and joins its like, while the acid moisture of the *stypteria* (στυπτηρία ὑγρά) and the fleecy salt *Trichitis*, love the clay of Melos. Everything in inanimate nature hastens to associate itself with its like. No earthly element (and who will dare to class light as such?) can therefore be found in a pure and virgin state. Everything as soon as formed hastens to enter into new combinations, and nought, save the disjoining art of man, can present in a separate state ingredients which ye would vainly seek in the interior of the earth, or in the moving oceans of air and water. In dead inorganic matter absolute repose prevails as long as the bonds of affinity remain unsevered, and as long as no third substance intrudes to blend itself with the others; but even after this disturbance unfruitful repose soon again succeeds.

"Different, however, is the blending of the same substances in animal and vegetable bodies. Here vital force imperatively asserts its rights, and, heedless of the affinity and antagonism of the atoms asserted by Democritus, unites substances which in inanimate nature ever flee from each other, and separates that which is incessantly striving to unite.

"Draw nearer to me, my disciples, and recognise in the 'Rhodian Genius,' in the expression of his youthful vigour, in the butterfly on his shoulder, in the commanding glance of his eye, the symbol of *vital force* as it animates every germ of organic creation. The earthly elements at his feet are striving to gratify their own desires and to mingle with one another. Imperiously the Genius threatens them with upraised and high-flaming torch, and compels them, regardless of their ancient rights, to obey his laws.

"Look now on the new work of art which the Tyrant has sent me to explain; and turn your eyes from the picture of life to the picture of death. The butterfly has soared upwards, the extinguished torch is reversed, and the head of the youth is drooping. The spirit has fled to other spheres, and the vital force is extinct. Now the youths and maidens join their hands in joyous accord. Earthly matter again resumes its rights. Released from all bonds they impetuously follow their sexual instincts, and the day of his death is to them a day of nuptials.—Thus dead matter, animated by vital force, passes through a countless series of races, and perchance enshrines in the very substance in which of old a miserable worm enjoyed its brief existence, the divine spirit of Pythagoras.*

"Go, Polycles, and tell the Tyrant what thou hast heard! And ye, my beloved, Euryphamos, Lysis, and Scopas, come nearer—and yet nearer to me! I feel that the faint vital force within me can no longer retain in subjection the earthly matter, which now reclaims its freedom. Lead me once more to the Poecile, and thence to the wide sea-shore. Soon will ye collect my ashes."

* The very same idea is expressed in Schiller's *Walk under the Linden Trees.*—Ed.

---

## ILLUSTRATION AND NOTE.

IN the Preface to the Second and Third Editions of this work (See preliminary pages of this translation) I have already noticed the republication of the preceding tale, which was first printed in Schiller's *Horen* (for the year 1795, part 5, pages 90—96). It embodies the development of a physiological idea in a semi-mythical garb. In the year 1793, in the Latin *Aphorisms from the Chemical Physiology of Plants*, appended to my *Subterranean Flora*, I had defined the *vital force* as the unknown cause which prevents the elements from following their original attractive forces. The first of my aphorisms ran thus:—

"Rerum naturam si totam consideres, magnum atque durabile, quod inter elementa intercedit, discrimen perspicies, quorum altera affinitatum legibus obtemperantia, altera, vinculis solutis, varie juncta apparent. Quod quidem discrimen in elementis ipsis eorumque indole neutiquam positum, quum ex sola distributione singulorum petendum esse videatur. Materiam segnem, brutam, inanimam eam vocamus, cujus stamina secundum leges chymicæ affinitatis mixta sunt. Animata atque organica ea potissimum corpora appellamus, quæ, licet in novas mutari formas perpetuo tendant, vi interna quadam continentur, quominus priscam sibique insitam formam relinquant.

"Vim internam, quæ chymicæ affinitatis vincula resolvit, atque obstat, quominus elementa corporum libere conjungantur, vitalem vocamus. Itaque nullum certius mortis criterium putredine datur, qua primæ partes vel stamina rerum, antiquis juribus revocatis, affinitatum legibus parent. Corporum inanimorum nulla putredo esse potest."*

* See *Aphorismi ex doctrina Physiologiæ chemicæ Plantarum*, in Humboldt, *Flora Fribergensis subterranea*, 1793, pp. 133—136. *Translation*;—"If you attentively consider the whole nature of things, you will discover a great and permanent difference amongst elements, some of which obeying the laws of affinity, others independent, appear in various combinations. This difference is by no means inherent in the elements themselves and in their nature, but seems to be derived solely from their particular distribution. We call that matter inert, brute, and

These opinions, against which the acute Vicq d'Azyr has protested in his *Traité d'Anatomie*, vol. i. p. 5, but which are still entertained by many eminent persons among my friends, I have placed in the mouth of Epicharmus. Reflection and prolonged study in the departments of physiology and chemistry have deeply shaken my earlier belief in peculiar, so-called vital forces. In the year 1797, at the conclusion of my *Versuche über die gereizte Muskel- und Nervenfaser, nebst Vermuthungen über den chemischen Process des Lebens in der Thier- und Pflanzenwelt* (vol. ii. pp. 430—436), I already declared that I by no means regarded the existence of these peculiar vital forces as established. Since that period I have not applied the term *peculiar forces* to that which may possibly be produced only by the combined action of the separate already long known substances and their material forces. We may, however, deduce a more certain definition of *animate* and *inanimate* substances from the chemical relations of the elements, than can be derived from the criteria of voluntary movement, the circulation of fluid in solid parts, and the inner appropriation and fibrous arrangement of the elements. I call that substance *animate* "whose voluntarily separated parts change their composition after separation has taken place, the former external relations still continuing the same." This definition is merely the expression of a fact. The equilibrium of the elements is maintained in animate matter by virtue of their being parts of one whole. One organ determines another, one gives to another the temperature, the tone as it were, in which these, and no other affinities operate. Thus in organisation all is reciprocal, means and end. The rapidity with which organic parts change their compound state, when separated from a complex of living organs, differs greatly according to the degree of

inanimate, the particles of which are combined according to the laws of chemical affinity. On the other hand, we call those bodies animate and organic, which, although constantly manifesting a tendency to assume new forms, are restrained by some internal force from relinquishing that originally assigned them. That internal force, which dissolves the bonds of chemical affinity, and prevents the elements of bodies from freely uniting, we call vital. Accordingly, the most certain criterion of death is putrescence, by which the first parts, or stamina of things, resume their pristine state, and obey the laws of affinity. In inanimate bodies there can be no putrescence."

their dependence, and the nature of the component materials. The blood of animals, which is variously modified in the different classes, undergoes a change earlier than the juices of plants. Fungi generally decompose more rapidly than the leaves of trees; and muscle more readily than the cutis.

Bone, the elementary structure of which has only been understood of late years, the hair of animals, the ligneous part of vegetable substances, the shells or husks of fruit, and the feathery calix (*pappus*) of plants, are not inorganic and devoid of life; but approximate, even in life, to the condition which they manifest after their separation from the rest of the organism. The higher the degree of vitality or irritability of an animate substance, the more striking or rapid will be the change in its compound state after separation. "The aggregate of the cells is an organism, and the organism *lives* as long as its parts continue actively subservient to the whole. Considered antithetically to inanimate nature, the organism *appears* to be self-determining."* The difficulty of satisfactorily referring the vital phenomena of organism to physical and chemical laws, depends chiefly (and almost in the same manner as the prediction of meteorological processes in the atmosphere) on the complication of the phenomena, and on the great number of the simultaneously acting forces, as well as the conditions of their activity.

I have faithfully adhered in the *Cosmos* to the same mode of representing and considering the so-called *vital forces*, and affinities,† the formative impulse and the principle of organising activity. I there wrote as follows:‡ "The mythical ideas long entertained of the imponderable substances, and vital forces, peculiar to each mode of organization, have complicated our views generally, and shed an uncertain light on the path we ought to pursue.

"The most various forms of intuition have thus, age after age, aided in augmenting the prodigious mass of empirical knowledge, which in our own day has been enlarged with ever-increasing rapidity. The investigating spirit of man

* Henle, *Allgemeine Anatomie*, 1841, pp. 216—219.

† Pulteney Alison, in the *Transact. of the Royal Soc. of Edinburgh*, vol. xvi. p. 305.

‡ *Cosmos*, vol. i. p. 58. (Bohn's Edition.)

strives, from time to time, with varying success, to break through those ancient forms and symbols invented to subject rebellious matter to rules of mechanical construction."

Further in the same work,* I have said, "It must, however, be remembered, that the inorganic crust of the earth contains within it the same elements that enter into the structure of animal and vegetable organs. A physical cosmography would therefore be incomplete, if it were to omit a consideration of these forces, and of the substances which enter into solid and fluid combinations in organic tissues, under conditions which, from our ignorance of their actual nature, we designate by the vague term of *vital forces*, and group into various systems, in accordance with more or less perfectly conceived analogies."†

* Vol. i. p. 349. (Bohn's Edition.)

† Compare also the critique on the acceptation of special vital forces in Schleiden's *Botanik als inductive Wissenschaft*, part i. pp. 60, and the lately published and admirable treatise of Emil du Bois-Reymond, *Untersuchungen über thierische Elektricität*, vol. i. pp. xxxiv—l.

## THE
# PLATEAU, OR TABLE-LAND,
OF
# CAXAMARCA,
## THE ANCIENT CAPITAL OF THE INCA ATAHUALLPA,
AND THE
## FIRST VIEW OF THE PACIFIC OCEAN,
*From the Ridge of the Andes.*

---

AFTER having sojourned for a whole year on the ridge of the Andes, or Antis, (1), between 4° north and 4° south latitude, amidst the table-lands of New Granada, Pastos, and Quito, and consequently at an elevation varying between 8500 and 13,000 feet above the level of the sea, it is delightful to descend gradually through the more genial climate of the Cinchona or Quina Woods of Loxa, into the plains of the Upper Amazon. There an unknown world unfolds itself, rich in magnificent vegetation. The little town of Loxa has given its name to the most efficacious of all fever barks,—the Quina, or the Cascarilla fina de Loxa. This bark is the precious produce of the tree, which we have botanically described as the Cinchona Condaminea; but which, (from the erroneous supposition that all the Cinchona known in commerce was obtained from one and the same tree,) had previously been called Cinchona officinalis. The fever bark first became known, in Europe, about the middle of the seventeenth century. Sebastian Badus affirms, that it was brought to Alcala de Henares in the year 1632; but according to other accounts, it was brought to Madrid in 1640, when the Countess de Chinchon (2), the wife of the Peruvian Viceroy, arrived from Lima, (where she had been cured of an intermittent fever,) accompanied by her physician, Juan del Vego. The finest kind of Cinchona is obtained at the distance of from eight to twelve miles southward of the town of Loxa, among the mountains of Uritusinga, Villonaco, and Rumisitana. The trees which yield this bark grow on mica slate and gneiss, at the moderate elevations of 5755 and 7673 feet above the level of the sea, nearly corresponding, respectively, with the heights of the Hospital on the Grimsel, and the Pass of the Great St. Bernard. The Cinchona Woods in these parts are bounded by the little rivulets Zamora and Cachyacu.

The tree is felled in its first flowering season, or about the fourth or seventh year of its growth, according as it may have been reared from a strong shoot or from seed. At the time of my journey in Peru we learned, with surprise, that the quantity of the Cinchona Condaminea annually obtained at Loxa by the Cascarilla gatherers, or Quina hunters (*Cascarilleros* and *Caçadores de Quina*), amounted only to 110 hundred weight. At that time none of this valuable product found its way into commerce; all that was obtained was shipped at Payta, a port of the Pacific, and conveyed round Cape Horn to Cadiz, for the use of the Spanish Court. To procure the small supply of 11,000 Spanish pounds, no less than 800 or 900 Cinchona trees were cut down every year. The older and thicker stems are becoming more and more scarce; but, such is the luxuriance of growth, that the younger trees, which now supply the demand, though measuring only six inches in diameter, frequently attain the height of from 53 to 64 feet. This beautiful tree, which is adorned with leaves five inches long and two broad, seems, when growing in the thick woods, as if striving to rise above its neighbours. The upper branches spread out, and when agitated by the wind the leaves have a peculiar reddish colour and glistening appearance which is distinguishable at a great distance. The mean temperature of the woods of the Cinchona Condaminea varies between 60° and 66° Fahrenheit; that is to say, about the mean annual temperature of Florence and the Island of Madeira: but the extremes of heat and cold experienced at those points of the temperate zone, are never felt in the vicinity of Loxa. However, comparisons between climates in very different degrees of latitude, and the climate of the table-lands of the tropical zone, must, from their very nature, be unsatisfactory.

Descending from the mountain node of Loxa, south-south-east, into the hot valley of the Amazon River, the traveller passes over the Paramos of Chulucanas, Guamani, and Yamoca. These Paramos are the mountainous deserts, which have been mentioned in another portion of the present work; and which, in the southern parts of the Andes, are known by the name of Puna, a word belonging to the Quichua language. In most places, their elevation is about 10,125 feet. They are stormy, frequently enveloped for several successive days in thick fogs, or visited by terrific hail-storms; the hail-stones being not only of different forms, generally much flattened by rotation, but also run together into thin floating plates of ice called papa-cara, which cut the face and hands in their fall. During this meteoric process, I have sometimes known the thermometer to sink to 48° and even 43° Fahrenheit, and the electric tension of the atmosphere, measured by the voltaic electrometer, has changed, in the space of a few minutes, from positive to negative. When the temperature is below 43° Fahrenheit, snow falls in large flakes, scattered widely apart; but it disappears after the lapse of a few hours. The short thin branches of the small leaved myrtle-like shrubs, the large size and luxuriance of the blossoms, and the perpetual freshness caused by the absorption of the moist atmosphere—all impart a peculiar aspect and character to the treeless vegetation of the Paramos. No zone of Alpine vegetation, whether in temperate or cold climates, can be compared with that of the Paramos in the tropical Andes.

The solemn impression which is felt on beholding the deserts of the Cordilleras, is increased in a remarkable and unexpected manner, by the circumstance that in these very regions there still exist wonderful remains of the great road of the Incas, that stupendous work by means of which, communication was maintained among all the provinces of the empire along an extent of upwards of 1000 geographical miles. On the sides of this road, and nearly at equal distances apart, there are small houses, built of well-cut freestone. These buildings, which answered the purpose of stations, or caravanseries, are called Tambos, and also Inca-Pilca, (from Pircca, the Wall). Some are surrounded by a sort of fortification; others were destined for baths, and had arrangements for the conveyance of warm water: the larger ones were intended exclusively for the family of the sovereign. At the foot of the volcano Cotopaxi, near Callo, I had previously seen buildings of the same kind in a good state of preservation. These I accurately measured, and made drawings from them. Pedro de Cieça, who wrote in the sixteenth century, calls these structures Aposentos de Mulalo (3). The pass of the Andes, lying between Alausi and Loxa, called the Paramo del Assuay, a much frequented route across the Ladera de Cadlud, is at the elevation of 15,526 feet above the level of the sea, and consequently almost at the height of Mont Blanc. As we were proceeding through this pass, we experienced considerable difficulty in guiding our heavily laden mules over the marshy ground on the level height of the Pullal; but whilst we journeyed onward for the distance of about four miles, our eyes were continually rivetted on the grand remains of the Inca Road, upwards of 20 feet in breadth. This road had a deep under-structure, and was paved with well-hewn blocks of black trap porphyry. None of the Roman roads which I have seen in Italy, in the south of France and in Spain, appeared to me more imposing than this work of the ancient Peruvians; and the Inca road is the more extraordinary, since, according to my barometrical calculations, it is situated at an elevation of 13,258 feet above the level of the sea, a height exceeding that of the summit of the Peak of Teneriffe by upwards of 1000 feet. At an equal elevation, are the ruins said to be those of the palace of the Inca Tupac Yupanqui, and known by the name of the Paredones del Inca, situated on the Assuay. From these ruins the Inca road, running southward in the direction of Cuenca, leads to the small but well-preserved fortress of the Cañar (4), probably belonging to the same period, viz.: the reign of Tupac Yupanqui, or that of his warlike son Huayna Capac.

We saw still grander remains of the ancient Peruvian Inca road, on our way between Loxa and the Amazon, near the baths of the Incas on the Paramo of Chulucanas, not far from Guancabamba, and also in the vicinity of Ingatambo, near Pomahuaca. The ruins at the latter place are situated so low, that I found the difference of level between the Inca road at Pomahuaca, and that in the Paramo del Assuay, to be upwards of 9700 feet. The distance in a direct line, as determined by astronomical latitudes, is precisely 184 miles; and the ascent of the road is about 3730 feet greater than the elevation of the Pass of Mont Cenis, above the Lake of Como. There are two great causeways, paved with flat stones, and in some places covered with cemented gravel (5), on Macadam's plan. One of these lines of road runs through the broad and barren plain lying between the sea-coast and the chain of the Andes, whilst the other passes along the ridge of the Cordilleras. Stones, marking the distances at equal intervals, are frequently seen. The rivulets and ravines were crossed by bridges of three kinds; some being of stone, some of wood, and others of rope. These bridges are called by the Peruvians, Puentes de Hamaca, or Puentes de Maroma. There were also aqueducts for conveying water to the Tambos and fortresses. Both lines of road were directed to Cuzco, the central point and capital of the great Peruvian empire, situated in 13° 31′ south lat., and according to Pentland's Map of Bolivia, at the elevation of 11,378 feet above the level of the sea. As the Peruvians had no wheeled carriages, these roads were constructed for the march of troops, for the conveyance of burthens borne by men, and for flocks of lightly laden Lamas; consequently, long flights of steps (6), with resting-places, were formed at intervals in the steep parts of the mountains. Francisco Pizarro and Diego Almagro, in their expeditions to remote parts of the country, availed themselves with much advantage of the military roads of the Incas; but the steps just mentioned were formidable impediments in the way of the Spanish cavalry, especially as in the early period of the Conquista, the Spaniards rode horses only, and did not make use of the sure-footed mule, which, in mountainous precipices, seems to reflect on every step he takes. It was only at a later period that the Spanish troops were mounted on mules.

Sarmiento, who saw the Inca roads whilst they were in a perfect state of preservation, mentions them in a *Relacion* which he wrote, and which long lay buried in the Library of the Escurial. "How," he asks, "could a people, unacquainted with the use of iron, have constructed such great and magnificent roads, (*caminos tan grandes, y tan sovervios*), and in regions so elevated as the countries between Cuzco and Quito, and between Cuzco and the coast of Chili?" "The Emperor Charles," he adds, "with all his power, could not have accomplished even a part of what was done by the well-directed Government of the Incas, and the obedient race of people under its rule." Hernando Pizarro, the most educated of the three brothers, who expiated his misdeeds by twenty years of captivity in Medina del Campo, and who died at

100 years of age, in the odour of sanctity (*en olor de Santidad*), observes, alluding to the Inca roads: "Throughout the whole of Christendom, no such roads are to be seen as those which we here admire." Cuzco and Quito, the two principal capitals of the Incas, are situated in a direct line south-south-east, north-north-west in reference the one to the other. Their distance apart, without calculating the many windings of the road, is 1000 miles; including the windings of the road, the distance is stated by Garcilaso de la Vega, and other Conquistadores, to be "500 Spanish leguas." Notwithstanding this vast distance, we are informed, on the unquestionable testimony of the Licentiate Polo de Ondegardo, that Huayna Capac, whose father conquered Quito, caused certain materials to be conveyed thither from Cuzco, for the erection of the royal buildings, (the Inca dwellings). In Quito, I found this tradition still current among the natives.

When, in the form of the earth, nature presents to man formidable difficulties to contend against, those very difficulties serve to stimulate the energy and courage of enterprizing races of people. Under the despotic centralizing system of the Inca Government, security and rapidity of communication, especially in relation to the movement of troops, were matters of urgent state necessity. Hence the construction of great roads, and the establishment of very excellent postal arrangements by the Peruvians. Among nations in the most various degrees of civilization, national energy is frequently observed to manifest itself, as it were by preference, in some special direction; but the advancement consequent on this sort of partial exertion, however strikingly exhibited, by no means affords a criterion of the general cultivation of a people. Egyptians, Greeks (7), Etruscans, and Romans, Chinese, Japanese, and Indians, present examples of these contrasts. It would be difficult to determine, what space of time may have been occupied in the execution of the

Peruvian roads. Those great works, in the northern part of the Inca Empire, on the table-land of Quito, must certainly have been completed in less than thirty or thirty-five years; that is to say, in the short interval between the defeat of the Ruler of Quito, and the death of the Inca Huayna Capac. With respect to the southern, or those specially styled the Peruvian roads, the period of their formation is involved in complete obscurity.

The date of the mysterious appearance of Manco Capac is usually fixed 400 years prior to the arrival of Francisco Pizarro, (who landed on the Island of Puná in the year 1532), consequently, about the middle of the twelfth century, and full 200 years before the foundation of the city of Mexico (Tenochtitlan); but instead of 400 years, some Spanish writers represent the interval between Manco Capac and Pizarro to have been 500, or even 550 years. However the history of the Peruvian empire records only thirteen reigning princes of the Inca dynasty, which, as Prescott justly observes, is not a number sufficient to fill up so long a period as 550, or even 400 years. Quezalcoatl, Botchia, and Manco Capac, are the three mythical beings, with whom are connected the earliest traces of cultivation among the Aztecs, the Muyscas, (properly Chibchas), and the Peruvians. Quezalcoatl, who is described as bearded and clothed in black, was High Priest of Tula, and afterwards a penitent, dwelling on a mountain near Tlaxapuchicalco. He is represented as having come from the coast of Panuco; and, therefore, from the eastern part of Anahuac, on the Mexican table-land. Botchia, or rather the bearded, long-robed Nemterequeteba (8), (literally messenger of God, a Buddha of the Muyscas), came from the grassy steppes eastward of the Andes chain, to the table-lands of Bogotá. Before the time of Manco Capac, some degree of civilization already existed on the picturesque shores of the Lake of Titicaca. The fortress of Cuzco, on the hill of Sacsahuaman, was built on the model of the more ancient structures of Tiahuanaco. In like manner, the Aztecs imitated the pyramidal buildings of the Toltecs, and the latter copied those of the Olmecs (Hulmecs); and thus, by degrees, we arrive at historic ground in Mexico as early as the sixth century of the Christian era. According to Siguença, the Toltecic Step Pyramid of Cholula, was copied from the Hulmecic Step Pyramid of Teotihuacan. Thus, through every stage of civilization, we pass into an earlier one, and as human intelligence was not aroused simultaneously in both continents, we find that in every nation the imaginative domain of mythology immediately preceded the period of historical knowledge.

The early Spanish Conquistadores were filled with admiration on first beholding the roads and aqueducts of the Peruvians; yet not only did they neglect the preservation of those great works, but they even wantonly destroyed them. As a natural consequence of the destruction of the aqueducts, the soil was rendered unfertile by the want of irrigation. Nevertheless, those works, as well as the roads, were demolished for the sake of obtaining stones ready hewn for the erection of new buildings; and the traces of this devastation are more observable near the sea-coast, than on the ridges of the Andes, or in the deeply cleft valleys with which that mountain-chain is intersected. During our long day's journey from the syenitic rocks of Zaulac to the valley of San Felipe, (rich in fossil remains and situated at the foot of the icy Paramo of Yamoca), we had no less than twenty-seven times to ford the Rio de Guancabamba, which falls into the Amazon. We were compelled to do this on account of the numerous sinuosities of the stream, whilst on the brow of a steep precipice near us, we had continually within our sight the vestiges of the rectilinear Inca road, with its Tambos. The little mountain stream, the Rio de Guancabamba, is not more than from 120 to 150 feet broad; yet so strong is the current, that our heavily laden mules were in continual danger of being swept away by it. The mules carried our manuscripts, our dried plants, and all the other objects which we had been a whole year engaged in collecting; therefore, every time that we crossed the stream, we stood on one of the banks in a state of anxious suspense until the long train of our beasts of burthen, eighteen or twenty in number, were fairly out of danger.

This same Rio de Guancabamba, which in the lower part of its course has many falls, is the channel for a curious mode of conveying correspondence from the coast of the Pacific. For the expeditious transmission of the few letters that are sent from Truxillo to the province of Jaen de Bracamoros, they are despatched by a swimming courier, or, as he is called by the people of the country, "*el correo que nada.*" This courier, who is usually a young Indian, swims in two days from Pomahuaca to Tomependa; first proceeding by the Rio de Chamaya, (the name given to the lower part of the Rio de Guancabamba) and then by the Amazon river. The few letters of which he is the bearer, he carefully wraps in a large cotton handkerchief, which he rolls round his head in the form of a turban. On arriving at those parts of the rivers in which there are falls or rapids, he lands, and goes by a circuitous route through the woods. When wearied by long-continued swimming, he rests by throwing one arm on a plank of a light kind of wood of the family of the Bombaceæ, called by the Peruvians *Ceiba*, or *Palo de balsa.* Sometimes the swimming courier takes with him a friend to bear him company. Neither troubles himself about provisions, as they are always sure of a hospitable reception in the huts which are surrounded by abundant fruit-trees in the beautiful Huertas of Pucara and Cavico.

Fortunately, the river is free from crocodiles, which are first met with in the upper course of the Amazon, below the cataract of Mayasi; for the slothful animal prefers to live in the more tranquil waters. According to my calculation, the Rio de Chamaya has a fall (9) of 1778 feet, in the short distance of 52 geographical miles; that is to say, measuring from the Ford (*Paso*) de Pucara, to the point where the Chamaya disembogues in the river Amazon, below the village of Choros. The Governor of the province Jaen de Bracamoros assured me, that letters sent by the singular water post conveyance just mentioned, are seldom either wetted or lost. After my return from Mexico, I myself received, when in Paris, letters from Tomependa, which had been transmitted in this manner. Many of the wild Indian tribes, who dwell on the shores of the Upper Amazon, perform their journeys in a similar manner; swimming sociably down the stream in parties. On one occasion, I saw the heads of thirty or forty individuals, men, women, and children, of the tribe of the Xibaros, as they floated down the stream on their way to Tomependa. The *Correo que nada* returns by land, taking the difficult route of the Paramo del Paredon.

On approaching the hot climate of the basin of the Amazon, the aspect of beautiful and occasionally very luxuriant vegetation delights the eye. Not even in the Canary Islands, nor on the warm coasts of Cumana and Caracas, had we beheld finer orange-trees than those which we met with in the Huertas de Pucara. They consisted chiefly of the sweet orange-tree (*Citrus aurantium*, Risso); the bitter orange-tree (*Citrus vulgaris*, Risso) was less numerous. These trees, laden with their golden fruit in thousands, attain there a height of between 60 and 70 feet; and their branches, instead of growing in such a way as to give the trees rounded tops or crowns, shoot straight up like those of the laurel. Near the ford of Cavico a very unexpected sight surprised us. We saw a grove of small trees, about 18 or 19 feet high, the leaves of which, instead of being green, appeared to be of a rose colour. This proved to be a new species of Bougainvillæa, a genus first determined by Jussieu the elder, from a Brazilian specimen in Commerson's *Herbarium*. But on a nearer approach we found that these trees were really without leaves, properly so called, and that what, from a distant view, we had mistaken for leaves, were bright rose-coloured bracts. Owing to the purity and freshness of the colour, the effect was totally different from that of the hue which so pleasingly clothes many of our forest-trees in autumn. The Rhopala ferruginea, a species of the South African family of the Proteaceæ, has found its way hither, having descended from the cool heights of the Paramo de Yamoca into the warm plains of the Chamaya. We likewise frequently saw here the beautifully pinnated Porlieria hygrometrica, one of the Zygophylleæ, which, by the closing of its leaves, indicates change of weather, generally the approach of rain. This plant is more certain in its tokens than any of the Mimosaceæ, and it very rarely deceived us.

At Chamaya we found rafts (*balsas*) in readiness to convey us to Tomependa, where we wished to determine the difference of longitude between Quito and the mouth of the Chinchipe; a point of some importance to the geography of South America on account of an old observation of La Condamine (10). We slept as usual in the open air, and our resting-place was on the sandy shore called the Playa de Guayanchi, at the confluence of the Rio de Chamaya and the Amazon. Next morning we proceeded down the latter river as far as the Cataract and the Narrows, or the Pongo of Rentema. Pongo, the name given to River Narrows by the natives, is a corruption of the word *Puncu*, which, in the Quichua language, signifies a door or gate. In the Pongo de Rentema huge masses of rock consisting of coarse-grained sandstone (conglomerate), rise up like towers and form a rocky dam across the stream. I measured a base line on the flat sandy shore, and found that the Amazon River, which, further eastwards, spreads into such mighty width, is, at Tomependa, scarcely 1400 feet broad. In the celebrated River Narrows, called the Pongo de Manseriche, between Santiago and San Borja, the breadth is less than 160 feet. The Pongo de Manseriche is formed by a mountain ravine, in some parts of which the overhanging rocks, roofed by a canopy of foliage, permit only a feeble light to penetrate, and by the force of the current all the drift-wood, consisting of trunks of trees in countless numbers, is broken and dashed to atoms. The rocks by which all these Pongos are formed, have, in the course of centuries, undergone many changes. The Pongo de Rentema, which I have mentioned above, was, a year before my visit to it, in part broken up by a high flood; indeed the inhabitants of the shores of the Amazon still preserve by tradition a lively recollection of the sudden fall of the once lofty masses of rock along the whole length of the Pongo. This fall took place in the early part of the last century, and the debris suddenly dammed up the river and impeded the current. The consequence was, that the inhabitants of the village of Puyaya, situated at the lower part of the Pongo de Rentema, were filled with alarm on beholding the dry bed of the river; but, after the lapse of a few hours, the waters recovered their usual course. There appears to be no reason for believing that these remarkable phenomena are occasioned by earthquakes. The river, which has a very strong current, seems, as it were, to be incessantly labouring to improve its bed. Of the force of its efforts some idea may be formed from the fact that, notwithstanding its vast breadth, it sometimes rises upwards of 26 feet above its ordinary level in the space of 20 or 30 hours.

We remained seventeen days in the hot valley of the Marañon or the Amazon River. To proceed from thence to the coast of the Pacific it is necessary to cross the chain of the Andes, between Micuipampa and Caxamarca (in 6° 57 S. lat., and 78° 34′ W. long.), at a point where, according to my observations, it is intersected by the magnetic equator. At a still higher elevation are situated the celebrated silver mines of

Chota. Then, after having passed the ancient Caxamarca (the scene, 316 years ago, of the most sanguinary drama in the history of the Spanish Conquista), and also Aroma and Gangamarca, the route descends, with some interruptions, into the Peruvian lowlands. Here, as in nearly all parts of the Andes, as well as of the Mexican Mountains, the highest points are picturesquely marked by tower-like masses of erupted porphyry and trachyte, the former frequently presenting the effect of immense columns. In some places these masses give a rugged cliff-like aspect to the mountain ridges; and in other places they assume the form of domes or cupolas. They have here broken through a formation, which, in South America, is extensively developed on both sides of the equator, and which Leopold von Buch, after profound research, has pronounced to be cretaceous. Between Guambos and Montan, nearly 12,800 feet above the level of the sea, we found marine fossils (11) (Ammonites about 15 inches in diameter, the large Pecten alatus, oyster shells, Echini, Isocardias, and Exogyra polygona). A species of Cidaris, which, in the opinion of Leopold von Buch, does not differ from one found by Brongniart in the old chalk at the Perte du Rhone, we collected in the basin of the Amazon at Tomependa, and likewise at Micuipampa; that is to say, at elevations differing the one from the other by no less than 10,550 feet. In like manner, in the Amuich chain of the Caucasian Daghestan, the chalk of the banks of the Sulak, scarcely 530 feet above the level of the sea, is again found on the Tchunum, at the elevation of full 9600 feet, whilst, on the summit of the Shadagh Mountain, 13,950 feet high, the Ostrea diluviana (Goldf.), and the same chalk, present themselves. Abich's admirable Caucasian observations furnish the most decided confirmation of Leopold von Buch's geognostic views respecting the cretaceous Alpine development.

From the solitary farm of Montan, surrounded with flocks of Lamas, we ascended further southward the eastern declivity of the Cordilleras, until we reached the level height in which is situated the argentiferous mountain Gualgayoc, the principal site of the far-famed mines of Chota. Night was just drawing in, and an extraordinary spectacle presented itself to our observation. The Cerro de Gualgayoc is separated by a deep cleft-like valley (Quebrada), from the limestone mountain Cormolache. The latter is an isolated hornstone rock, presenting, on the northern and western sides, almost perpendicular precipices, and containing innumerable veins of silver, which frequently intersect and run into each other. The highest shafts are 1540 feet above the floor of the stoll or ground-work, called the Socabon de Espinachi. The outline of the mountain is broken by numerous tower-like points and pyramidal notches; and hence the summit of the Cerro de Gualgayoc bears the name of Las Puntas. This mountain presents a most decided contrast to that smoothness of surface which miners are accustomed to regard as characteristic of metalliferous districts. "Our mountain," said a wealthy mine-owner whom we visited, "looks like an enchanted castle (*como si fuese un castillo encantado*)." The Gualgayoc bears some resemblance to a cone of dolomite, but it is still more like the notched ridges of the Mountain of Monserrat in Catalonia, which I have also visited, and which has been so pleasingly described by my brother. Not only is the silver mountain Gualgayoc perforated on every side, and to its very summit, by many hundred large shafts, but the mass of the siliceous rock is cleft by natural openings, through which the dark blue sky of these elevated regions is visible to the observer standing at the foot of the mountain. The people of the country call these openings windows (*Las ventanillas de Gualgayoc*). On the trachytic walls of the volcano of Pichincha similar openings were pointed out to us, and there, likewise, they were called windows, (*Ventanillas de Pichincha.*) The singular aspect of the Gualgayoc is not a little increased by numerous sheds and habitations, which lie scattered like nests over the fortress-looking mountain wherever a level spot admits of their erection. The miners carry the ore in baskets, down steep and dangerous footpaths, to the places where it is submitted to the process of amalgamation.

The value of the silver obtained from the mines of Gualgayoc during the first thirty years of their being worked, from 1771 to 1802, is supposed to have amounted to upwards of thirty-two millions of piastres. Notwithstanding the hardness of the quartzose rock, the Peruvians, even before the arrival of the Spaniards, extracted rich argentiferous galena from the Cerro de la Lin, and also from the Chupiquiyacu; of this fact many old shafts and galleries bear evidence. The Peruvians also obtained gold from the Curimayo, where also natural sulphur is found in the quartz rock as well as in the Brazilian Itacolumite. We took up our temporary abode, in the vicinity of the mines, in the little mountain town of Micuipampa, situated at an elevation of 11,873 feet above the sea, and where, though only 6° 43′ from the equator, water freezes within doors, at night, during a great part of the year. This wilderness, almost devoid of vegetation, is inhabited by 3000 or 4000 persons, who are supplied with articles of food from the warm valleys, as they themselves can grow nothing but some kinds of cabbage and salad, the latter exceedingly good. Here, as in all the mining towns of Peru, *ennui* drives the richer inhabitants, who, however, are not the best informed class, to the dangerous diversions of cards and dice. The consequence is, that the wealth thus quickly won is still more quickly spent. Here one is continually reminded of the anecdote related of one of the soldiers of Pizarro's army, who complained that he had lost in one night's play, "a large piece of the sun," meaning a plate of gold which he had obtained at the plunder of the Temple of Cuzco. At Micuipampa the thermometer, at eight in the morning, stood at 34°.2, and at noon, at 47°.8 Fahrenheit. Among the thin Ichhu-grass (possibly our Stipa eriostachya), we found a beautiful Calceolaria (*C. Sibthorpioides*), which we should not have expected to see at such an elevation.

Near the town of Micuipampa there is a high plain called the Llano or the Pampa de Navar. In this plain there have been found, extending over a surface of more than four English square miles, and immediately under the turf, immense masses of red gold ore and wire-like threads of pure silver. These are called by the Peruvian miners *remolinos*, *clavos*, and *vetas manteadas*, and they are overgrown by the roots of the Alpine grasses. Another level plain, to the west of the Purgatorio, and near the Quebrada de Chiquera, is called the Choropampa (the Muscle-Shell Plain), the word *churu* signifying in the Quichua language a muscle or cockle, particularly a small eatable kind, which the people of the country now distinguish by their Spanish names *hostion* or *mexillon*. The name Choropampa refers to fossils of the cretaceous formation, which in this plain are found in such immense numbers that at an early period they attracted the attention of the natives. In the Choropampa there has been found near the surface of the earth, a rich mass of pure gold, spun round, as it were, with threads of silver. This fact proves how slight may be the affinity between many of the ores upheaved from the interior of the earth, through fissures and veins, and the nature of the adjacent rock, and how little relative antiquity exists between them and that of the formation they have broken through. The rock of the Gualgayoc, as well as that of the Fuentestiana, is very watery, whilst in the Purgatorio perfect dryness prevails. In the Purgatorio, notwithstanding the height of the strata above the sea-level, I found to my astonishment, that the temperature in the mine was 67°.4 Fahr., whilst in the neighbouring Mina de Guadalupe the water in the mine was about 52°.2 Fahr. In the open air the thermometer indicates only 42°.1 Fahr., and the miners, who labour very hard, and who work almost without clothing, say that the subterranean heat in the Purgatorio is stifling.

The narrow path from Micuipampa to the ancient Inca city Caxamarca is difficult even for mules. The original name of the town was Cassamarca or Kazamarca, that is to say, the City of Frost. Marca, in the signification of a district or town, belongs to the northern dialect of the Chinchaysuyo, or the Chinchasuyu, whilst in the common Quichua language the word means the story of a house, and also a fortress and place of defence. For the space of five or six miles, the road led us through a succession of Paramos, where we were without intermission exposed to the fury of a boisterous wind and the sharp angular hail peculiar to the ridges of the Andes. The height of the road is for the most part between 9600 and 10,700 feet above the sea-level. There I had the opportunity of making a magnetic observation of general interest, viz., for determining the point where the north inclination of the needle passes into the south inclination, and also the point at which the traveller has to cross the magnetic equator (12).

Having at length reached the last of these mountain wildernesses, the Paramo de Yanaguanga, the traveller joyfully looks down into the fertile valley of Caxamarca. It presents a charming prospect, for the valley, through which winds a little serpentine rivulet, is an elevated plain of an oval form, in extent from 96 to 112 square miles. The plain bears a resemblance to that of Bogota, and like it is probably the bed of an ancient lake; but in Caxamarca there is wanting the myth of the miracle-working Botchia, or Idacanzas, the High Priest of Iraca, who opened a passage for the waters through the rocks of Tequendama. Caxamarca lies 640 feet higher than Santa Fé de Bogota, and consequently its elevation is equal to that of the city of Quito; but being sheltered by surrounding mountains, its climate is much more mild and agreeable. The soil of

Caxamarca is extraordinarily fertile. In every direction are seen cultivated fields and gardens, intersected by avenues of willows, varieties of the Datura (bearing large red, white, and yellow flowers), Mimosas, and beautiful Quinuar trees (our Polylepsis villosa, a Rosacea approximating to the Alchemilla and Sanguisorba). The wheat harvest in the Pampa de Caxamarca is, on the average, from fifteen to twenty-fold; but the prospect of abundant crops is sometimes blighted by night frosts, caused by the radiation of heat towards the cloudless sky, in the strata of dry and rarefied mountain air. These night frosts are not felt within the roofed dwellings.

Small mounds, or hillocks, of porphyry (once perhaps islands in the ancient lake) are studded over the northern part of the plain, and break the wide expanse of smooth sandstone. From the summit of one of these porphyry hillocks, we enjoyed a most beautiful prospect of the Cerro de Santa Polonia. The ancient residence of Atahuallpa is on this side, surrounded by fruit gardens, and irrigated fields of lucern (Medicago sativa), called by the people here *Campos de alfalfa*. In the distance are seen columns of smoke, rising from the warm baths of Pultamarca, which still hear the name of Baños del Inca. I found the temperature of these sulphuric springs to be 156°.2 Fahr. Atahuallpa was accustomed to spend a portion of each year at these batl.s, where some slight remains of his palace have survived the ravages of the Conquistadores. The large deep basin or reservoir (*el tragadero*) for supplying these baths with water, appeared to me, judging from its regular circular form, to have been artificially cut in the sandstone rock, over one of the fissures whence the spring flows. Tradition records that one of the Inca's sedan-chairs, made of gold, was sunk in this basin, and that all endeavours to recover it have proved vain.

Of the fortress and palace of Atahuallpa, there also remain but few vestiges in the town, which now contains some beautiful churches. Even before the close of the sixteenth century, the thirst for gold accelerated the work of destruction, for, with the view of discovering hidden treasures, walls were demolished and the foundations of buildings recklessly undermined. The Inca's palace is situated on a hill of porphyry, which was originally cut and hollowed out from the surface, completely through the rock, so that the latter surrounds the main building like a wall. Portions of the ruins have been converted to the purposes of a town jail and a Municipal Hall (Casa del Cabildo). The most curious parts of these ruins, which however are not more than between 13 and 16 feet in height, are those opposite to the monastery of San Francisco. These vestiges, like the remains of the dwelling of the Caciques, consist of finely-hewn blocks of free-stone, two or three feet long, laid one upon another without cement, as in the Inca-Pilca, or fortress of the Cañar, in the high plain of Quito.

In the porphyritic rock there is a shaft which once led to subterraneous chambers and into a gallery, (by miners called a stoll,) from which, it is alleged, there was a communication with the other porphyritic rocks already mentioned;—those situated at Santa Polonia. These arrangements bear evidence of having been made as precautions against the events of war, and for the security of flight. The burying of treasure was a custom very generally practised among the Peruvians in former times; and subterraneous chambers still exist beneath many private dwellings in Caxamarca.

We were shown some steps cut in the rock, and the foot-bath used by the Inca (*el lavatorio de los pies*). The operation of washing the sovereign's feet was performed amidst tedious court ceremonies (13). Several lateral structures, which, according to tradition, were allotted to the attendants of the Inca, are built some of free-stone with gable roofs, and others of regularly shaped bricks, alternating with layers of siliceous cement. The buildings constructed in this last-mentioned style, to which the Peruvians give the name of *Muros y obra de tapia*, have little arched niches or recesses. Of their antiquity I was for a long time doubtful, though I am now convinced that my doubts were not well-grounded.

In the principal building, the room is still shown in which the unfortunate Atahuallpa was confined for the space of nine months, from the date of November, 1532 (14). The notice of the traveller is still directed to the wall, on which he made a mark to denote to what height he would fill the room with gold, on condition of his being set free. This height is variously described. Xerez in the *Conquista del Peru* (which Barcia has preserved to us), Hernando Pizarro in his letters, and other writers, all give different accounts of it. The captive monarch said, "that gold in bars, plates, and vessels should be piled up as high as he could reach with his hand." The dimensions of the room, as given by Xerez, are equivalent to 23 feet in length and 18 in breadth. Garcilaso de la Vega, who quitted Peru in 1560, in his twentieth year, estimates that the treasures brought from the temples of the Sun in Cuzco, Huaylas, Huamachuco, and Pachacamac, up to the fatal 29th of August, 1533, the day of the Inca's death, amounted to 3,838,000 ducados de oro (15).

In the chapel of the town jail, which, as I have mentioned above, is erected on the ruins of the Inca Palace, a stone, stained, as it is alleged, with "indelible spots of blood," is viewed with horror by the credulous. It is placed in front of the altar, and consists of an extremely thin slab, about 13 feet in length, probably a portion of the porphyry or trachyte of the vicinity. To make an accurate examination of this stone, by chipping a piece off, would not be permitted. The three or four spots, said to be blood stains, appear in reality to be nothing but hornblende and pyroxide run together in the fundamental mass of the rock. The Licentiate Fernando Montesinos, though he visited Peru scarcely a hundred years after the taking of Caxamarca, gave currency to the fabulous story that Atahuallpa was beheaded in prison, and that traces of blood were still visible on a stone on which the execution had taken place. There appears no reason to question the fact, since it is borne out by the testimony of many eye-witnesses, that the Inca willingly allowed himself to be baptized by his cruel and fanatical persecutor, the Dominican monk, Vicente de Valverde. He received the name of Juan de Atahuallpa, and submitted to the erecmony of baptism to avoid being burnt alive. He was put to death by strangulation (*el garrote*), and his execution took place publicly in the open air. Another tradition relates that a chapel was erected above the stone on which Atahuallpa was strangled, and that the remains of the Inca repose beneath that stone. Supposing this to be correct, the alleged spots of blood are not accounted for. The fact is, however, that the body was never deposited under the stone in question. After the performance of a mass for the dead and other solemn funeral ceremonies, at which the brothers Pizarro were present in deep mourning (!), the body was conveyed first to the cemetery of the Convento de San Francisco, and afterwards to Quito, Atahuallpa's birthplace. This removal to Quito was in compliance with the wish expressed by the Inca prior to his death. His personal enemy, the crafty Rumiñavi, from artful political motives, caused the body to be interred in Quito with great solemnity. Rumiñavi (literally the stone-eye) received this name from a defect in one of his eyes, occasioned by a wart. (In the Quichua language *rumi* signifies stone, and *ñavi* eye.)

Descendants of the Inca still dwell in Caxamarca, amidst the dreary architectural ruins of departed splendour. These descendants are the family of the Indian Cacique, or, as he is called in the Quichua language, the Curaca Astorpilca. They live in great poverty, but nevertheless contented and resigned to their hard and unmerited fate. Their descent from Atahuallpa, through the female line, has never been a doubtful question in Caxamarca; but traces of beard would seem to indicate some admixture of Spanish blood. Huascar and Atahuallpa, two sons of the great Huayna Capac (who for a child of the Sun was somewhat disposed to free-thinking) (16), reigned in succession before the invasion of the Spaniards. Neither of these two princes left any acknowledged male heirs. In the plains of Quipaypan, Huascar was made prisoner by Atahuallpa, by whose order he was shortly after secretly put to death. Atahuallpa had two other brothers. One was the insignificant youth Toparca, who in the autumn of 1533 Pizarro caused to be crowned as Inca; and the other was the enterprising Manco Capac, who was likewise crowned, but who afterwards rebelled: neither of these two princes left any known male issue. Atahuallpa indeed left two children; one a son, who received in Christian baptism the name of Don Francisco, and who died young; the other a daughter, Doña Angelina, who became the mistress of Francisco Pizarro, with whom she led a wild camp life. Doña Angelina had a son by Pizarro, and to this grandson of the slaughtered monarch the Conqueror was fondly attached. Besides the family of Astorpilca, with whom I became acquainted in Caxamarca, the families of Carguaraicos and Titu-Buscamayca were, at the time I visited Peru, regarded as descendants of the Inca dynasty. The race of Buscamayca has since that time become extinct.

The son of the Cacique Astorpilca, an interesting and amiable youth of seventeen, conducted us over the ruins of the ancient palace. Though living in the utmost poverty, his imagination was filled with images of the subterranean splendour and the golden treasures which, he assured us, lay hidden beneath the heaps of rubbish over which we were treading. He told us that one of his ancestors once blindfolded the eyes of his wife, and then, through many intricate passages cut in the rock, led her down into the subterranean gardens of the Inca. There the lady beheld, skilfully imitated in the purest gold, trees laden with leaves and fruit, with birds perched on their branches. Among other things, she saw Atahuallpa's gold sedan-chair (*una de las andas*) which had been so long searched for in vain, and which is alleged to have sunk in the basin at the Baths of Pultamarca. The husband commanded his wife not to touch any of these enchanted treasures, reminding her that the period fixed for the restoration of the Inca empire had not yet arrived, and that whosoever should touch any of the treasures would perish that same night. These golden dreams and fancies of the youth were founded on recollections and traditions transmitted from remote times. Golden gardens, such as those alluded to (*Jardines ó huertas de oro*), have been described by various writers who allege that they actually saw them; viz., by Cieza de Leon, Parmento, Garcilaso, and other early historians of the Conquista. They are said to have existed beneath the Temple of the Sun at Cuzco, at Caxamarca, and in the lovely valley of Yucay, which was a favourite seat of the sovereign family. In places in which the golden Huertas were not under ground, but in the open air, living plants were mingled with the artificial ones. Among the latter, particular mention is always made of the high shoots of maize and the maize-cobs (*mazorcas*) as having been most successfully imitated.

The son of Astorpilca assured me that underground, a little to the right of the spot on which I then stood, there was a large Datura tree, or Guanto, in full flower, exquisitely made of gold wire and plates of gold, and that its branches overspread the Inca's chair. The morbid faith with which the youth asserted his belief in this fabulous story, made a profound and melancholy impression on me. These illusions are cherished among the people here, as affording them consolation amidst great privation and earthly suffering. I said to the lad, "Since you and your parents so firmly believe in the existence of these gardens, do you not, in your poverty, sometimes feel a wish to dig for the treasures that lie so near you?" The young Peruvian's answer was so simple and so expressive of the quiet resignation peculiar to the aboriginal inhabitants of the country, that I noted it down in Spanish in my Journal. "Such a desire (*tal antojo*)," said he, "never comes to us. My father says that it would be sinful (*que fuese pecado*). If we had the golden branches, with all their golden fruits, our white neighbours would hate us and injure us. We have a little field and good wheat (*buen trigo*)." Few of my readers will I trust be displeased that I have recalled here the words of young Astorpilca and his golden dreams.

An idea generally spread and firmly believed among the natives is, that it would be criminal to dig up and take possession of treasures which may have belonged to the Incas, and that such a proceeding would bring misfortune upon the whole Peruvian race. This idea is closely connected with that of the restoration of the Inca dynasty, an event which is still expected, and which in the sixteenth and seventeenth centuries was looked forward to with especial confidence. Oppressed nations always fondly hope for the day of their emancipation, and for the re-establishment of their old forms of government. The flight of Manco Inca, the brother of Atahuallpa, who retreated into the forests of Vilcapampa, on the declivity of the Eastern Cordillera; and the abode of Sayri Tapac and Inca Tupac Amaru in those wildernesses, are events which have left lasting recollections in the minds of the people. It is believed that descendants of the dethroned dynasty settled still further eastward in Guiana, between the rivers Apurimac and Beni. These notions were strengthened by the myth of *el Dorado* and the golden city of Manoa, which popular credulity carried from the west and propagated eastward. So greatly was the imagination of Sir Walter Raleigh inflamed by these dreams, that he raised an expedition in the hope of conquering "the imperial and golden city." There he proposed to establish a garrison of three or four thousand English, and to levy from "the Emperor of Guiana, a descendant of Huayna Capac, and who holds his Court with the same magnificence, an annual tribute of £300,000 sterling, as the price of the promised restoration to the throne in Cuzco and Caxamarca." Wherever the Peruvian Quichua language prevails, traces of the expected restoration of the Inca rule (17) exist in the minds of many of the natives possessing any knowledge of their national history.

We remained five days in the capital of the Inca Atahuallpa, which, at that time, numbered only 7000 or 8000 inhabitants. Our departure was delayed by the necessity of obtaining a great number of mules to convey our collections, and of selecting careful guides to conduct us across the chain of the Andes to the entrance of the long but narrow Peruvian sandy desert called the *Desierto de Sechura.* Our route across the Cordilleras lay from north-east to south-west. Having passed over the old bed of the lake, on the pleasant level height of Caxamarca, we ascended an eminence at an elevation of scarcely 10,230 feet: and we were then surprised by the sight of two strangely-shaped porphyritic mounds called the Aroma and the Cunturcaga. The latter is a favourite haunt of the gigantic vulture, which we call the Condor; *kacca*, in the Quichua language, signifying *the rocks.* The porphyritic heights just mentioned are in the form of columns having five, six, or seven sides, from 37 to 42 feet in height, and some of them are crooked and bent as if in joints. Those which crown the Cerro Aroma are remarkably picturesque. The peculiar distribution of the columns, which are ranged in rows one above another, and frequently converging, presents the appearance of a two-storied building, roofed by a dome of massive rock, which is not columnar. These erupted masses of porphyry and trachyte are, as I have on a former occasion remarked, characteristic of the ridges of the Andes, to which they impart a physiognomy totally different from that of the Swiss Alps, the Pyrenees, and the Siberian Altai.

From Cunturcaga and Aroma we descended, by a zigzag route, a steep declivity of 6400 feet into the cleft-like valley of the Magdalena, the lowest part of which is 4260 feet above the sea level. Here there is an Indian village consisting of a few miserable huts, surrounded by the same species of cotton-trees (*Bombax discolor*), which we first observed on the banks of the Amazon. The scanty vegetation of the valley of Magdalena somewhat resembles that of the province of Jaen de Bracamoros, but we missed, with regret, the red groves of Bougainvillæa. Magdalena is one of the deepest valleys I have seen in the chain of the Andes. It is a decided cleft, running transversely from east to west, and bounded on each side by the Altos of Aroma and Guangamarca. Here recommences the same quartz formation which was so long enigmatical to me. We had previously observed it in the Paramo de Yanaguanga, between Micuipampa and Caxamarca, at an elevation of 11,722 feet, and on the western declivity of the Cordillera it attains the thickness of many thousand feet. Since Leopold von Buch has proved that the cretaceous formation is widely extended, even in the highest chains of the Andes, and on both sides of the isthmus of Panama, it may be concluded that the quartz formation, of which I have just made mention (perhaps transformed in its texture by the action of volcanic power), belongs to the free sandstone intervening between the inner chalk and the gault and greensand. From the genial valley of the Magdalena we again proceeded westward, and, for the space of two hours and a half, we ascended a steep wall of rock 5116 feet high, which rises opposite to the porphyritic groups of the Alto de Aroma. In this ascent we felt the change of temperature the more sensibly, as the rocky acclivity was frequently overhung with cold mist.

After having travelled for eighteen months without inter

mission, within the restricted boundaries of the interior of a mountainous country, we felt an ardent desire to enjoy a view of the open sea, a desire which was heightened by repeated disappointments. Looking from the summit of the volcano of Pichincha, over the thick forests of the Provincia de las Esmeraldas, no sea horizon is distinctly discernible owing to the great distance and the height of the point of view. It is like looking down from a balloon into empty space; the fancy divines objects which the eye cannot distinguish. Afterwards, when, between Loxa and Guancabamba, we arrived at the Paramo de Guamani (where there are many ruins of buildings of the times of the Incas), our mule-drivers confidently assured us that, beyond the plain, on the other side of the low districts of Piura and Lambajeque, we should have a view of the sea. But a thick mist overhung the plain and obscured the distant coast. We beheld only variously-shaped masses of rock, now rising like islands above the waving sea of mist, and now vanishing. It was a view similar to that which we had from the Peak of Teneriffe. We experienced a similar disappointment whilst proceeding through the Andes Pass of Guangamarca, which I am now describing. Whilst we toiled along the ridges of the mighty mountain, with expectation on the stretch, our guides, who were not very well acquainted with the way, repeatedly assured us that, after proceeding another mile, our hopes would be fulfilled. The stratum of mist, in which we were enveloped, seemed sometimes to disperse for a moment, but whenever that happened, our view was bounded by intervening heights.

The desire which we feel to behold certain objects is not excited solely by their grandeur, their beauty, or their importance. In each individual this desire is interwoven with pleasing impressions of youth, with early predilections for particular pursuits, with the inclination for travelling, and the love of an active life. In proportion as the fulfilment of a wish may have appeared improbable, its realization affords the greater pleasure. The traveller enjoys, in anticipation, the happy moment when he shall first behold the constellation of the Cross, and the Magellanic clouds circling over the South Pole; when he shall come in sight of the snow of the Chimborazo, and of the column of smoke ascending from the volcano of Quito; when, for the first time, he shall gaze on a grove of tree-ferns, or on the wide expanse of the Pacific Ocean. The days on which such wishes are fulfilled mark epochs in life, and create indelible impressions; exciting feelings which require not to be accounted for by any process of reasoning. The longing wish I felt to behold the Pacific from the lofty ridges of the Andes was mingled with recollections of the interest with which, as a boy, I had dwelt on the narrative of the adventurous expedition of Vasco Nunez de Balboa (18). That happy man, whose track Pizarro followed, was the first to behold, from the heights of Quarequa, on the isthmus of Panama, the eastern part of the great "South Sea." The reedy shores of the Caspian, viewed from the point whence I first beheld them, viz., from the Delta formed by the mouths of the Volga, cannot certainly be called picturesque, yet the delight I felt on first beholding them, was enhanced by the recollection that, in my very earliest childhood, I had been taught to observe, on the map, the form of the Asiatic inland sea. The impressions aroused within us in early childhood, or excited by the accidental circumstances of life (19), frequently, in after years, take a graver direction, and become stimulants to scientific labours and great enterprises.

After passing over many undulations of ground, on the rugged mountain ridges, we at length reached the highest point of the Alto de Guangamarca. The sky, which had so long been obscured, now suddenly brightened. A sharp southwest breeze dispersed the veil of mist; and the dark blue canopy of heaven was seen between the narrow lines of the highest feathery clouds. The whole western declivity of the

Cordillera (adjacent to Chorillos and Cascas), covered with huge blocks of quartz 13 or 15 feet long; and the plains of Chala and Molinos, as far as the sea coast near Truxillo, lay extended before our eyes, with a wonderful effect of apparent proximity. We now, for the first time, commanded a view of the Pacific. We saw it distinctly; reflecting along the line of the coast an immense mass of light, and rising in immeasurable expanse until bounded by the clearly-defined horizon. The delight which my companions, Bonpland and Carlos Montufar, shared with me in viewing this prospect, caused us to forget to open the barometer on the Alto de Guangamarca. According to a calculation which we made at a place somewhat lower down (an isolated farm called the Hato de Guangamarca), the point at which we first gained a view of the ocean, must have been at no greater an elevation than between 9380 and 9600 feet.

The view of the Pacific was solemnly impressive to one, who, like myself, was greatly indebted for the formation of his mind, and the direction given to his tastes and aspirations, to one of the companions of Captain Cook. I made known the general outline of my travelling schemes to John Forster, when I had the advantage of visiting England under his guidance, now more than half a century ago. Forster's charming pictures of Otaheite had awakened throughout Northern Europe a deep interest (mingled with a sort of romantic longing), in favour of the islands of the Pacific Ocean. At that period, when but few Europeans had been fortunate enough to visit those islands, I cherished the hope of seeing them, at least in part; for the object of my visit to Lima was twofold: first, to observe the transit of Mercury over the solar disc, and secondly, to fulfil a promise I had made to Captain Baudin, on my departure from Paris. This promise was to join him in the circumnavigatory voyage which he was to undertake as soon as the French Republic could furnish the necessary funds.

American papers circulated in the Antilles announced that the two French corvettes, *Le Géographe* and *Le Naturaliste*, were to sail round Cape Horn, and to touch at Callao de Lima. This information, which I received when in the Havannah, after having completed my Orinoco journey, caused me to relinquish my original plan of proceeding through Mexico to the Philippines. I lost no time in engaging a ship to convey me from Cuba to Cartagena de Indias. But Captain Baudin's expedition took quite a different course from that which had been expected and announced. Instead of proceeding by the way of Cape Horn, as had been intended at the time when it was agreed that Bonpland and I should join it, the expedition sailed round the Cape of Good Hope. One of the objects of my visit to Peru, and of my last journey across the chain of the Andes, was thus thwarted; but I had the singular good fortune, at a very unfavourable season of the year, in the misty regions of Lower Peru, to enjoy a clear bright day. In Callao I observed the passage of Mercury over the sun's disc, an observation of some importance in aiding the accurate determination of the longitude of Lima (20), and of the south-western part of the new continent. Thus, amidst the serious troubles and disappointments of life, there may often be found a grain of consolation.

---

## ILLUSTRATIONS AND ADDITIONS.

### (1) p. 390—"*On the Ridge of the Andes or Antis.*"

The Inca Garcilaso, who was well acquainted with the native language of his country, and who loved to trace etymologies, invariably calls the chain of the Andes, "las Montañas de los Antis." He states positively that the great mountain-chain, eastward of Cuzco, derives its name from the race of the Antis and from the province Anti, which was situated to the east of the capital of the Incas. The quaternary divisions of the Peruvian empire, according to the four cardinal points, reckoning from Cuzco, did not derive their names from the very circumstantial words (having reference to the sun) which in the Quichua language signify east, west, north, and south (intip llucsinanpata, intip yaucunanpata, intip chaututa chayananpata, intip chaupunchau chayananpata). Those divisions were named from provinces and races of people (Provincias llamadas Anti, Cunti, Chincha y Colla) situated to the east, west, north, and south, with reference to the city of Cuzco, which was the centre of the empire. The four divisions of the Inca theocracy were accordingly named Antisuyu, Cuntisuyu, Chinchasuyu, and Collasuyu; the word *Suyu* signifying *strip* or *part.* Notwithstanding the great distance between them, Quito belonged to Chinchasuyu; and in proportion as the Incas, by their religious wars, extended their faith, their language, and their despotic government, these Suyus acquired greater dimensions and became more unequal in magnitude. With the names of the provinces was thus associated an indication of their position; and "to name those provinces," observes Garcilaso, "was the same as to say to the east or to the west." (Nombrar aquellos Partidos era lo mismo que decir al Oriente, ó al Poniente.) The snow-chain of the Andes was regarded as an eastern chain. "La Provincia Anti da nombra á las Montañas de los Antis. Llamáron à la parte del Oriente Antisuyu, por la qual tambien llaman Anti á toda aquella gran Cordillera de Sierra Nevada que pasa al Oriente del Peru, por dar á entender, que está al

Oriente." (*Commentarios Reales*, p. i. pp. 47, 122.)* Later writers have supposed the name of the Andes chain to be derived from the word Anta, which, in the Quichua language, signifies copper. That metal was indeed of the highest importance to a people who for their edged-tools or cutting instruments, employed not iron, but a sort of copper mixed with tin; but still the name of copper mountains would scarcely have been extended over so vast a chain. Professor Buschmann has justly observed, that the final "*a*" is retained in the word *anta* when it forms part of a compound; and Garcilaso expressly adduces as an example *anta*, copper, and *antamarca*, province of copper. Moreover in the ancient language of the Inca empire (the Quichua), words and their compounds are so simple in formation that the conversion of "*a*" into "*i*" is out of the question; so that *Anta*, copper, and *Anti* or *Ante* (the country or an inhabitant of the Andes or the mountain-chain itself) must be regarded as words totally distinct from each other. In dictionaries of the Quichua language, with explanations in Spanish, the word *Anti* or *Ante* has the following interpretations: *la tierra de los Andes;—el Indio, hombre de los Andes;—la Sierra de los Andes.* The original signification or derivation of the word is buried in the darkness of past ages. Besides Antisuyu, some other compounds of which Anti or Ante forms a part, are, Anteruna (the native inhabitant of the Andes), Anteunccuy or Antionccoy (the sickness of the Andes; *mal de los Andes pestifero.*)

(2) p. 390—"*The Countess de Chinchon.*"

This lady was the wife of the Viceroy Don Geronimo Fernandez de Cabrera, Bobadilla y Mendoza, Conde de Chinchon, who governed Peru from 1629 to 1639. The cure of the Vice-Queen took place in the year 1638. A tradition which is current in Spain, but which I have frequently heard contradicted in Loxa, names Juan Lopez de Cañizares, Corregidor of the Cabildo de Loxa, as the person by whom the

* *Translation.*—"From the Province Anti the Montañas of the Antis received their name. Antisuyu signified the eastern direction, and for that reason the name Anti was given to all that part of the great Cordillera of Sierra Nevada which runs along the east of Peru, to denote that it was situated in the east." (*Commentarios Reales*, pt. i. pp. 47, 122.)—Ed.

Quina (Cinchona) bark was first brought to Lima, and universally recommended as a medicine. In Loxa, I have heard it affirmed that the salutary properties of the tree were long previously, though not generally, known in the mountainous regions. Immediately after my return to Europe, I expressed doubts whether the discovery had really been made by the natives in the vicinity of Loxa, for the Indians in the neighbouring valleys, where intermittent fevers are very prevalent, have an aversion to the Quina bark.* The story which sets forth that the natives learned the virtues of the Cinchona from the lions, "who cure themselves of intermittent fever by gnawing the bark of the Quina tree,"† appears to be merely a monkish fiction, and wholly of European origin. No such disease as the lion's fever is known in the New Continent; for the so-called great American lion (*Felis concolor*) and the small mountain lion (the *Puma*, whose footmarks I have seen on the snow) are never tamed, consequently never become the subjects of observation. Nor are the various species of the feline race, in either continent, accustomed to gnaw the bark of trees. The name "Countess's Powder" (*Pulvis Comitissæ*) originated in the circumstance of the bark having been dealt out as a medicine by the Countess de Chinchon. But this name was subsequently metamorphosed into "Cardinal's" or "Jesuit's" Powder, because Cardinal de Lugo, Procurator-General of the Order of the Jesuits, made known the medicine, whilst he was on a journey through France, and recommended it the more urgently to Cardinal Mazarin, as the brethren of the Order were beginning to carry on a profitable trade in the South American Quina bark, which they contrived to obtain through their missionaries. It is scarcely necessary to mention that Protestant physicians suffered themselves sometimes to be influenced by religious intolerance and hatred of the Jesuits, in the long controversy that was maintained, respecting the good or evil effects of the fever bark.

### (3) p. 393—"*Aposentos de Mulalo.*"

The Aposentos are dwellings or inns. They are called in

* See my *Treatise on the Quina Woods*, inserted in the *Magazin der Gesellschaft naturforschender Freunde zu Berlin, Jahrg.* i. 1807, s. 59.

† *Histoire de l'Acad. des Sciences, année* 1738. Paris, 1740, p. 233.

the Quichua language *Tampu*, whence the Spanish term *Tambo* (an inn). On the subject of these Aposentos see Cieça's *Chronica del Peru* (cap. 41 ed. de 1544, p. 108), and my *Vues des Cordillères* (Pl. xxiv).

(4) p. 394—" *The fortress of the Cañar.*"

This fortress is situated near Turche, and at an elevation of about 10,640 feet.* Not far distant from the Fortaleza del Cañar is situated the celebrated ravine of the sun, called the Inti Guaycu (in the Quichua language *huaycco*). In this ravine there are some rocks on which the natives imagine they see the image of the sun, and a bench called the Inga-Chungana (Incachuncana), the Inca's play. I made drawings of both. (*Vues des Cord.*, pl. xviii. et xix.)

(5) p. 394—" *Causeways covered with cemented gravel.*"

See Velasco's *Historia de Quito*, 1844, (t. i. p. 126—128), and Prescott's *History of the Conquest of Peru*, (vol. i. p. 157.)

(6) p. 395—" *Flights of Steps.*"

See Pedro Sancho in Ramusio, vol. iii. fol. 404, and the Extracts from Manuscript Letters of Hernando Pizarro, of which Mr. Prescott, the great historical writer, now at Boston, has so advantageously availed himself (vol. i. p. 444). "El camino de las sierras es cosa de ver, porque en verdad en tierra tan fragosa en la cristiandad no se han visto tan hermosos caminos, toda la mayor parte de calzada."†

(7) p. 396—" *Greeks, Romans, &c., present examples of these contrasts.*"

"The Greeks," says Strabo, (lib. v. p. 235, Casaub,) "in building their cities sought to produce a happy result by aiming at the union of beauty and solidity; but, on the other hand, the Romans directed particular attention to objects which the Greeks neglected; paving the streets with stone,

* I have given a drawing of it in the *Vues des Cordillères*, pl. xvii.; see also Cieça, cap. 44, P. i. p. 120.

† *Translation.*—"The road of the Sierras is wonderful to behold; for truly, throughout all Christendom, there are not to be seen such beautiful roads on such rugged ground, and, for the most part they are paved.'

building aqueducts to provide a plentiful supply of water, and constructing drainage for carrying all the uncleanliness of the city into the Tiber. They likewise paved all the roads in the country, so that the merchandize brought by trading vessels might be conveniently transported from place to place."

(8) p. 397—"*Nemterequeteba, the messenger of God.*"

Civilization in Mexico (the Aztec country of Anahuac), and in that country which, in the Peruvian theocracy, was called the Empire of the Sun, has so rivetted the attention of Europe, that a third point of dawning civilization, the mountainous regions of New Granada, was long totally lost sight of. I have already treated this subject in some detail.* The government of the Muyscas of New Granada bore some resemblance to the constitution of Japan: the temporal ruler corresponded with the Cubo or Seogun at Jeddo, and the spiritual ruler was like the sacred Daïri at Meaco. The table-land of Bogota was called by the natives of the country Bacata, *i. e.*, the utmost limit of the cultivated plains considered with reference to the mountain wall. When Gonzalo Ximenez de Quesada advanced thither he found the country ruled by three powers, whose relative subordination one to another is not now clearly understood. The spiritual chief was the electoral high priest of Iraca or Sogamoso (Sugamuxi, the place at which Nemterequeteba is said to have disappeared), the temporal princes were the Zake (Zaque of Hunsa or Tunja), and the Zipa of Funza. The last-named prince seems to have been, in the feudal constitution, originally subordinate to the Zake.

The Muyscas had a regular system of computing time, with intercalation for the amendment of the lunar year. For money they made use of small circular gold plates, cast, and all equal in diameter, (a circumstance worthy of remark, as traces of coinage even among the ancient and highly civilized Egyptians have hitherto been sought in vain). Their temples of the Sun were built with stone columns, some vestiges of which have recently been discovered in Leiva.† The race of

* See *Vues des Cordillères et Monumens des peuples indigènes de l'Amérique*, ed. in 8vo. t. ii. pp. 220—267.

† Joaquin Acosta, *Compendio historico del Descubrimiento de la*

the Muyscas should properly be distinguished by the denomination Chibchas; for Muysca, in the Chibcha language, merely signifies *men* or *people*. The origin and the elements of civilization, introduced among the Muyscas, were attributed to two mythical beings, Bochica and Nemterequeteba, who are frequently confounded one with another. Bochica was the most mythical of the two; having been in some degree regarded as divine and even equal to the Sun. His fair companion Chia or Huythaca occasioned, through her magical art, the submersion of the beautiful valley of Bogota, and for that reason she was banished from the earth by Bochica, and made to revolve round it as the moon. Bochica struck the rocks of Tequendama, and thereby opened a passage through which the waters flowed off, in the neighbourhood of the Giants' Field (Campo de Gigantes), where, at the elevation of 8792 feet above the level of the sea, the bones of elephant-like Mastodons have been discovered. It is stated by Captain Cochrane,* and by Mr. John Ranking,† that animals like the Mastodon still live in the Andes, and that they cast their teeth. Nemterequeteba, surnamed Chinzapogua, (*el enviado de Dios*, the envoy of God,) was regarded as a human being. He is represented as a bearded man, who came from the East, from Pasca, and who disappeared at Sogamoso. The foundation of the sanctuary of Iraca is sometimes ascribed to Nemterequeteba and sometimes to Bochica. The latter, it would appear, also bore the name of Nemterequeteba, and, therefore, that the one should have been confounded with the other, on such unhistoric ground, is a circumstance easily accounted for.

My old friend Colonel Acosta, in his admirable work entitled *Compendio de la Historia de la Nueva Granada*, endeavours to show, through the evidence of the Quichua language, that New Granada is the native land of the potato plant. In the *Compendio* (p. 185), he observes, "that as the potato (*Solanum tuberosum*) is known in Usmè by the indigenous name *Yomi*, and not by the Peruvian name, and as it was found by Quesada, cultivated in the province of Velez in

*Nueva Granada*, 1848, pp. 188, 196, 206, and 208; *Bulletin de la Société de Géographie de Paris*, 1847, p. 114.

* *Journal of a Residence in Columbia*, 1825, vol. ii. p. 390.

† *Historical Researches on the Conquest of Peru*, 1827, p. 397.

1537, a period when its introduction from Chile, Peru, and Quito must have been improbable, the plant may be regarded as indigenous to New Granada." It must, however, be borne in mind that the Peruvians had invaded Quito, and made themselves completely masters of it before 1525, in which year the death of the Inca Huayna Capac occurred. Indeed, the southern provinces of Quito fell under the dominion of Tupac Inca Yupanqui at the close of the fifteenth century.* The history of the first introduction of the potato into Europe is, unfortunately, involved in much obscurity, but the merit of the introduction is still very generally supposed to be due to Sir John Hawkins, who is said to have brought the plant from Santa Fe in the year 1563 or 1565. But a fact, which appears to be better authenticated, is, that the first potatoes grown in Europe were those planted by Sir Walter Raleigh on his estate at Youghal in Ireland, from whence they were conveyed to Lancashire. The Banana-tree (*Musa*), which, since the arrival of the Spaniards, has been cultivated in all the warmer parts of New Granada, is believed, by Colonel Acosta (p. 205), to have been known only in Choco before the Conquista. The name Cundinamarca, which by affected erudition was applied to the young republic of New Granada in the year 1811, a name suggestive of golden dreams (sueños dorados), would properly be Cundirumarca, not Cunturmarca.† Luis Daza, who accompanied the small invading army commanded by the Conquistador Sebastian de Belalcazar, who advanced from the south, mentions having heard of a distant country, rich in gold, and inhabited by the race of the Chicas. This country, Daza states, was called Cundirumarca, and its prince solicited auxiliary troops from Atahuallpa in Caxamarca. The Chichas have been confounded with the Chibchas or Muyscas of New Granada; and by a similar mistake the name of the unknown more southerly region has been transferred to this country.

(9) p. 400—"*Fall of the Rio de Chamaya.*"

See my *Recueil des Observ. Astron.*, vol. i. p. 304; Nivellement Barométrique, No. 236–242. I made a drawing of the

* Prescott's *Conquest of Peru,* vol. i. p. 332.

† See Garcilaso, lib. viii. cap. 2; also Joaquin Acosta, p. 189.

swimming courier, representing him in the act of winding round his head the handkerchief containing the letters. See *Vues des Cordillères*, pl. xxxi.

(10) p. 401—"*A point of some importance to the geography of South America, on account of an old observation of La Condamine.*"

My object was to connect chronometrically, Tomependa, (the starting-point of La Condamine's journey) and other places on the Amazon river, geographically determined by him, with the town of Quito. La Condamine was in Tomependa in June, 1743; consequently, 59 years before I visited that place, which I found, after astronomical observations made during three consecutive nights, to be situated in south lat. 5° 31′ 28″, and west long. 78° 34′ 55″). By my observations, and a laborious recalculation of all those previously made, Oltmanns has shewn that until the time of my return to France the longitude of Quito had been erroneously determined, and that the error made a difference of full 50½ arc-minutes.* Jupiter's satellites, lunar distances, and occultations afford a satisfactory accordance, and all the elements of the calculation are before the public. The too easterly longitude which had been determined for Quito was, by La Condamine, carried to Cuenca and the Amazon river. "Je fis," says La Condamine, "mon premier essai de navigation sur un radeau (*balsa*) en descendant la rivière de Chinchipe jusqu'à Tomependa. Il fallut me contenter d'en déterminer la latitude et de conclure la longitude par les routes. J'y fis mon testament politique en rédigeant l'extrait de mes observations les plus importantes."†

(11) p. 403—"*At the elevation of nearly* 12,800 *feet above the sea, we found marine fossils.*"

See my *Essai géognostique sur le Gisement des Roches*, 1823, p. 236; and for the first zoological determination of the fossils contained in the cretaceous formation of the Andes chain, see Leopold de Buch, *Pétrifications recueillies en Amérique* par Alex. de Humboldt et Charles Degenhardt, 1839 (in fol.), pp. 2, 3, 5, 7, 9, 11, 18, 22. Pentland found

* See Humboldt, *Recueil des Observ. Astron.* vol. ii. pp. 309—359.

† *Journal du Voyage fait à l'Equateur*, 1751, p. 186.

fossil shells of the Silurian formation in Bolivia, and on the Nevado of Antakana at the elevation of 17,480 feet. (See Mary Somerville's *Physical Geography*, 1849, vol. i. p. 185.)

(12) p. 407—"*The point at which the Andes-chain is intersected by the magnetic equator.*"

See my *Rélation Hist. du Voyage aux Régions Equinoxiales*, t. iii. p. 622; and *Cosmos*, vol. i. pp. 191, 432; where, through errors of the press, the longitude is in one place marked 48° 40′, and in another 80° 40′, whereas it ought to be 80° 54′.

(13) p. 409—"*Tedious court ceremonies.*"

Conformably with an ancient ceremonial, Atahuallpa spat, not on the ground, but into the hand of a distinguished lady of the Court circle. "This was done," observes Garcilaso, "by reason of his majesty." "El Inca nunca escupia en el suelo, sino en la mano de una Señora mui principal, por Magestad." (Garcilaso, *Comment. Reales*, p. ii. p. 46.)

(14) p. 410—"*Captivity of Atahuallpa.*"

The captive Inca was, at his own desire, a short time before he was put to death, conducted into the open air, for the purpose of seeing a large comet, described to have been of a greenish black hue, and nearly as thick as a man's body; ("*una cometa verdinegra, poco menos gruesa que el cuerpo de un hombre,*" Garcilaso, p. ii. p. 44). This comet, which Atahuallpa saw shortly before his death, (therefore, in July or August, 1533), he supposed to be the same comet of evil omen, which had appeared at the death of his father Huayna Capac, and was certainly identical with that observed by Appian.* The comet was seen by Appian, on the 21st of July, standing high in the north, near the constellation of Perseus; and it appeared like a sword held by Perseus, in his right hand.† The year in which the Inca Huayna Capac

* Pingré, *Cométographie*, t. i. p. 496; and Galle's *Verzeichniss aller bisher berechneten Cometenbahnen*, in Olbers' *Easiest method of calculating the course of a Comet*, 1847, p. 206.

† Mädler's *Astronomie*, 1846, p. 307; also Schnurrer's *Chronik der Seuchen in Verbindung mit gleichzeitigen Erscheinungen,*" 1825, part ii. p. 82.

died, is considered by Robertson not to be satisfactorily determined; but the investigations of Balboa and Velasco shew, that the event must have occurred about the end of 1525. The statements of Hevelius (*Cométographie*, p. 844), and of Pingré (vol. i. p. 485), obtain additional confirmation from the testimony of Garcilaso, (p. i. p. 321,) and the traditions preserved among the Amautas ("que son los filosofos de aquella republica"). I may here add the remark, that Oviedo is certainly incorrect in stating in the yet unpublished continuation of his "*Historia de las Indias*," that the name of the Inca was not Atahuallpa, but Atabaliva. See Prescott's *Conquest of Peru*, vol. i. p. 498.

(15) p. 410—"*Ducados de Oro*," (3,838,000 *golden ducats.*)

The sum mentioned in the text is that stated by Garcilaso de la Vega.* On this subject, however, Padre Blas Valera and Gomera give different accounts.† Moreover, it is difficult to ascertain the precise value of the Ducado Castellano or Peso de Oro.‡ The intelligent historian, Prescott, has had the opportunity of consulting a manuscript, bearing the promising title of "*Acta de Reparticion del Rescate de Atahuallpa*," (Act of assessment for the ransom of Atahuallpa). The Peruvian booty shared by the brothers Pizarro and by Almagro, appears to be too highly estimated by Prescott, who says it amounted to 3,500,000*l.* sterling, but the ransom money, the treasures of the different temples of the Sun, and of the Huertas de Oro, were all included in that amount.§

(16) p. 412—"*The great Huayna Capac, who, for a Child of the Sun, was somewhat disposed to free-thinking.*"

The nightly disappearance of the sun excited, in the mind of the Inca, many philosophic doubts respecting the government of the world by that luminary. Among the Inca's remarks on this subject, as recorded by Padre Blas Valera, are the following:—"Many maintain that the sun lives and is the

* *Commentarios reales de las Incas*, parte ii. 1722, pp. 27, 51.

† *Historia de las Indias*, 1533, p. 67. See my *Essai Politique sur la Nouvelle Espagne*, ed. 2, t. iii. p. 424.

‡ See the *Essai politique*, t. iii. p. 371, 377; and also Joaquin Acosta's *Descubrimiento de la Nueva Granada*, 1848, p. 14.

§ Prescott's *Conquest of Peru*, vol. i. pp. 464—477.

creator and maker of all things (*el hacedor de todas las cosas*); but whosoever desires to do a thing completely must continue at his task without intermission. Now many things are done when the sun is absent, therefore, he cannot be the creator of all. It may also be doubted whether the sun be really living, for, though always moving round in a circle, he is never weary (*no se cansa*). If the sun were a living thing he would, like ourselves, become weary; and if he were free, he would, doubtless, sometimes move into parts of the heavens in which we never see him. The sun is like an ox bound by a rope, being obliged always to move in the same circle (*como una Res atada que siempre hace un mismo cerco*), or like an arrow which can only go where it is sent, and not where it may itself wish to go." (Garcilaso, *Comment. Reales*, p. i. lib. viii. cap. 8, p. 276.) The Inca's simple comparison of the circling movement of a heavenly body to that of an ox fastened by a rope is very curious, owing to a circumstance which may be explained here. Huayna Capac died at Quito in 1525 (seven years prior to the invasion of the Spaniards), and his empire was divided between Huascar and Atahuallpa. Now, in the native language of Peru, the name Huascar signifies rope, and Atahuallpa means a cock or a fowl. Instead of *res* Huayna Capac probably used the word signifying, in his native language, *animal* generally; but, even in Spanish, the word *res* is not applied exclusively to oxen, but is employed to denote cattle of all kinds. How far the Padre, with the view of weaning the natives from the dynastic service of the Inca, may have mingled passages from his own sermons with the heresies of the Inca, we need not here inquire. That it was deemed very important to keep these doubts from the knowledge of the lower classes of the people is evident, from the very conservative policy and the state maxims of the Inca Roca, the conqueror of the province of Charcas. This Inca founded schools exclusively for the higher classes, and, under heavy penalties, prohibited instruction being given to the common people, lest it should render them presumptuous, and cause them to disturb the State. (No es licito que enseñen á los hijos de los Plebeios las Ciencias, porque la gente baja no se eleve y ensobervezca y menoscabe la Republica; Garcilaso, p. i. p. 276.) Thus the theocracy of the Incas may be said to have resembled the Slave States in the free land of the North American Union

(17) p. 415—"*Expected restoration of the Inca rule.*"

I have treated this subject at length in another work. Sir Walter Raleigh had heard of an old prophecy current in Peru, which foretold "that from Inglaterra those Ingas shoulde be againe in time to come restored and deliuered from the seruitude of the said conquerors. I am resolued that if there were but a small army afoote in Guiana marching towards Manoa, the chiefe citie of Inga, he would yield her Majesty by composition, so many hundred thousand pounds yearely, as should both defend all enemies abroad and defray all expenses at home, and that he woulde besides pay a garrison of 3000 or 4000 soldiers very royally to defend him against other nations. The Inca will be brought to tribute with great gladnes." A restoration project, which promised to be highly satisfactory to both parties, but, unfortunately for the success of the scheme, the dynasty which was to be restored and which was to pay for the restoration was wanting.

(18) p. 418—"*The adventurous expedition of Vasco Nuñez de Balboa.*"

I have, in another work, mentioned the fact that Columbus, long before his death, full ten years prior to Balboa's expedition, was aware of the existence of the South Sea, and its near proximity to the eastern coast of Veragua.* Columbus was led to the knowledge of this fact, not by theoretical speculations on the configuration of Eastern Asia, but by positive and local information obtained from the inhabitants themselves, information which he collected on his fourth voyage (11th May, 1502, to the 7th November, 1504). This fourth voyage led the Admiral from the coast of Honduras to the Puerto de Mosquitos, and even as far as the western extremity of the Isthmus of Panama. The natives reported (and Columbus commented on their reports in the *Carta rarissima* of the 7th of July, 1503), "that not far from the Rio de Belen, the other sea (the South Sea), turns

* *Relation hist.*, t. iii. pp. 703, 705, 713.

† Raleigh, *The Discovery of the large, rich, and beautiful Empire of Guiana, performed in* 1595. Edition published by Sir Robert Schomburgk, 1848, pp. 119 and 137.

‡ *Examen critique de l'histoire de la Géographie du Nouveau Continent et des progrès de l'Astronomie nautique aux* 15*me et* 16*me siècles*, t. i. p. 349.

(boxa) to the mouths of the Ganges; so that the countries of the Aurea (*i.e.*, the Chersonesus Aurea of Ptolemy) are situated, in relation to the eastern shores of Veragua, as Tortosa (at the mouth of the Ebro) is in relation to Fuentarabia (on the Bidassoa) in Biscay, or as Venice in respect to Pisa." But, although Balboa first saw the South Sea from the heights of the Sierra de Quarequa, on the 25th of September,* it was several days later before Alonzo Martin de Don Benito, who had discovered a passage from the mountains of Quarequa to the gulf of San Miguel, embarked on the South Sea in a canoe.†

The recent acquisition of the western coast of the New Continent by the United States of North America, and the fame of the golden treasures of New (now called Upper) California, have rendered the question of forming a direct communication between the shores of the Atlantic and the western regions, by the isthmus of Panama, more urgent than ever. I, therefore, consider it my duty here once more to direct attention to the fact, that the shortest route to the shores of the Pacific, as pointed out by the natives to Alonzo Martin de Don Benito, is in the eastern part of the Isthmus, and led to the Golfo de San Miguel. We know that Columbus‡ sought for a narrow pass (estrecho de tierra firme); and in the official documents extant, of the dates of 1505, 1507, and especially in that of 1514, mention is made of the sought-for opening (abertura), and of the pass (passo), which, in this district, should lead directly to the "Indian Land of Spices." A channel of communication between the Atlantic and the Pacific, is a subject which has more or less occupied my attention for the space of forty years; and in my published works, as well as in the several memoirs which, with honourable confidence, the Free States of Spanish America have requested me to write, I have constantly recommended a hypsometrical survey of the Isthmus throughout its whole length, but more especially at two points, viz., where at Darien and what was formerly the deserted province of Biruquete, it joins the South American Continent, and where,

* Peter Martyr's *Epist.* dxl. p. 296.

† Joaquin Acosta, *Compendio hist. del Descubrimiento de la Nueva Granada*, p. 49.

‡ *Vida del Almirante por Don Fernando Colon*, cap. 90.

between Atrato and the Bay of Cupica, on the shore of the Pacific, the mountain chain of the Isthmus almost entirely disappears.*

In the year 1828 and 1829, General Bolivar, at my request, caused the Isthmus between Panama and the mouth of the Rio Chagres to be accurately levelled by Lloyd and Falmarc.† Since that time, other measurements have been executed by intelligent and experienced French engineers, and plans have been drawn out for canals and railways with locks and tunnels. But these measurements have invariably been made in the meridian direction between Porto-bello and Panama, or westward from thence, towards Chagres and Cruces. The most important points of the eastern and south-eastern parts of the Isthmus, on both shores, have in the meantime been overlooked. Until those parts shall be described geographically, according to accurate (but easily obtained) chronometrical determinations of latitude and longitude; and hypsometrically, with reference to their superficial conformation, by barometrical measurements and elevations, I see no reason to alter the views I have always entertained on this subject. Accordingly, at the present time (1849), I here repeat the opinion I have often before expressed; viz., *that the assertion is groundess and altogether premature*, that the Isthmus of Panama is unsuited to the formation of an Oceanic Canal—one with fewer sluices than the Caledonian Canal—capable of affording an unimpeded passage, at all seasons of the year, to vessels of that class which sail between New York and Liverpool, and between Chili and California.

According to examinations, the results of which the Directors of the Deposito Hidrografico of Madrid have caused to be inserted in all their maps since 1809, it appears that on the Antillean shore of the Isthmus, the creek called the Ensenada de Mandinga, stretches so far to the south that its distance from the Pacific shore, eastward of Panama, appears to be only between 4 and 5 German geographical miles (15 to

* See my *Atlas géographique et physique de la Nouv. Espagne*, pl. iv. and *Atlas de la Relation historique*, pl.xxii. xxiii.; also my *Voyage aux regiones équinoxiales du Nouveau Continent*, t. iii. pp. 117—154, and *Essai politique sur la royaume de la Nouvelle Espagne*, t. i. 2nde ed. 1825, pp. 202—248.

† *Philosophical Transactions of the Royal Soc. of London for the year* 1830, pp. 59—68.

an equatorial degree) or 16 to 20 English geographical miles. On the Pacific coast also, the deep Golfo de San Miguel, into which falls the Rio Tuyra, with its tributary the river Chuchunque (Chucunaque), runs far into the Isthmus. The river Chuchunque too, in the upper part of its course, runs within 16 geographical miles of the Antillean shore of the Isthmus, westward of Cape Tiburon. For upwards of twenty years I have been repeatedly consulted on the problem of the Isthmus of Panama, by companies having ample pecuniary means at their disposal; but in no instance has the simple advice I have given been followed. Every engineer who has been scientifically educated knows the fact that between the tropics, even without corresponding observations, good barometrical measurements (horary variations being taken into account) may be relied on as correct, within from 75 to 96 feet. Besides it would be easy to establish, for the space of a few months, one on each shore, two fixed barometric stations; and frequently to compare the portable instruments used in the preliminary levelling with each other, and with those at the fixed stations. The point demanding the most attentive examination is that where the range of mountains between the Isthmus and the main continent of South America sinks into hills. Considering the importance of this subject to the commercial interests of the whole world, the examination should not, as heretofore, be restricted within narrow bounds. A complete comprehensive survey, including the whole eastern part of the Isthmus—the results of which would be alike useful in facilitating every possible scheme, whether of canals or railroads—can alone decide the much discussed problem, either affirmatively or negatively. This work will in the end be undertaken, but had my advice been adopted, it would have been done at first.

(19) p. 418—"*Impressions excited by the accidental circumstances of life.*"

In *Cosmos* I have adverted to the incitements to the Study of Nature. (Vol. ii. p. 371, Bohn's edition.)

(20) p. 420—"*Of importance in determining the longitude of Lima.*"

At the time of my expedition the longitude of Lima,

as determined by Malaspina and marked in the maps published by the Deposito Hidrografico de Madrid, was $5^h 16' 53''$. The transit of Mercury over the Sun's disc, on the 9th of November, 1802 (which I observed at Callao, the port of Lima, from the Round Tower of the Fort of San Felipe), gave for Callao, by the mean of the contact of both limbs, $5^h 18' 16'' 5$; by the external contact only, $5^h 18' 18''$ (79° 34′ 30″). This result, obtained from the transit of Mercury, has been confirmed by Lartigue and Duperrey; and by observations made during Capt. Fitzroy's expeditions of the "Adventurer" and the "Beagle." Lartigue fixed the longitude of Callao at $5^h 17' 58''$; Duperrey made it $5^h 18' 16''$; and Capt. Fitzroy $5^h 18' 15''$. After having calculated the longitudinal difference between Callao and the Convent of San Juan de Dios at Lima, by carrying chronometers from the one place to the other during four journeys, I found that the observations of the transit of Mercury determined the longitude of Lima to be $5^h 17' 51''$ (79° 27′ 45″ W. from Paris, or 77° 6′ 3″ W. from Greenwich.) See my *Recueil d'observations astron.*, vol. ii. p. 397, and *Relation hist.*, t. iii. p. 592.

POTSDAM, *June*, 1849.

THE END.

# INDEX.

LONDON: PRINTED BY HARRISON AND SON, ST. MARTIN'S LANE.

Printed in the United States
By Bookmasters